B4

Food Science

The Biochemistry of Food and Nutrition

Fourth Edition

Kay Yockey Mehas
Family and Consumer Sciences Teacher
Principal/Director of Schools
Eugene, Oregon

Sharon Lesley Rodgers
Chemistry/Physics Teacher
Henry D. Sheldon High School
Eugene, Oregon

Glencoe
McGraw-Hill

New York, New York   Columbus, Ohio   Woodland Hills, California   Peoria, Illinois

Safety Notice

The reader is expressly advised to consider and use all safety precautions described in this book or that might also be indicated by undertaking the activities described herein. In addition, common sense should be exercised to help avoid all potential hazards and, in particular, to take relevant safety precautions concerning any known or likely hazards involved in food preparation, or in use of the procedures described in the Title, such as the risk of knife cuts or burns.

Publisher and Authors assume no responsibility for the activities of the reader or for the subject matter experts who prepared this book. Publisher and Author make no representation or warranties of any kind, including but not limited to, the warranties of fitness for particular purpose or merchantability, nor for any implied warranties related thereto, or otherwise. Publisher and Author will not be liable for damages of any type, including any consequential, special or exemplary damages resulting, in whole or in part, from reader's use or reliance upon the information, instructions, warnings or other matter contained in this book.

Brand Disclaimer

Publisher does not necessarily recommend or endorse any particular company or brand name product that may be discussed or pictured in this text. Brand name products are used because they are readily available, likely to be known to the reader, and their use may aid in the understanding of the text. Publisher recognizes that other brand name or generic products may be substituted and work as well or better than those featured in the text.

Glencoe/McGraw-Hill

A Division of The **McGraw·Hill** *Companies*

Copyright © 2002, 1997, 1994, 1989 by Glencoe/McGraw-Hill. All rights reserved. Except as permitted under the United States Copyright Act, no part of this publication may be reproduced or distributed in any form or by any means, or stored in a database or retrieval system, without prior written permission from the publisher.

Send all inquiries to:
Glencoe/McGraw-Hill
3008 W. Willow Knolls Drive
Peoria, IL 61614-1083

ISBN 0-07-822603-1

Printed in the United States of America

1 2 3 4 5 6 7 8 9 10 071 05 04 03 02 01 00

Technical Reviewers

Robert G. Brannan, MS
Researcher
University of Massachusetts
Amherst, Massachusetts

Carol A. Costello, PhD
Associate Professor
University of Tennessee
Knoxville, Tennessee

Eileen M. Ferguson, PhD
Environmental Consultant
Carson City, Nevada

Sarah T. Hawkins, PhD, RD, CFCS
Professor
Indiana State University
Terre Haute, Indiana

Teacher Reviewers

Lynn Beard, MEd
Life Management Instructor
Everett High School
Lansing, Michigan

Irma Bode
Family and Consumer Sciences Teacher
Southeast Raleigh High School
Raleigh, North Carolina

Mark E. Cahn
Science Teacher
Willow Glen High School
San Jose, California

Barbara Cerotsky, MS
Family and Consumer Sciences Teacher
North Eugene High School
Eugene, Oregon

Carol Atkinson Dunn, MEd
Family and Consumer Sciences Teacher
Proviso East High School
Maywood, Illinois

Nancy L. Glasgow
Food Science Instructor
W.G. Enloe Magnet High School
Raleigh, North Carolina

Susan P. Green
Family and Consumer Sciences Instructor
Heber Springs High School
Heber Springs, Arkansas

Lola Holl, MS
Family and Consumer Sciences Instructor
Minneapolis High School
Minneapolis, Kansas

Martha Maddox, MS, CFCS
Family and Consumer Sciences Teacher
Davidson High School
Mobile, Alabama

Jaclyn Stockton-Sooy, MEd
Family and Consumer Sciences Teacher
Arundel High School
Gambrills, Maryland

Aaron T. Warner, MEd
Teacher of Biology and Chemistry
Terre Haute South Vigo High School
Terre Haute, Indiana

Contents in Brief

Unit 5 The Chemistry of Food..292

Unit 6 The Microbiology of Food Processing.....................398

Contents

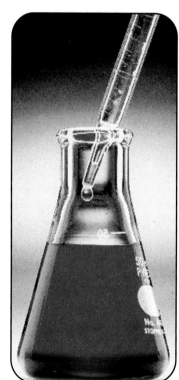

Unit ③ Chemistry Fundamentals 100

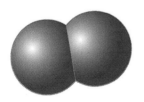

Unit 4   The Science of Nutrition 172

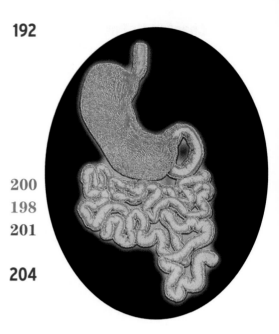

Chapter 17: Protein 256

Hemoglobin

Chapter 18: Vitamins and Minerals 274

Unit ⑤ The Chemistry of Food.........................292

Chapter 24: Food Additives   378

Unit ⑥ The Microbiology of Food Processing.....................398

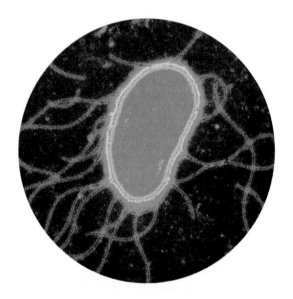

FOOD SCIENCE Careers

EXPERIMENTS

Nutrition*Link*

Charts

TechWatch

SIX BILLION AND COUNTING

With the earth's population predicted to reach ten billion in less than two hundred years, feeding the world will be an increasing challenge. How are food scientists using biotechnology to help? By developing disease-resistant plants, they help farmers increase crop yields. An experimental potato hybrid contains genes resistant to a new, more deadly strain of the disease that caused the Irish potato famine in the 1840s. A particular gene may enable wheat to grow where it couldn't before. A vaccine for "shipping fever," the biggest killer of beef cattle in feedlots, has also been developed. With progress through biotechnology, food scientists are finding answers for the future.

UNIT 1

The World of Food Science

CHAPTER 1
What Is Food Science?

CHAPTER 2
Why Study Food Science?

TechWatch Activity
Research information on disease-resistant plants. Then write a paper on how this type of biotechnology is helping the agriculture industry.

What Is Food Science?

Objectives

● Trace the development of the scientific study of food.

● Describe areas included in the field of food science.

● Identify different types of work that food scientists do.

Terms to remember

biotechnology
food science

What do these three situations have in common? A manufacturer creates a popular candy with appealing colors. A nutrition researcher explains why eating broccoli may reduce cancer risk. A health organization ships packages of high-protein drink to malnourished children in a remote part of the world. These situations all share a link with food science. Without the work of food scientists, none of them would have been possible.

Food science is *the study of producing, processing, preparing, evaluating, and using food*. The field crosses many branches of science, including biology, botany, physiology, zoology, and bacteriology. Organic chemistry and physics, however, are used most often.

From the trivial—like a colorful candy coating—to matters of life and death—like hunger—research in food science leads to new discoveries every day. Daily life is filled with events that demonstrate the close relationship between food and the scientific world. Until you do some exploring, you can't imagine all that goes on in the field of food science.

History of Food Science

On a snowy March day in 1626, the noted English scientist, writer, and philosopher Sir Francis Bacon stopped to buy a chicken on his carriage ride home. He wanted to see how stuffing the bird with snow affected the "conservation . . . of bodies." In other words, did freezing preserve chicken? Sadly, Bacon developed bronchitis as a result of his experiment and died soon after. A potential food scientist was lost.

Bacon's method and curiosity were typical of food science research for many years. Investigating how physical laws affected food was part of chemistry and biology, but not a separate study.

The first food science "textbooks" may have been cookbooks. A recipe, after all, is based on facts that cooks have learned by trial and error, just as a chemistry text presents facts that scientists have learned from years of experiments.

In the mid-1800s, German chemist Justus von Liebig tried to explain some things that cooks achieve in his book *Researches on the Chemistry of Food.* In 1896 in the United States, Fannie Farmer continued the idea with her *Boston Cooking School Cook Book.* Farmer wrote that she hoped the book's "condensed scientific knowledge" would "awaken an interest … which will lead to deeper thought and broader study of what to eat." That goal still serves as a basic definition of food science.

During the same time period, scientist George Washington Carver, shown on page 24, was finding new uses for traditional foods of the American South. Remarkably, Carver developed over 100 products from peanuts, pecans, and sweet potatoes.

In the early 1900s, advances in technology gave rise to commercial food processing. For instance, in 1925 Clarence Birdseye helped the frozen foods industry get started with a refined method for quickly freezing fish. The food industry began supporting scientific studies of foods as important to its success. Such research has been largely industry-driven ever since.

Food Science Today

Like other sciences, food science has advanced rapidly in recent decades. New technology continues to open the doors to discovery.

In your lifetime alone, food scientists have been behind many new products. The diet-conscious can now choose from many fat-free products, including ice cream and cookies. Many developments are not as well known. Poultry producers can spray chickens with a substance that the birds consume as they clean it from their feathers. The substance then kills disease-

Nutrition *Link*

Food Scientists and Nutrition

Food science and nutrition make good partners. As you learn more about food science, you'll discover the remarkable link between food science and people's health.

Have you noticed the "veggie burgers" now on many restaurant menus? By replacing some or all of the meat with soy meal, food scientists have created a high-protein, low-fat "hamburger" that costs less to produce than the tra-

ditional patty. New foods such as these increase your options for nutritious eating.

Food technology has other benefits too. Developing methods for freezing vegetables within hours of harvest retains more nutrients for your meals at home. New techniques in preservation allow nutritious foods to be transported quickly wherever they're needed—all over the world.

Applying Your Knowledge

1. Some products of food science also make it easy to eat less nutritiously. How is this possible?
2. Besides veggie burgers, what other nutritious food products have been developed by food scientists in recent years? How willing are you to try them?

Food production and marketing have changed over the years. How does this scene compare to the typical supermarket of today?

causing germs in the chicken's intestines. Developments in food science just keep coming, making people wonder what the future holds.

Areas of Food Science

Production, processing, evaluation, and use— each area of food science contains a wide range of responsibilities. Because all areas are inter-related, concerns of different areas often over-lap. Cooperation is essential. A food scientist who researches sugar substitutes may need to exchange ideas with people who market sugar-free products.

Food Production

Food science is part of food production "from the ground up." For example, a scientist might research ways to treat the soil in order to pro-duce the most nutritious crop.

Through **biotechnology**, *scientists use the tools of modern genetics in the age-old process of improving plants, animals, and microor-ganisms for food production*. Throughout his-tory, humans have been working with foods and changing gene structures without realizing. Modern biotechnology allows food producers to do the same, but with greater understanding and selectivity.

To get results, scientists used to breed an organism for several generations. Now they can alter the organism's genetic material to speed this natural process.

Through biotechnology, scientists improve agricultural products by enhancing or eliminat-ing traits that a plant or animal inherits. In one recent achievement, a way to turn on a plant's natural defenses against disease and insects was found. Thus, an alternative to chemical pesticides was created.

Food scientists involved in production typi-cally specialize in one food group and its com-ponents, such as cereal grains, fish and seafood, or beverages. Others focus on a partic-ular food ingredient, such as sugar or fat substi-tutes. Still others, called microbiologists, research molds, bacteria, and other organisms that are needed to produce foods or that threaten food quality and safety.

Food Processing

Processing takes food that has been produced and puts it through steps to create the final marketable result. Industries and government agencies employ those with food science ex-pertise for work in processing.

Such products as turkey "bacon" and "meats" made from vegetable protein can be part of a healthful diet. Can you name other similar products developed by food scientists?

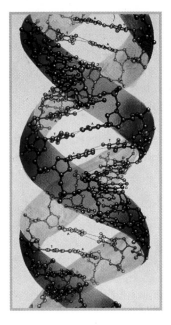

DNA (deoxyribonucleic acid) exists in the cells of all organisms. The complex chemical code on DNA strands provides the genetic map for each organism. Scientists can work with DNA to alter certain features of an organism. That's how a food scientist makes some plants disease-resistant.

Quality control specialists also use food science principles. By monitoring each step in processing, these professionals help ensure that final products meet government, industry, and company standards. They determine whether raw products meet purchasing specifications. Selling an apple for eating or for applesauce depends partly on them. They check warehouses and other storage facilities for sanitary conditions and to be sure the temperature and humidity are appropriate to preserve the foods kept there. The nutrition information on a food label comes from scientists who test foods for nutrient content.

Food Preparation

When you bake a new brand of frozen pizza, you trust that following the package directions will

Researchers in food science may help develop more economical processing methods that retain greater food quality. The work of a food scientist often takes place in the laboratory, where experimentation and research dig deeply into the chemical properties of substances and the way they react with each other. They often work with engineers, microbiologists, flavor experts, sensory evaluation experts, packaging specialists, and marketing analysts. Any of these individuals may be a food scientist with expertise in a particular area.

Scientific investigation is the foundation for work in the food science field. Scientists study everything from the composition of foods to how packaging affects food. All that they learn helps them make advances and improvements in food processing.

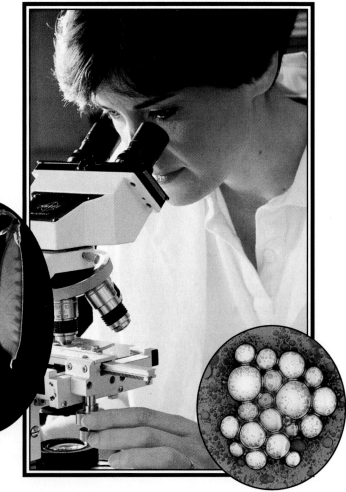

Careers

R & D Specialist

Why aren't some potato chips salt-free? Ask an . . . R & D Specialist

A s an R & D specialist (research and development) for a large snack foods company, Joel Krider works with people throughout the organization to develop new products. "We come up with plenty of ideas, but they don't always work.

"With today's concerns about consuming too much salt, I thought salt-free potato chips might sell. But they've been tried again and again for the past twenty years, and people still don't buy them. Other products sell well—like rice cakes. This crunchy snack food is cost-effective to make and so popular that our company produces several flavors. Sometimes we're surprised by what sells and what doesn't."

Education Needed: Specialists in food or nutrition research and development need a bachelor's degree in food science, chemistry, or human nutrition. Significant laboratory experience is also helpful. With coursework in statistics, specialists can evaluate the results of product development experiments and the data gathered from consumer tasting panels.

KEY SKILLS
- Organizational skills
- Creativity
- Decision-making skills
- Communication skills
- Problem-solving skills
- Keen senses

Life on the Job: R & D specialists participate in all aspects of product development. They do market research and test prototypes. During production, they often troubleshoot problems. Participating in quality control and even marketing and sales can go along with this specialist's work. Giving presentations and travel may be expected in some positions.

Your Challenge

Suppose you're an R & D specialist helping to determine a new product. The product under consideration has been tried before. Opinions are mixed about whether it will work this time.

1. What societal situations might cause a particular snack food to fail or succeed?
2. What strategies could your company use to promote a product?

yield the promised results. Your confidence comes courtesy of those who help decide the best preparation methods for the products you buy. Determining times, temperatures, and other factors begins with scientific knowledge about how different food components react to various conditions. Not only your enjoyment, but also food safety, depends on this understanding.

The same is true of foods that you order in a restaurant or create from a cookbook. Food science explains the whys behind so many kitchen processes. Why won't an egg white whip if a drop of yolk is present? Why does a sauce thicken? Why do some potatoes have green spots? Food science provides answers to questions like these.

Evaluation of Food

"Will people buy this product?" To answer that crucial question, food technologists send new items to evaluation laboratories and test kitchens. In these facilities, food scientists work with other experts to refine a product's taste, texture, appearance, and other qualities. They systematically vary ingredients and preparation methods to design the most pleasing and profitable food product they can. A food design consultant might suggest changing the sweetener used in a recipe, while a flavor chemist can explain how that change would chemically alter the product and affect its taste.

Evaluation is a before-and-after process. Initially it's used during the production phase to decide whether a food can be created at a reasonable cost. Later it helps refine a product before test marketing with consumers. A food may be reevaluated any time a manufacturer looks closely at consumer preferences and profitability. For example, frozen dinners are in larger supply than ever and have improved greatly over the earliest products. Why do you think that has happened?

Utilization of Food

Can you name ten things made from soybeans? You might list soy milk and soybean oil, but what about newsprint? Also, don't be surprised to find corn products in your next tank of gasoline. The search to find new uses for food products may lead anywhere, even to products that are not food for humans. As growing consumer demand strains the earth's resources, getting the most from every resource gains urgency.

Studying how foods can be used completes—or begins, if you like—a cycle of science in the food industry. This investigation may be the first step in producing a new food item.

Nutrition Science

Nutrition and science have a productive history together that dates back to about the early 1900s. Questioning minds began to wonder why people eat food. They studied the complex structures of food and began to connect foods to the way the body functions. Although much has been learned over the years, many questions still await answers. Some questions may not have even been asked.

Scientists who study nutrition explore chemistry and biology, as well as certain physi-

Imagine trying to prepare daily meals for hundreds of people. Knowledge of food science builds confidence and expertise in such situations. Could this knowledge be helpful at home too?

The capsaicin (kap-SAY-uh-sun) in chili peppers makes them hot. Some research shows it may also interfere with cancer development.

Food manufacturers increase interest in a product by offering samples to consumers. *How does taste testing at an early stage of product development help a manufacturer?*

cal laws that govern life. They deal with the atoms, molecules, and cells that make up both human beings and the food they eat. From their research comes new information and developments. Some scientists create nutritious food products. Others refine foods to make them better for you. Still others draw conclusions about what you should be eating and give recommendations. The work of all these people combines to give you the hope of living a long and healthy life.

Science in Real Life

The quick "walk" through food science that you've just taken only skims the surface. Science principles are applied to food everywhere—in farm fields, in food processing plants, in home and restaurant kitchens, and in research laboratories. Until this course, you might not have noticed the interesting connection between food and science.

Science has always been a tool for explaining the mysteries of the world. If you've never thought of food as mysterious, you may begin to rethink that attitude in the weeks ahead. Science will gain new relevance as you make discoveries about food. Why do people put eggs in yeast breads? Why do fresh banana slices turn brown? Discovering answers may just lead you to many new questions.

Food scientists have developed many uses for corn. Field corn feeds livestock. Corn is used to make cornmeal, breakfast foods, cornstarch, and corn syrups. As a raw material, this versatile food is used by many industries. With an Internet search, you can discover the amazing list of products linked to corn.

CHAPTER 1 Review and Activities

Chapter Summary

- Food science is the study of producing, processing, preparing, evaluating, and using food.
- People have long been interested in understanding food scientifically.
- Advancements in technology have lead to many discoveries in the field of food science.
- Food science is part of every stage in the production, processing, evaluation, and use of food.
- Food scientists contribute to a safe and enjoyable food supply.
- Knowledge of food science has aided the study of nutrition, and both are increasingly important.
- This course will help you see the interesting connection between food and science.

Using Your Knowledge

1. What branches of science are most strongly connected with food science?
2. How did Sir Francis Bacon display the qualities of a scientist?
3. Describe the impact of two people on the advancement of food science.
4. How have food science and the food industry been important to each other?
5. What is biotechnology? Give two examples.
6. What are three responsibilities of a quality control specialist?
7. How does food science help people who cook in restaurants or at home?
8. Why do food manufacturers need to evaluate their products?
9. Why do people need to make thorough use of the foods grown on earth?
10. How is food science related to nutrition?

Real-World Impact

Biotechnology

Biotechnology has produced new food products and also some concerns. Gene transfer among organisms has raised such questions as these: "How will new products affect the environment? Will people with food allergies be harmed by products that contain some of a food's genes? How will new foods fit with personal or religious dietary laws? What labeling should be required?" More complex questions are bound to arise as more sophisticated, and controversial, techniques are developed.

Thinking It Through

1. Suppose researchers developed a new, more nutritious variety of pea. What concerns about this product might a food scientist have in each area of the food industry?
2. Give some arguments for and against labeling foods that are genetically altered.

Skill-Building Activities

1. **Communication.** Locate an article relating to food science in a newspaper or magazine. Present an oral report summarizing the article's contents.
2. **History.** Conduct further research on how one of the people mentioned in this chapter influenced food science. If you wish, choose another person, such as Henry Kellogg, Charles Post, Dr. Harvey Wiley, or Ellen Richards. Share your findings with the class.
3. **Critical Thinking.** Identify a food that you enjoy. How do you think a food scientist would be involved in the production, processing, and marketing of that food product?
4. **Teamwork.** With a partner, make a list of questions you have about the science of foods. For example, you might wonder why carrots are orange or why gravy thickens. Save your questions and search for answers as you complete this course.

Understanding Vocabulary

Use the Internet to find sites that relate to "food science." Compare results with other students and create a composite list of useful sites.

Thinking Lab

Food Production

In 1922 U.S. farmers used the first hybrid seed corn, created by crossbreeding two corn plants. Hybrid corn accounted for a 600 percent increase in U.S. corn production between 1930 and 1985. Through the work of food scientists, farmers have continued to grow more food and better quality crops. Some biotech crops can produce more per acre. Some can be raised with less tilling and less fuel.

Analysis

Today the world contains little remaining land to use for farming. Taking rain forests and wetland habitats for farming would have serious consequences. Advances like those described above help farmers grow more bountiful harvests on existing farmland. At the same time, negative impacts on the environment can be reduced. Less tilling means preservation of precious topsoil. Reduced farm runoff helps protect streams and rivers.

Thinking Critically

1. Search the Internet to find information on rain forests. Why are they so critical to the world? Report your findings to the class.
2. What other solutions to feeding the world's population are possible besides making better use of existing farmland?

CHAPTER 2   Why Study Food Science?

Objectives

- Describe personal benefits of studying topics in food science.
- Describe contributions of food science to increasing food supplies.
- Explain the role of food science in preserving the environment.
- Explain contributions of food science to nutrition and food safety.
- Relate food science to social change and technological advances.

Terms to remember

biodiversity
entry-level jobs
integrated pest management (IPM)
sustainable farming

Suppose you want to make a fruit salad, but the peaches aren't ripe. If you put them in a paper bag, close it securely, and let the bag sit for a day or two, the peaches will ripen and sweeten. Peaches naturally produce ethylene gas, which speeds up ripening. By trapping the gas in a bag, ripening occurs more quickly. Knowing this simple principle of food science can come in handy.

Food science principles have many impacts. It's nice to know how to ripen a peach quickly, but it's essential to know how to prepare and store food so that it won't cause illness. Society uses food science information to get food to those who need it around the world and to keep food supplies safe. From simple applications in the kitchen to the benefits experienced by a community, food science offers a huge body of useful information. That alone makes it worth studying.

Personal Benefits

On a personal level, your study of food science will be helpful to you. The benefits can start at your next meal and continue well into the future.

Protecting Your Health

Suppose you overheard this comment from a teen in the school lunch line: "My parents keep telling me to eat right, but I know what I like. I'd rather eat pie for lunch any day than vegetables." The advantage you have over this student is that you're taking food science. People who learn the chemistry behind eating right don't need to be told how to eat. They have a better way to make decisions.

You've probably heard that you need vitamins in your diet, but do you know just how they function? Can you explain scientifically why experts advise against eating too much saturated fat? After studying the science of foods and nutrition, you'll have answers to questions like these. Chances are, you won't have to be coaxed to make smart food choices because your own knowledge of the facts will be much more convincing.

The foods of different cultures vary nutritionally. Beans, which are a mainstay in Hispanic cultures, supply good protein, especially when combined with rice. A diet that relies heavily on rice and fish, as in Japan, tends to be low in fat. Information from nutrition scientists helps people from each culture determine what else is needed to make their diet healthful.

Understanding Food Safety

Suppose you're at a picnic with friends. You notice that the potato salad has been sitting out for a while, and everyone is headed for a game of volleyball. You take the salad and put it back on ice in the cooler for storage until the group eats again. No one will ever know for sure, but you might have saved them all from illness.

Careless handling can make a food more harmful than healthful. People who study food science learn what it takes to keep the food supply safe for themselves and others. You can use such knowledge now and in your own family someday.

Gaining Practical Skills

Understanding the chemical processes at work in the kitchen can be quite useful. Even if you never become a professional chef, knowledge of food science can increase your cooking skills.

You'll know why certain food combinations produce the reactions they do. You might decide to put a little raw pineapple with some meat in the refrigerator overnight to tenderize the meat before making stew. You might store green bananas on the counter after learning that cold air in the refrigerator causes cell damage that prevents ripening.

An understanding of food science may encourage you to expand your skills in the kitchen. Knowing how and why egg whites whip into a foam, for instance, can inspire you to make a soufflé. As knowledge builds your confidence, you become more willing to try something new.

Exploring Careers

Does the idea of learning how microwaves affect food sound interesting to you? Would you rather

Storing food correctly has economic, health, and safety advantages. What would those advantages be?

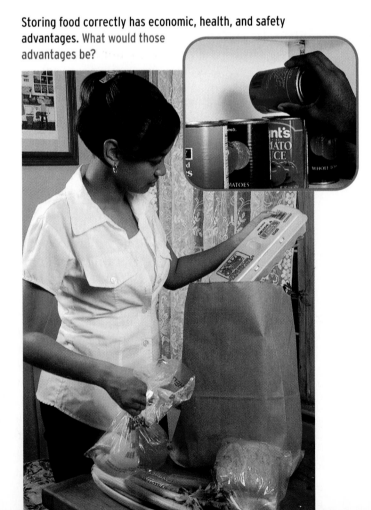

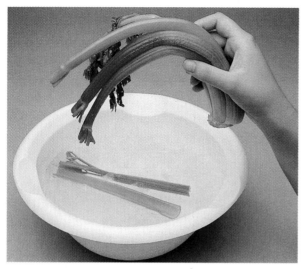

Celery that has lost its crispness won't be pleasing on a relish tray. If soaked in ice water, however, the cells of the celery absorb water and become crisp again. You and your classmates might like to create a booklet of practical tips that you collect from family, friends, and research. Look for the scientific principles that explain each idea.

You can find food scientists in many settings. In the space program, a scientist experiments to find food products and packaging that can go into space. A food scientist who works for an airline makes decisions about what foods to serve in flight and how to preserve and transport them. Some scientists test foods for contaminants in their work for food manufacturers.

Whether your interests lie in chemistry, nutrition, athletics, or baking, they might be applied in a food-related career. After all, whatever else people do, they still have to eat.

Social Impacts

Food science is a kind of "current event." With its many applications, food science is a major

plan a menu for residents of a nursing home or advise professional athletes on how to eat to reach top performance? Perhaps you'd like to be a scientist who develops new food products that protect against heart disease. Because the field of food science is so varied, many different careers are possible. Throughout this book, you'll be introduced to people who use food science in their work. You may even find a career that's right for you.

Some careers don't require a college degree. **Entry-level jobs**, *those that don't require experience or training*, are often found in the food service industry. You could work in a restaurant or a hospital kitchen, for example. Which chapters in this text do you think would be most useful in food service careers?

Many careers in food science require a college degree. Scientists, family and consumer scientists, food editors, and many others have at least a bachelor's degree (four years after high school). People who do scientific research often take additional years of coursework to get an advanced degree, a master's or a Ph.D.

Many careers spring from entry-level positions. You can learn about foods by working where food is prepared. Some people go on to school to get the education and training they need for a specific career.

Many colleges and universities today offer programs in food science. By contacting them, you can find out what these programs are like. Try looking for information on the Internet by searching for food science or particular universities.

force that shapes the world today. Learning how food science can influence, and especially improve, the world situation is critical to the survival of a healthy world population. By studying food science, you may learn what you can do—now as a student and perhaps later in a career—to help bring about the solutions that food science has to offer.

Food for the World's People

Did you know that the world's population is growing by about 250,000 people every day? Think about this: over one-quarter of the world's children under age five are malnourished. To food scientists, these statistics translate to concerns about the world's food supply.

In addressing global food issues, scientists and others point to several problems. For one thing, much food is wasted in more prosperous areas of the world. Better preservation and distribution methods could help get food to where it's needed. Also, as populations grow, fertile farmland is lost to housing. In some countries, farmland isn't put to the best use. Until solutions are found for these complicated problems, hunger in the world will continue. The work of food scientists can make a difference.

In some areas of the world, food is scarce. People in these areas need information and resources that will help them get the most from the land they use to grow crops and raise animals for food.

Housing developments and apartment complexes keep spring-ing up in order to accommodate an ever-growing population. How does this affect the food supply? What other effects does an increasing population have on people and communities?

studying the nutritional needs of specific popu-lations, nutrition scientists can determine which vitamins and minerals should be added to foods in those areas.

The safety of the food supply has had more attention lately, especially as food trade grows among nations with different safety standards. That's one reason you'll find food scientists in processing plants on large fishing vessels today. They help ensure the safety and quality of the fish caught even though the ship may remain at sea for weeks. This and many other efforts are made to ensure safety as foods travel around the world to reach new destinations.

Environmental Impacts

Imagine using disposable utensils to eat a meal—and then eating the utensils. These ex-perimental utensils, made from easily degrad-able plant starches, are one example of applying food science to protect the environment.

Environmental protection goes hand in hand with agricultural improvement. Breeding hardier plants reduces the need for chemical fertilizers and pesticides. This is an economical benefit as well, especially where chemical addi-tives are expensive or unavailable. Also, raising plants that need less water reduces the amount of salt residue left by irrigation.

Concerned about the environmental effects of using fossil fuels, scientists stress energy effi-ciency in all areas of food use. Research contin-ues into energy- and cost-saving technologies for processing, storing, and transporting foods. With newer packaging, foods can be stored without refrigeration. This is an asset in places where refrigeration is too costly to use. Likewise, because dried foods are very light-weight, they cost much less to transport long distances. You'll learn more about these and other types of processing in later chapters.

Advances in biotechnology are having a pos-itive effect on the food supply. Scientists have developed tougher strains of traditional food crops. Some are more resistant to disease and pests. Others have a shorter growing season or tolerate poor soils or irregular rainfall better.

Biodiversity

Some genetic engineers concentrate on **biodi-versity**, which is *cultivating a variety of plants and animals.* Biodiversity offers pro-tection against "putting all your eggs in one basket." In other words, there is less reliance on a few food sources.

Technologists promote biodiversity by pre-serving established plant and animal sources as well as breeding new ones. When creating new strains of organisms for a few selected traits, scientists try to retain the hardiness that a plant or animal has developed over the generations.

Public Health and Safety

While expanding the food supply is an impor-tant goal, food scientists also aim to make the food supply the best it can be nutritionally. Raising cattle instead of crops, for instance, adds protein and minerals to the local diet. By

FOOD SCIENCE Careers

Food Scientist

What careers are in food science? Ask a . . . Food Scientist

I didn't even know food scientists existed," Lauren Ingersoll says, "until I met someone in the field. Combining my interest in science with my lifelong love of food seemed like a good match for me. Now in my work for a food producer, I experiment with different vitamin mixes to get the right blends for different food products.

"Not all food science careers are alike. Many food scientists use scientific research and engineering to create new and improved foods. Some develop ways to process and distribute foods. Working with engineers, flavor experts, and packaging and marketing experts is part of the job. As the field continues to grow, the range of careers will just keep expanding."

Education Needed: Food scientists typically need a bachelor's degree in food science or human nutrition, with coursework in biochemistry and statistics. Companies need qualified professionals in such areas as biotechnology, chemistry, chemical engineering, medical technology, and data analysis.

KEY SKILLS
- Detail-oriented
- Organizational skills
- Communication skills
- Laboratory skills

College programs may have internships that provide on-the-job experiences.

Life on the Job: The daily routine of a food scientist is as varied as the many positions. Laboratory work, often managed alone, is common. Detailed research and experimentation involves the physical, chemical, and microbiological composition of foods. Teamwork and contact with the public come with some positions.

Your Challenge

Suppose you're exploring careers in hopes of deciding what work you would like in the future. A career in food science sounds interesting.

1. What qualities do you have that might fit a food science career?
2. Write five questions about food science careers that you want to answer during this course.

Since hunger is a problem in the world, food waste deeply concerns many people. Waste can be managed on both a large and small scale—by the food industry and in personal lives. What steps could be taken in each?

tainable farming practices are old-fashioned techniques gaining new favor. Other examples include composting plant waste and rotating crops from one growing season to the next.

Your Study of Food Science

If you had looked for a high school course in food science ten years ago, you would have been less likely to find one than you are today. More and more, educators are seeing the need for students to learn what a scientific approach to the study of foods can teach.

This text offers an overview of the topics and issues at the heart of food science. As you study, you'll learn what makes one food different from another. You'll explore why people form food preferences. You'll see science at work in everyday life.

Newer, more efficient processing also reduces food waste. More food can be preserved longer and stored under challenging conditions. Thus, it can be transported greater distances without spoiling, for access to more people.

Sustainable Farming

Some food scientists develop techniques for **sustainable farming**, or *producing food by natural methods that fit with local needs and conditions.* Instead of heavily working and eroding the land, a farmer might raise native plants that grow easily. Likewise, a farmer might graze dairy cows on land that can't support crops. In both situations, food scientists research new uses for the products.

One practice central to sustainable agriculture is **integrated pest management (IPM)**, which *controls pests with nonchemical deterrents.* As the name suggests, IPM is a system-wide approach. A variety of natural controls helps to manage the land, not overwhelm it. Instead of spraying pesticides, harmful insects live along with their natural enemies, keeping the "bad bugs" in check. That and other sus-

Just as farmers use techniques to protect the land, home gardeners do too. Some people create a compost pile. Once established, this mixture of decaying organic matter can be used to fertilize and condition garden soil.

Nutrition Link

Nutrition and the Internet

As you study nutrition, you'll investigate a huge amount of information—and misinformation. To judge the worth of nutrition news, from the Internet or elsewhere, use the following guidelines:

- Learn the names of respected authorities, such as the American Dietetic Association (www.eatright.org) and the *New England Journal of Medicine* (www.nejm.org).
- Accept limited claims. Good science is careful, detailed work. The claim, "whole grains are linked to lower cancer rates," is less exciting but more likely to be true than "whole grains prevent cancer."
- Look for verification. The more studies and sources that support a claim, the more you can trust it. Examine a variety of sources. Look for nutrition journals that are peer reviewed, meaning the researchers' colleagues accept the soundness of the study and its results.
- Value the test of time. Findings from 20 years ago may be more reliable than recent studies, especially when accepted after retesting with today's technology.

Applying Your Knowledge

1. Why do you think the field of nutrition is so prone to misinformation?
2. An Internet "nutritionist" claims that a "secret" herbal supplement is proven to cure a variety of minor complaints. What is your impression?

As part of your study, you'll carry out laboratory experiments. These experiences will show you firsthand how foods and substances in them react under certain conditions. You may be the family member "in the know" at home when future meals are prepared.

Why study food science? The information you gain will serve you well throughout life. If that's not enough, you may go much farther than this course alone. A volunteer at a food bank? A community leader who cares? A career in food science? So much is possible.

Studying food science is fun. Many students enjoy the experimentation that brings science principles to life.

Food Labels and Nutrition

Nearly all packaged foods sold in the United States are required by law to display nutritional information on their containers for the benefit of consumers. You can find this information on the Nutrition Facts panel typically printed on the side or back of a package. Food scientists have contributed to the formulation of this information.

Although you'll learn more in Unit 4 about the meaning of terms on a Nutrition Facts panel, you can start to become familiar with the data now. To begin evaluating the contributions of food science to your diet, compare the labels on two products that appear regularly at many breakfast tables.

Equipment and Materials

Nutrition Facts panels from labels on two different food packages

Procedure

1. Obtain one Nutrition Facts panel from a container of oatmeal and one from a container of ready-to-eat cereal from your teacher. (Use other products for comparison if your teacher provides different ones.)
2. Looking at the sample data table below, create a similar table of your own on paper or on the computer.
3. Examine the Nutrition Facts panels on each product and record the information called for in the sample data table.

Analyzing Results

1. How do the foods compare in calories, total fat, and cholesterol?
2. Which food was higher in sodium and sugar?
3. Which of the two products was higher in vitamins and minerals?
4. How do you account for differences in nutritional value?
5. Based on your current understanding of nutrition, what advantages and disadvantages do you see in eating each food item?

SAMPLE DATA TABLE

Nutrient	Oats		Ready-to-Eat Cereal	
	Amount	% Daily Value	Amount	% Daily Value
Calories				
Total Fat				
Cholesterol				
Sodium				
Dietary Fiber				
Sugars				
Protein				
Vitamins*				
Minerals*				

*List separately.

Chapter Summary

- Understanding concepts in food science helps you take responsibility for your own eating habits and nutrition.
- A study of food science may add to your choices when considering careers.
- Because food is a universal need, food science is a view to understanding some global situations, problems, and possible solutions.
- Techniques in food science can be used to increase the quantity and quality of food available.
- Food scientists can provide the means to help sustain food production and the environment at the same time.
- Food scientists play a role in promoting food safety and good nutrition.

Using Your Knowledge

1. How can studying food science help you manage your physical health?
2. Besides health, what other personal benefits might you get from studying food science?
3. Do food science careers require a college education? Explain your answer.
4. Describe what the career of a food scientist is like?
5. What global issues might find solutions in food science?
6. Why is biodiversity helpful?
7. Why do the food safety procedures one country follows make a difference to the people in other countries?
8. Give three examples of how food science promotes environmental protection.
9. What is sustainable farming?
10. What guidelines can help you conduct accurate research on the Internet?

Real-World Impact

Focus on Hunger

World hunger is often blamed on "too many people." Another reason cited is "too much poverty." About 95 percent of all births each year occur in developing nations, which are often environmentally and politically unstable. Natural disasters, from flood to draught, hit these regions hardest. Inexperienced or unpopular governments cannot or will not help their people. Some even refuse aid, including the technology people need. Until governments and individuals commit to deep-seated change, the cycle of poverty will continue.

Thinking It Through

1. In some countries people lack the resources and expertise to use technology that could help them. How can this problem be solved?
2. If people in a nation receive assistance without requiring political and economic reform, what positive and negative results might result?

Skill-Building Activities

1. **Critical Thinking.** Amber and Renaldo are both interested in food science careers. They want to learn more before making a decision. What volunteer activities and part-time employment ideas would you suggest to them? How else might they gain information?

2. **Science.** Research the development of one food through food science. How has it been changed to be more available, appealing, nutritious, or profitable to produce? Examples include tomatoes, meats, and popcorn.

3. **Critical Thinking.** Locate a newspaper, magazine, or Internet article concerning a current issue in health or nutrition. Analyze the situation from a food scientist's point of view. What concerns, questions, or solutions might a food scientist bring to the discussion? Share your ideas with classmates.

4. **Teamwork.** With a partner, identify five foods that you think are unappealing. If you were food scientists, how would you make these foods more acceptable to consumers? List two ideas for each food item.

5. **Social Studies.** In small groups, discuss the possible economic, environmental, and social consequences of these situations: a) A large company sells genetically engineered seed corn that cannot be saved for replanting, but must be bought every year; b) Concerned about the unknown effects of biotechnology, citizens of some countries persuade their governments to ban the importing of genetically altered foods.

Understanding Vocabulary

Discuss with your class why sustainable farming techniques are especially important in underdeveloped countries.

Thinking Lab

Food Science Careers

Food science reaches into every aspect of the food supply, from food service, to food processing, to nutrition. Each application of food science represents a variety of careers. Many of these careers may be unfamiliar to you at this point in your life.

Analysis

Choosing a career can be difficult when you don't know what positions exist in the world of work. Food science may surprise you. Every day people in this field are involved in work of all kinds. By studying food science and the careers it offers, you may find a career goal that you had never considered before.

Organizing Information

1. Use the Internet to research jobs in food science. Select one and write a job description. Include annual salary, typical responsibilities, and employment opportunities. Share this information with your class.

2. Visit the web site of a university with a food science and technology department. Learn what courses are needed to major in food science. Also find out what courses are needed to major in nutrition.

THE ELECTRONIC NOSE

In food science the human nose has commonly been used to evaluate the odors and flavors of new food products. The invention of the electronic nose, however, provides an alternative. This device uses sensors that respond differently to certain chemicals in a food sample, much the way the human nose does—but the electronic nose is more objective. The "nose" allows for fast analysis and reliable results. Food companies use the "nose" to check deterioration during shelf-life studies, monitor products during transport, and ensure that packaging odors don't contaminate products. Electronic noses can verify that Parmesan cheese is authentic and precisely control the roasting and blending of coffee. They can also test the freshness of fruit from harvest to sale.

UNIT 2 The Food Science Lab

TechWatch Activity
Use the Internet and other resources to learn more about the electronic nose. Prepare a sensory demonstration on the use of this technology in the food industry.

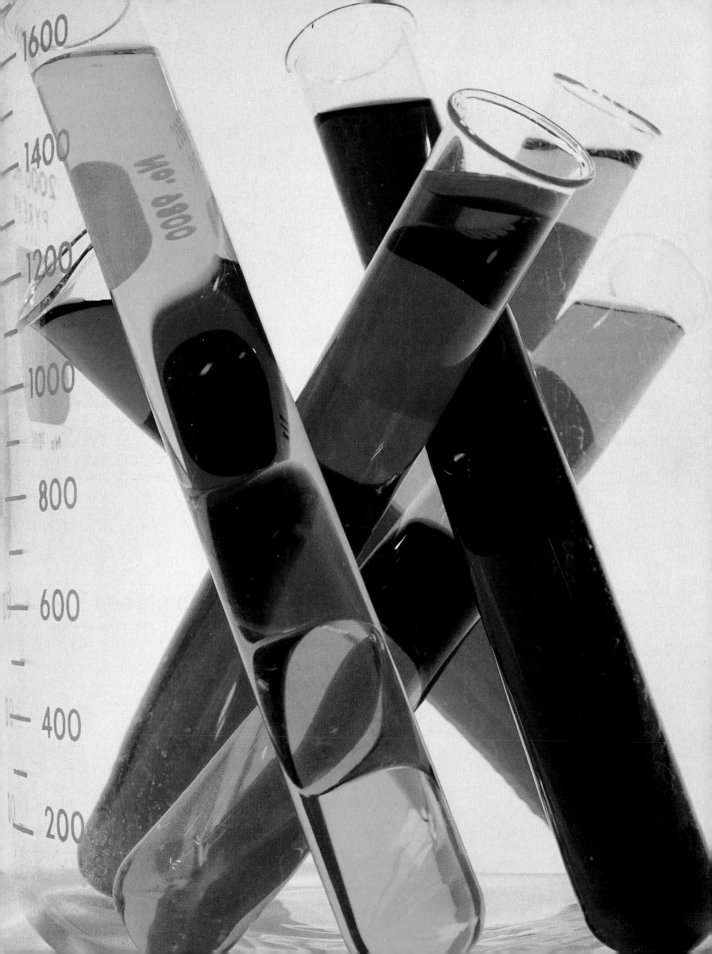

Using Laboratory Equipment

Objectives

- Choose laboratory equipment that is suited for specific tasks.
- Demonstrate proper use and maintenance of laboratory equipment.
- Demonstrate techniques for working safely in a food science laboratory.

Terms to remember

balance
beaker
buret
calibrate
Erlenmeyer flask
graduated cylinder
insoluble
meniscus
tare

Sewing a shirt is no simple task if you don't know how to thread a sewing machine. Making an oak table isn't easy either if you can't operate a saw. In any specialized field, success—and often safety—depend on how well you manage "the tools of the trade."

The "tools" you need to be familiar with in food science are the same ones you would find in any chemistry or biology lab. Most of them are found in actual industrial food science labs too. To make your efforts worthwhile, you need to know how to use this equipment properly. Equally important, you need to learn the rules for equipment care and your safety. This chapter serves as a handbook to help you manage successfully in the food science lab.

Laboratory Equipment

Every experiment in the laboratory requires a certain set of supplies and equipment. Some of these are fragile and expensive to replace. With each experiment, your instructor will show you techniques for using equipment correctly.

Many laboratory items have specific purposes. Using them as intended may take varying degrees of expertise. Mixing substances with a stirring rod is a basic skill. Using a funnel to pour without spilling takes a bit of care. With practice using a microscope, you'll identify things the eye can't see alone. You'll want to pay close attention when heating substances on a laboratory burner like the one in **Figure 3-1** on page 50.

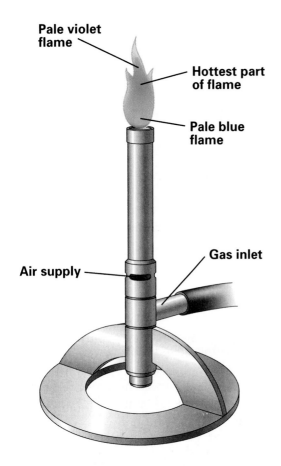

Figure 3-1
In a laboratory burner the incoming air and gas must both be adjusted to get the correct flame.

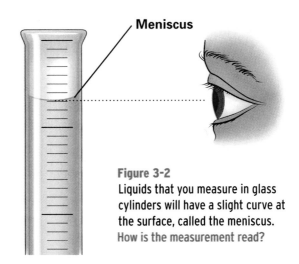

Figure 3-2
Liquids that you measure in glass cylinders will have a slight curve at the surface, called the meniscus. How is the measurement read?

Containers and Holding Devices

As you can imagine, many substances used in experiments are not meant to be hand-held. Different types of containers are basic to lab work. Common containers, some of which are displayed in **Figure 3-3**, include:

- **Beakers.** A **beaker** is *a glass container that has a wide mouth and holds solids and liquids.* Beakers come in different sizes. A beaker isn't intended for the most exact measurements; use it only for measuring approximate amounts.
- **Graduated cylinders.** A **graduated cylinder** is *a tall, cylindrical container used for measuring the volume of liquids.* Use a graduated cylinder for exact measure-

ments. The volume of liquid is read from the **meniscus** (muh-NIS-kus), which is *the bottom of the curve a liquid forms in a container.* To read, place the cylinder on a flat surface at eye level as illustrated in **Figure 3-2**.

- **Test tubes.** A test tube is a cylindrical container that holds a small quantity of a solid or liquid. Several test tubes are conveniently held in a test tube rack. Test tube brushes are designed to wash test tubes and other narrow equipment.
- **Erlenmeyer flasks.** An **Erlenmeyer flask** (UR-lun-my-er) is *a cone-shaped container with a narrow neck and a broad, flat bottom.* Use it to swirl liquid without spilling.
- **Burets.** A **buret** (byur-ET) is *a long, thin cylinder, which is marked to 0.1 of a milliliter.* Use it to transfer exact amounts of liquid in experiments.
- **Petri dishes.** A petri dish (PEA-tree) is a shallow dish with a loose-fitting cover. In this course, you'll use a petri dish to grow bacteria.
- **Holding devices.** To hold equipment, you'll use ring stands, iron rings, and clamps. A ring stand is a metal pole mounted in a base to which a clamp or iron ring can be attached. Iron rings often hold beakers and flasks, while clamps support test tubes, burets, and also thermometers.

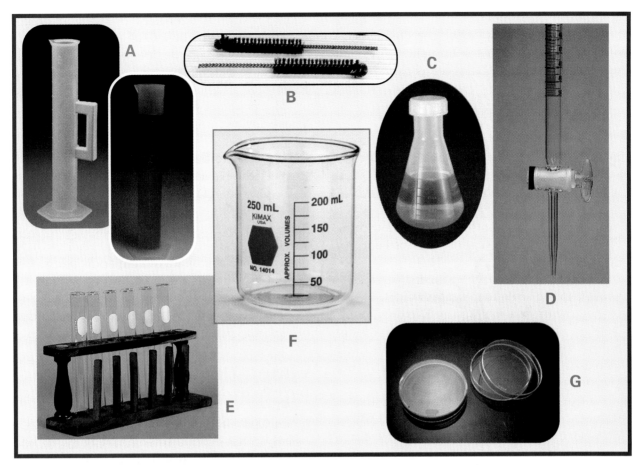

Figure 3-3
Here you see the following equipment: A. graduated cylinders; B. test tube brushes; C. flask; D. buret; E. test tubes and rack; F. beaker; and G. petri dishes. How would you use these items?

Thermometers

You've probably used a thermometer at home to see whether your freezer was cold enough or if you were running a fever. During experiments, you'll use a different thermometer to find the temperature of substances, as shown in **Figure 3-4**.

Many home thermometers, such as those used in cooking appliances and in candy making, are marked with the Fahrenheit scale. Laboratory thermometers use the Celsius scale, which will be described in a later chapter.

Medical thermometers are intended to record temperatures within a very limited range; 106°F is as high as they need to go. A

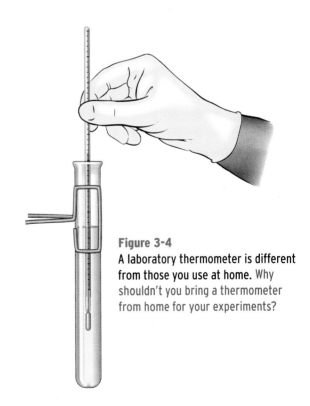

Figure 3-4
A laboratory thermometer is different from those you use at home. Why shouldn't you bring a thermometer from home for your experiments?

laboratory thermometer must handle much higher temperatures—above the boiling point of water. A thermometer brought from home will quickly explode at the temperatures a lab thermometer routinely handles. Also unlike a medical thermometer, a laboratory thermometer is never shaken to lower the reading.

Balances

A **balance** is *a scientific instrument that determines the mass of materials.* Mass is a measure of the amount of matter in a sample. Finding the mass of a substance placed on a balance is known as "massing" it.

In contrast, the term "weighing" applies to using a spring scale to measure weight. Weight is the gravitational pull of the Earth on an object. An object's weight decreases as the distance between it and the center of the Earth increases. That's why astronauts are weightless in space.

Unlike weight, mass is constant. No matter where a particular object is measured, it will always have the same mass value. While you may see and hear the terms weight and mass used interchangeably, "mass" and "massing" are used in this text when referring to the use of a balance.

To mass a substance, you need a container to hold it. How do you find the mass of a substance but not its container? You use a procedure called **taring**. *First mass the empty container; then subtract that value from the mass of the substance and the container together.*

Triple-Beam Balances

A triple-beam balance, shown in **Figure 3-5**, has a platform on the left. This platform, called a pan, holds the object or material to be massed. To the right are three arms, or beams, that join together at a pointer. The beams hold riders, which are weights of various sizes that begin at zero. The pointer swings freely up and down in front of a scale that has a zero line in the middle. When the pan is empty and all the riders on the beams are set to zero, the pointer

should rest on the zero line or swing an equal distance above and below the line.

Electronic Balances

Electronic balances that give digital readouts are also commonly used in science laboratories. These devices are described on page 57.

Caring for the Balance

Respect the balance as a delicate—and expensive—instrument. Never attempt to mass items heavier than the balance can measure. Massing liquids obviously requires a container; solids too should be massed either in a container or on a piece of massing or waxed paper. *Never mass chemicals directly on the pan of the balance.* If you spill any solid or liquid chemical on the balance, clean it off immediately. This will help prevent the pan from rusting or corroding.

Calibrating Measuring Equipment

When you use equipment for measuring, how can you be sure the measurements are accurate? To ensure correct readings, instruments are calibrated. To **calibrate** (KAL-uh-brayt) means *to check, adjust, or standardize the marks on a measuring instrument.*

The finer an instrument's calibration is, the better it measures. For example, a ruler marked in centimeters can measure length accurately only to that point. A ruler calibrated in millime-

Figure 3-5

This is a triple-beam balance. When carrying a balance, place one hand under the balance and the other hand on the beams' support.

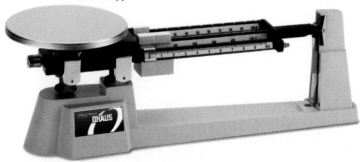

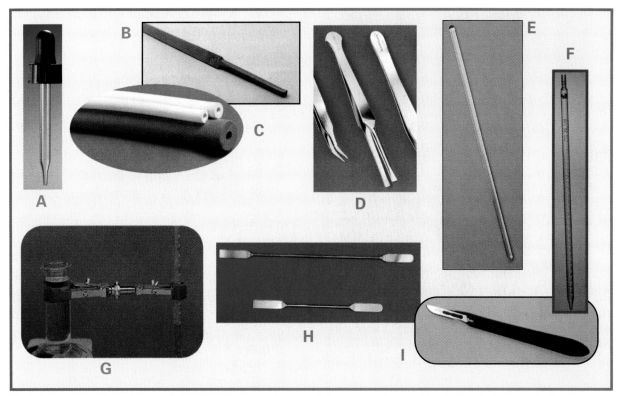

The lab equipment shown here includes these items: A. dropper; B. triangular file; C. rubber tubing; D. forceps; E. stirring rod; F. thermometer; G. clamp; H. spatulas; and I. scalpel.

ters, a smaller division of centimeters, gives a greater degree of measuring accuracy.

Some equipment needs calibration each time it's used. Because exact temperature is very important in making candy, the thermometer must be calibrated before you begin. To do this, first place the thermometer in boiling water and read the temperature. Since water typically boils at 100°C, note how far above or below 100°C the reading is. This is the adjustment you'll need to make when using the thermometer for candy making. If your thermometer shows water boils at 102°C, for example, you'll need to add two degrees to any temperatures in your candy recipe.

Laboratory Safety

In any science laboratory, people work with potentially dangerous substances, procedures, and equipment. Chemicals, especially, require careful handling. You can prevent harm to yourself and others by following the guidelines in the rest of this chapter.

Preparing for an Experiment

When an experiment is scheduled, prepare in the following ways:

- Read the procedure carefully *before* coming to class. Ask questions before you begin the experiment rather than hoping to "figure it out as you go along."
- Dress with safety in mind. Don't wear a coat, jacket, or sweater with bulky sleeves that could knock over equipment. Tie back long hair. Wear safety goggles or an apron if instructed, and for as long as instructed.
- Check that you have all the needed equipment and supplies. You'll save time and achieve better results.
- Wash your hands before beginning any food experiment.

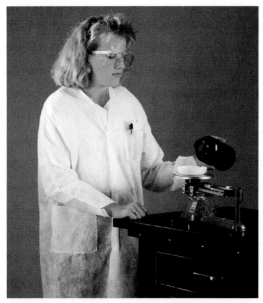

"It's better to be safe than sorry." This warning has been around for years, and it's still just as meaningful. What safety rules would you set for working in a laboratory?

- If you sneeze, cough, or touch your face or hair during an experiment, wash your hands before continuing.
- Use a cutting board for all cutting to prevent sliced fingers and scratched counters and tables.

Flames and other heat sources are an added element of risk. Remember the following precautions when using heat:

- Handle hot glass with tongs, a potholder, or heat-resistant gloves. Use potholders for hot pans.
- Don't reach over lighted gas burners or containers on a hot plate or range.
- Use a metal trivet when you heat a beaker on a range element.
- Place a wire gauze between glass and an open flame when heating a beaker or flask over a laboratory burner, as shown in **Figure 3-6** on page 56.

Working in the Laboratory

Laboratory work should be interesting and enjoyable. To make it that way, take the work seriously and avoid actions that could lead to accident or injury.

A logical, organized manner makes laboratory work easier and more efficient. You're less apt to feel rushed and risk carelessness when you take a systematic approach. Potential safety hazards are eliminated. The guidelines below will help you create a well-ordered environment.

- Keep your workspace uncluttered.
- Wash items and put them away as you finish using them.
- Dispose of paper materials and garbage promptly and properly.
- Keep drawers and cabinet doors closed except to remove or return items.
- Stay at your assigned location as much as possible.
- Read instructions and measure substances carefully for safety and accuracy.

The following rules stress personal hygiene:

- Wash any substances from your hands immediately; never lick your fingers.

Sometimes you will separate a substance from a liquid by filtering. Filter paper is folded into a cone and then placed in a funnel. The substance catches in the paper when you pour the liquid through the funnel.

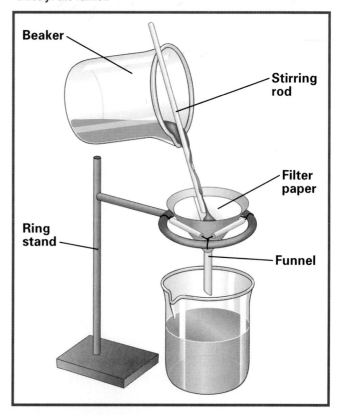

Lab Technologist

How is food quality maintained?
Ask a . . . Lab Technologist

A s a lab technologist for a juice producer, Dean Mobley tests products for quality. "In the lab we run chemical tests to check for everything from manufacturing mistakes to contamination," Dean explains. "Careful work is critical in the laboratory. The lab is my second home. I know the equipment there better than some things in my own home. To do the job right, we have to keep our equipment in good condition.

"For us, juice production isn't just a matter of making a good-tasting drink. The product has to be safe too. An inferior product affects sales, but a tainted product can also affect health. We can't allow that to happen."

Education Needed: Many food science laboratory technologists have a bachelor's degree. However, positions exist for high school graduates who do well in math, chemistry, and biological sciences. Employers are especially interested in people who have laboratory experience and know how to conduct experiments and use the equipment.

KEY SKILLS
- Laboratory skills
- Record keeping
- Analytical skills
- Mathematics
- Keen senses
- Communication

Life on the Job: Laboratory technologists carry out experiments using standard laboratory equipment and a PC. In food science they often use test cultures to ensure that foods and additives meet government regulations and company standards. They also test foods the old-fashioned way—by observing, sniffing, and tasting.

Your Challenge

Suppose your boss hasn't asked for your opinion about what type of microscope to buy. As a technologist, you use the equipment every day, and you'd like to be included in the decision.

1. Would you try to become involved in the decision? If so, how?
2. If you don't agree with the model selected, how would you respond?

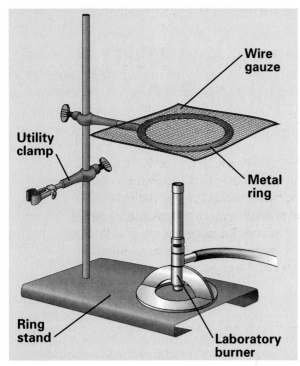

Figure 3-6
Depending on how a ring stand is set up, it can support many different experimental procedures. Here, wire gauze sits on an o-ring above the laboratory burner.

Labels: Wire gauze, Utility clamp, Metal ring, Ring stand, Laboratory burner

- Never heat an empty beaker or allow the liquid in one to completely boil away; the beaker will break.
- Never leave anything that is being heated unattended.
- When heating substances in a pan, turn the handle toward the back or center of the counter or range to reduce the risk of knocking over the pan.
- To prevent steam burns, remove the lid from a hot pan by tilting it away from you.

After the Experiment

As with working in a kitchen, a good cleanup promotes safety in a laboratory. After completing an experiment, follow these guidelines:

- Return all equipment to its proper place.
- Turn off all burners, hot plates, and range elements.
- Thoroughly clean your work area.
- Wash your hands well.
- Place paper towels and other disposable items in the designated receptacle.
- Never discard any water-insoluble substances in a sink. **Insoluble** (in-SAHL-yuh-bul) materials *don't dissolve*. Fat and egg shells are two materials that are insoluble in water. Wrap such substances in paper towels and place them carefully in the wastebasket.
- Clean up spills immediately.
- Wash any utensil dropped on the floor before reusing it.
- Sweep broken glass into a dustpan, preferably with a brush or broom. If you must use your hands, protect them with a paper towel. Every laboratory should have a special container for broken glass.
- Immediately report all spills, fires, cuts, and any other accidents to your teacher. Follow the first-aid and clean-up instructions you're given.

At all times during any experiment, stay alert. Suppose you pick up a beaker, not knowing that your laboratory partner just heated it. Hot glass looks just like cold glass. Unless you are aware of what's going on, you could get a burn. A serious approach to every experiment keeps you safe and leads to reliable results.

Accidents can happen in the laboratory. If glass breaks, you need to clean up the broken pieces immediately. What are the safe ways to do this?

Using an Electronic Balance

SAFETY FIRST

Review these safety guidelines before you begin this experiment.

One of the first skills you need to learn in food science is how to use a balance, the instrument used to mass materials. An electronic balance has a pan similar to those found on a triple-beam balance, but no arms or beams. On the front of most electronic balances are two buttons called sensors. One is labeled "function" and the other is labeled "tare." Don't press the function button unless instructed to do so by your teacher, since it will change the units in which the mass of your sample will be reported.

The tare button will be very useful when you mass a sample on paper or in a container. Pressing this button sets the balance to zero. To use an electronic balance, tare it when empty and again after massing the paper or container you'll use. Then place your sample on the paper or in the container and read its mass. You will learn to use the tare button in conducting this experiment.

Equipment and Materials

electronic balance
3 items to mass

5-mL measuring spoon
straightedge spatula

table salt

Procedure

1. From your teacher, obtain three items to mass.
2. Place one of the items on the balance pan or platform. Read the mass of the item on the screen at the front of the balance.
3. In a data table similar to the sample data table shown below, record the name of the item and its mass.
4. Mass each of the other items and record the information in your data table.
5. Place a piece of massing or waxed paper on the balance pan or platform. Press the tare button. The balance will cancel the mass of the paper so that the screen will read zero.
6. Fill a 5-mL measuring spoon with table salt. Level it with a straightedge spatula.
7. Pour the salt onto the paper. Read the mass of the salt on the screen.
8. Record the mass of the salt in your data table.

Analyzing Results

1. How does the mass of each item you used compare to its size?
2. In what other aspects of food production do you think taring is used?

SAMPLE DATA TABLE

Item	Mass
5 mL table salt	

Chapter Summary

- To work in a food science laboratory, you need to be able to identify, use, and maintain laboratory equipment.
- Containers used in the laboratory are often chosen for the task and for their degree of accuracy.
- Laboratory thermometers are specially designed for laboratory use.
- Substances may be massed using a triple-beam or electronic balance.
- Laboratory equipment may need to be calibrated in order to yield correct results.
- Planning and organization set the stage for safety in laboratory work.
- Using good safety habits is important in conducting successful food science experiments.
- Following an experiment, equipment should be returned to its proper storage location and waste materials disposed of properly.

Using Your Knowledge

1. Compare a graduated cylinder with a buret.
2. How is the meniscus used when measuring substances?
3. Give two reasons why a home thermometer may not be suited for laboratory use.
4. What three suggestions would you give for taking care of a balance?
5. How does the measurement of mass differ from the measurement of weight?
6. What will result if you forget to tare a scale before massing a substance?
7. What might happen over time if an instrument is not calibrated as needed?
8. What does it mean to be well prepared to start a food science experiment?
9. How does an orderly environment promote safety? How can you create such a setting?
10. Give some safety tips for boiling water in a beaker over a laboratory burner.
11. Describe proper disposal of three different types of materials in the food science laboratory.

Real-World Impact

Revolution in Glass

Tempering is a 150-year-old process that strengthens glass by heating and cooling the molten material. In the early 1900s, scientists learned that the outer surface of glass cools more quickly than the inner one, leading to compression and strength. Heating and chilling in air, however, created uneven strength, which could cause a product to shatter explosively under the temperature extremes of laboratory use. After much experimentation, scientists found that molten chemical salts heat and cool glass uniformly. In 1936, Corning, Inc., patented its tempered glassware under the trademark Pyrex®.

Thinking It Through

1. How do you think tempered glass has advanced scientific research?
2. For what other purposes might tempered glass be used?

Skill-Building Activities

1. **Science.** How would you mass 10.0 g of table salt on a balance?
2. **Communication.** Develop a slogan to promote safe laboratory practices or proper use of equipment. Create a poster or flyer and post it in a helpful spot in the food science laboratory.
3. **Critical Thinking.** According to an old saying, "A stitch in time saves nine." In other words, taking action now can avoid problems in the future. Give three specific examples of how this adage can be applied to working in a food science laboratory.
4. **Problem Solving.** Suppose you have been assigned a partner for work in the food science laboratory. After doing two experiments together, you are concerned. Your partner always wants to rush through the experiment in order to be the first done. Safety practices are skipped, and you don't think your results are as accurate as they should be. What will you do?
5. **Problem Solving.** Suppose you notice a crack in a test tube that you used in the experiment you just finished. You don't see how you could have been responsible. What will you do and why?

Understanding Vocabulary

Write fill-in-the-blank statements using the "Terms to Remember." Exchange papers with a classmate and complete the sentences.

Thinking Lab

X Marks the Spot

For student and professional alike, working in the food science laboratory is essential to gaining a fuller understanding of the chemistry of food.

Analysis

At any level, working safely in a laboratory involves familiarizing yourself with the work setting as well as with experimental procedures.

Representing and Applying Data

Using your lab station as a vantage point, draw a map of your food science laboratory. Identify the location of the following: exits; fire extinguisher; fire blanket; disposal container for broken glass; wastebaskets; safety-goggle storage; apron storage, and any other important locations. Post or store this map where your teacher designates.

Objectives

- Demonstrate how to make accurate and precise laboratory measurements.

- Distinguish between metric units of length, mass, and volume, and the prefixes used with them.

- Compare temperatures on the Celsius and Fahrenheit temperature scales.

- Demonstrate techniques for taking length, volume, mass, and temperature readings.

Terms to remember

accuracy
decimal system
metric system
precision

I f an experimental procedure instructed you to prepare a salt-water solution, is it likely that you would add the same amount of salt as other students in your class? Would you even prepare the same volume of solution? This same sort of confusion would occur if a recipe instructed you to add sugar to whipped cream until sweet. How sweet? Whose definition of sweet would you use?

Scientific study, like cooking, relies on definite measurements. The very word *science* implies exactness. To obtain meaningful experimental results and to develop a procedure that can be duplicated, measurements must be both precise and accurate.

Accuracy and Precision

Accuracy and precision mean the same thing to most people, but to a scientist, they are two separate concepts. Both are vital to successful measuring.

Accuracy is *how close a single measurement comes to the actual or true value of the quantity measured*. If you need 10 grams of table salt, and the amount that your balance registers as 10 grams truly has that mass, then the measurement is considered accurate.

Precision is *how close several measurements are to the same value*. To attain precision, you need to make more than one measurement and compare the results. If you mass a quantity of table salt several times with similar results, you have measured precisely.

Your readings might be precise but not accurate if your balance isn't calibrated. Scientists need to work with calibrated equipment to be accurate.

Highway signs show distances in kilometers, and outdoor advertising signs flash the current temperature in degrees Celsius. Because these measurements are appearing more often in the United States, you need to become familiar with them. As a student of food science, you'll use them regularly in this course.

Metric Units

When measuring a quantity, a unit of measure is needed. In the English system, a unit that measures length is the inch. In the metric system, the unit for measuring length is the meter

The Metric System

What if a talented musician had some great song ideas but couldn't write music? Think how many good tunes might be lost because the musical ideas couldn't be written in the language designed for them.

All around the world a common language is used for measurement. It's called the **metric system**. Scientists routinely use the metric system in their work. Citizens worldwide also use this system in daily life. Beverages are sold by the liter. Food labels indicate mass in grams.

Figure 4-1

Metric Prefixes				
Prefix	Meaning	Numerical Value	Unit Example	Relationship to Unit
Kilo (k)	thousand	1000	Kilometer (km)	1 km = 1000 m
Deci (d)	tenth	0.1	Decimeter (dm)	10 dm = 1 m
Centi (c)	hundredth	0.01	Centimeter (cm)	100 cm = 1 m
Milli (m)	thousandth	0.001	Millimeter (mm)	1000 mm = 1 m

(m). Other quantities to be measured have specific metric units of their own. In this class you'll also use the gram (g) to measure mass and the liter (L) to measure volume.

Metric Prefixes

The metric system is based on the **decimal system**, in which *numbers are expressed in units of ten*. Prefixes are used to indicate what multiple or fraction of the base unit is used in a given situation. These prefixes can be combined with any unit of measure. For example, the prefix *kilo-* (KEY-low) means 1000. Just as 1000 meters is equal to 1 kilometer, 1000 grams equals 1 kilogram and 1000 liters equals 1 kiloliter.

You can use kilometers to describe a large number of meters, but how do you describe measurements that are smaller than 1 meter? The prefixes *milli-*, *centi-*, and *deci-* describe fractions of a meter. For example, a centimeter, is one one-hundredth (0.01) of a meter. **Figure 4-1** summarizes how these common prefixes are used.

Comparing Systems

Once you are comfortable with the metric system, you'll find it easy to use. Comparing the units used in the United States with metric units can help build your understanding. **Figure 4-2** shows you how some common metric units compare to the English system.

Figure 4-2

Comparison of Metric and English Systems			
Metric Unit	Symbol	Measurement Visualizations	English Equivalent
gram	g	a little heavier than a paper clip	.035 ounces
kilogram	kg	almost nine sticks of butter	2.2 pounds
meter	m	from the floor to about the doorknob	1 yard
kilometer	km	about two-thirds of a mile	62 miles
decimeter	dm	about the diameter of an average grapefruit	3.94 inches
centimeter	cm	length of a staple	0.39 inches
millimeter	mm	width of a dime	.039 inches
liter	L	just over a quart of milk	1.057 quarts
milliliter	mL	a little less than 1/4 teaspoon	0.03 fluid ounces

How Much Is a Serving?

To use the information in nutrition charts, you need to compare what you eat to the serving sizes listed. Measuring everything you eat is seldom practical. How could you visualize the serving sizes of foods? Practice thinking of the following comparisons:

- 125 mL (1/2 cup) of fruit, cereal, pasta, or rice = cupcake
- 85 g (3 ounces) of cooked meat, fish, or poultry = deck of cards
- 30 mL (2 tablespoons) of peanut butter = golf ball
- 42 g (1.5 ounces) natural cheese = 9-volt battery
- 1, 10-cm (4-inch) pancake = CD
- 1 medium apple or orange = baseball

Applying Your Knowledge

1. Why might a person overestimate the size of a serving of one food yet underestimate the same size serving of another food?
2. You can develop your own visual comparisons. Perhaps 250 mL of milk reaches the top of the letters on your personalized mug. A medium potato is about 12 cm long and weighs about 200 g. Name three items you might use for size comparison.

Recording Metric Data

As you conduct laboratory experiments, you'll be using the metric system to record data. By following certain principles, your results will be more reliable.

Estimating

A scientist must know not only how to measure correctly but also how to report findings precisely and accurately. The scales on laboratory measuring devices vary. A beaker may have markings for every 10 mL of volume, yet a 100-mL graduated cylinder is marked to the milliliter. As you use the scales on laboratory equipment, make it a habit to estimate to one decimal place past the unit of measure marked on the scale.

Suppose you're using a 100-mL graduated cylinder marked to whole milliliters, and the bottom of the meniscus is between 24 and 25 mL. You might report this as 24.3 mL or 24.8 mL, depending on how far above 24 mL the volume is. Anyone who reads your data table will know that the last digit is just an estimate, but it will tell them whether the volume was closer to 24 mL or 25 mL.

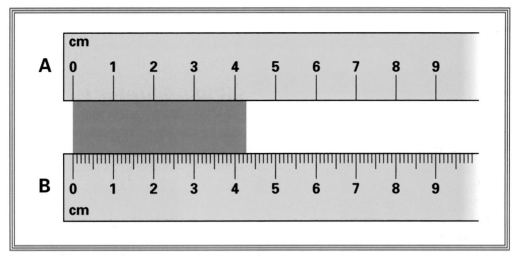

Figure 4-3
Ruler B has finer calibrations than ruler A. You might estimate the measurement on ruler A as 4.3, but another person might say 4.2 or 4.4. On ruler B the measurement is between 4.2 and 4.3. A good estimate might be 4.27. Why is one estimate to tenths and the other to hundredths?

Recording Final Zeros

What if the bottom of the meniscus is exactly on the 24-mL line? If you simply write 24 mL in your data table, it will suggest the volume is closer to 24 mL than to either 23 mL or 25 mL. If you record the data as 24.0 mL, however, you make it clear that the volume is exactly 24 mL. By recording the final zero, you report your data more precisely. This rule of recording data also applies to using an electronic balance.

Measuring in Metric

Whether you're measuring length, volume, or another quantity, the same principles of recording data apply. With each quantity, of course, you'll use different instruments for measuring. Not all instruments measure with the same degree of accuracy.

Measuring Length

When measuring length, you'll use a metric ruler or meter stick, as shown in **Figure 4-3**. Millimeters are generally the smallest divisions represented on these tools. Since metric uses the decimal system, you know without counting that there are 10 mm in one cm on a metric ruler. Each mm line, then, represents a tenth of a centimeter.

Since centimeters are usually the most practical unit of length measurement in food science, cm and mm marks will be most useful to you. If the height of a foam that you're measuring is even with the fourth line past 5 cm, the value you would report is 5.40 cm. If the height falls between 5.40 and 5.50 cm, estimate it as best you can.

Measuring Volume

You're becoming familiar with several different containers that measure volume. Each measuring device may be marked with a different scale, resulting in varying degrees of accuracy. Laboratory instructions will tell you to use different pieces of equipment to measure volume, depending on how accurate your measurement needs to be. Eventually you may be able to choose the appropriate device, just as a cook learns whether to use a tablespoon or a one-cup measure.

Careers

Technical Writer

What's new in food science and nutrition? Ask a . . . Technical Writer

A question that freelance technical writer Mia Nelson often hears is, "Are you a writer or a foods expert?" "Actually I have to be both," Mia explains. "My writing skills allow me to communicate clearly and accurately, and my background in foods and nutrition gives me a good understanding of the subjects I write about. Since information in the field changes daily, I still have to do research. I stay current by reading magazines and scientific publications. I also pay attention to what's on the Internet. Whether I'm reporting, "teaching," or doing public relations, the facts must always be right."

Education Needed: Because technical writers deal with expert-level information, they need to be well versed in their subject. Most writers in scientific fields have a degree in that area. In addition, technical writers need an excellent command of the language. A college internship in writing user's manuals, training material, or consumer education bulletins can teach about "writing to" a specific audience.

KEY SKILLS
- Written communication
- Teamwork
- Organization
- Computer literacy
- Technical knowledge
- Creativity

Life on the Job: Technical writers create a wide range of written materials, from reference and instruction manuals, to pamphlets for young children. They work with editors, artists, and other experts in the field to define each publication's goals and style. They may be employed by a company or work independently as freelancers.

Your Challenge

Suppose the American Sweet Potato Growers ask you, a technical writer, to create a promotional brochure that explains the nutritional benefits of sweet potatoes and promotes buying them.

1. What facts, topics, and format will make the brochure helpful and appealing?
2. What writing style and reading level will you use? Why?

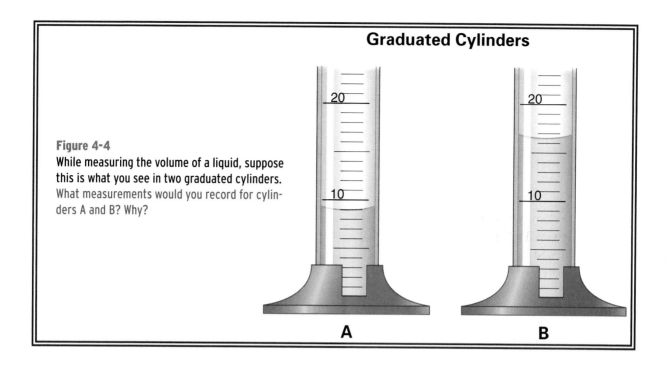

Graduated Cylinders

Figure 4-4
While measuring the volume of a liquid, suppose this is what you see in two graduated cylinders. What measurements would you record for cylinders A and B? Why?

A   B

The experiment at the end of this chapter shows the importance of using the right measuring device. Estimates can be faulty, not necessarily from carelessness, but because the scale on a device cannot measure with the accuracy needed. **Figure 4-4** shows how to measure with a graduated cylinder.

Measuring Mass

Depending on the type of balance in your food science lab, you may need to estimate the final digit for reporting mass in your data table. This is the case if you are using a triple-beam balance.

If you're using an electronic balance with a digital readout, however, you simply report all digits that appear on the scale. When using a decigram balance, which registers tenths of a gram, your readings will include one digit past the decimal—6.4 g, for example. If you're massing a substance using a centigram balance, two digits will appear after the decimal. You should report both of these, even if the last digit is a zero.

Metric Conversions

When you need to convert measurements, you can do so quickly with the help of a metric conversion chart. The chart shown in **Figure 4-5** on page 68 tells you what to do to convert from English to metric and vice versa.

The Celsius Temperature Scale

The Fahrenheit scale for measuring temperature, which is used in the English system, was invented in the early 1700s by a German physicist, Gabriel Fahrenheit. He is noted for making the thermometer more accurate by using mercury instead of alcohol.

Temperature in the metric system is measured with the Celsius scale. This gauge was named for Anders Celsius, the Swedish astronomer who developed it in 1742. This scale was formerly called the centigrade scale, which means divided into 100 parts.

Figure 4-5

Metric Conversion Chart

	When you want to convert:	Multiply by:	To find:
Length	inches	2.54	centimeters
	centimeters	0.39	inches
	feet	0.30	meters
	meters	3.28	feet
	yards	0.91	meters
	meters	1.09	yards
	miles	1.61	kilometers
	kilometers	0.62	miles
Volume	cubic inches	16.39	cubic centimeters
	cubic centimeters	0.06	cubic inches
	cubic feet	0.03	cubic meters
	cubic meters	35.31	cubic feet
	liters	1.06	quarts
	liters	0.26	gallons
	gallons	3.78	liters
Mass	ounces	28.35	grams
	grams	0.04	ounces
	pounds	0.45	kilograms
	kilograms	2.20	pounds
Area	square inches	6.45	square centimeters
	square centimeters	0.16	square inches
	square feet	0.09	square meters
	square meters	10.76	square feet
	square miles	2.59	square kilometers
	square kilometers	0.39	square miles
	hectares	2.47	acres
	acres	0.40	hectares

Figure 4-6 compares some common temperatures on the Fahrenheit and Celsius scales. Note that 100 degrees separate freezing and boiling points on the Celsius scale. On the Fahrenheit scale, however, the difference is 180 degrees. Therefore, a Celsius degree represents almost twice the temperature change as a Fahrenheit degree.

At times you may need to make conversions from Fahrenheit to Celsius or vice versa. **Figure 4-7** shows you how to do this.

Measuring Temperature

Most laboratory thermometers are marked in whole degrees, so temperature readings should be estimated to 0.1°. When measuring the temperature of a substance, be sure the thermometer bulb is completely immersed in the container but not touching the sides or bottom. Give the thermometer time to reach the temperature. You'll know this has occurred when the reading hasn't changed after ten seconds.

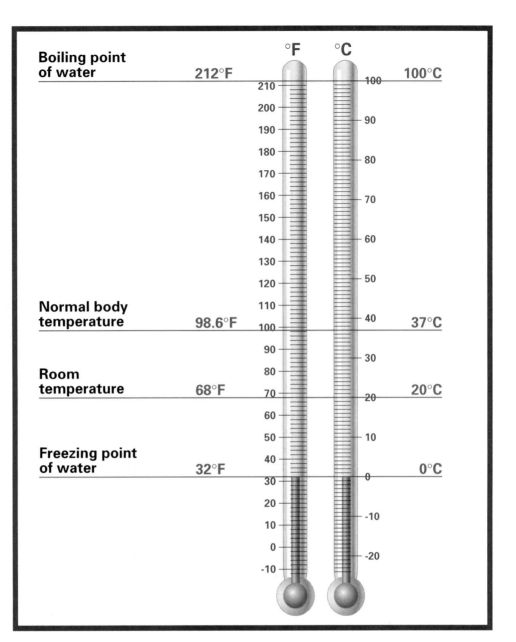

Figure 4-6
Here you see common temperatures in both Fahrenheit and Celsius scales. Suppose the temperature on a warm day registered 84ºF. What would that be on the Celsius scale?

Figure 4-7

Converting Temperatures		
Conversion	**Formula**	**Example**
Fahrenheit to Celsius	ºC = 5/9(ºF − 32)	200ºF − 32 = 168 168 x 5/9 = 93.3ºC
Celsius to Fahrenheit	ºF = 9/5(ºC) + 32	50ºC x 9/5 = 90 90 + 32 = 122ºF

EXPERIMENT 4

Measuring the Volume of a Liquid

Using the proper equipment for experiments in the food science laboratory makes tasks easier and results more accurate. Liquid volumes can be measured using a beaker, graduated cylinder, or buret. In this experiment you'll compare the degree of accuracy allowed by these pieces of equipment.

Equipment and Materials

water	10-mL graduated cylinder	50-mL buret	utility clamp
150-mL beaker	100-mL graduated cylinder	ring stand	

Procedure

1. Go to the station assigned to you by your teacher. Read as precisely as you can the volume of liquid in the beaker, graduated cylinder, and the two burets at the station. Record your readings in your data table. Remember to read volumes from the bottom of the meniscus if you can see one.

2. Return to your own lab station. Using the gradations on the side of the 150-mL beaker, add exactly 45 mL of tap water to the beaker. Without spilling any, pour this water into your 100-mL graduated cylinder and read the volume to the nearest 0.1 mL. Record in your data table. Empty the graduated cylinder.

3. Using the gradations on the side of the beaker, add exactly 7 mL of tap water to the beaker. Without spilling any, pour this water into your 10-mL graduated cylinder and read the volume to the nearest 0.01

mL. Record in your data table. Empty the graduated cylinder.

4. Clamp a clean 50-mL buret to a ring stand, using a utility clamp. Fill the buret with tap water. It is best to add water above the zero line and then release water slowly into your beaker until the liquid in the buret is at or below the zero line. Discard the extra water.

5. Release exactly 22.00 mL of water from the buret into the 100-mL graduated cylinder. Read the volume of liquid in your graduated cylinder as precisely as you can and record in your data table.

6. Release 6.55 mL of water from the same buret into the 10-mL graduated cylinder. Read and record this amount as accurately as possible.

Analyzing Results

1. Based on the reading from your graduated cylinder, is a beaker suitable for measuring 45 mL of water? For measuring 7 mL of water?

2. When, if ever, would you use a beaker to measure volumes of liquid in an experiment?

3. Compare your readings for the water released into each graduated cylinder. Which cylinder measured more accurately?

4. Do you think these same general results apply to measuring volumes of other types of substances, such as sand or oil? Why or why not?

SAMPLE DATA TABLES

Equipment	Volume	Station Number
Beaker		
Graduated cylinder		
Buret #1		
Buret #2		

VOLUME READINGS

Amount of Water	100-mL Graduated Cylinder	10-mL Graduated Cylinder
45 mL		
7 mL		
22 mL		
6.55 mL		

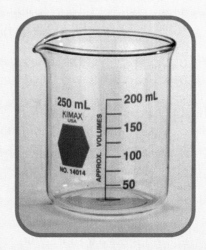

Chapter Summary

- Accurate measurements depend on the quality of the equipment used, while precise measurements depend on the skill of the experimenter.
- The metric units of length, volume, and mass are the meter, liter, and gram, respectively.
- The metric system, the international scientific measuring system, is based on multiples of ten.
- Metric prefixes can be used with any base unit of measure and indicate the relationship of a measure to the base unit.
- The Celsius scale is used to measure temperature in the metric system.
- Readings made on linear scales should be estimated to one decimal place past the smallest gradation shown. Digital readouts should be reported to as many decimal places as appear on the equipment.
- Different types of measuring equipment are used to measure different qualities.

Using Your Knowledge

1. Can you get precise results when using a faulty thermometer?
2. How are units of measurement related to the base unit in the metric system?
3. What would the measurement "6 dl" tell you about a substance?
4. Should you expect to pay more for a quart of milk than for a liter? Why or why not?
5. How many 115-g servings of fish can you get from a 0.75-kg package of fish fillets?
6. Would you wear a sweater in a classroom that was 32°C?
7. If a graduated cylinder is marked in whole milliliters, why wouldn't you report a volume as 25.55 mL?
8. What is the significance of the zero in the measurement 15.60 g?
9. Generally, how do you choose a measuring device for a given task?
10. Why might you have trouble accurately measuring the temperature of a solid with an English laboratory thermometer?

Real-World Impact

Gaining Skills and Confidence

With measuring, as with other areas of life, confidence in your skill can translate into confidence in your judgment. In a well-known experiment, the subject was asked to identify the longer of two straws. After doing so correctly, this person was joined by several others, who were working with the scientists running the experiment. These people deliberately named the shorter straw as the longer one. In most cases, the subject of the experiment eventually agreed, rather than defy the group's judgment.

Thinking It Through
1. What do you think was the purpose of the experiment?
2. When might a scientist experience a similar situation in real life?

Skill-Building Activities

1. **Research.** Find information on the effort in the United States to convert to metric. Write a report that summarizes this effort.
2. **Teamwork.** In a small group make a list of ways that metric measurements are commonly used in the United States. How would your team explain why metric is used in some areas and not in others?
3. **Problem Solving.** Suppose the only device you have to measure the length of an object is a meter stick that has broken off between the 2-cm and 3-cm lines. How could you use this stick to measure accurately?
4. **Critical Thinking.** If your lab partner reports the volume of a liquid as 25 mL, what do you presume are the smallest gradations on the measuring device? Why?
5. **Mathematics.** Convert these Fahrenheit temperatures to Celsius: 43°F, 74°F, and 170°F. Convert these Celsius temperatures to Fahrenheit: 20°C, 50°C, and 85°C.
6. **Critical Thinking.** Degrees in Celsius are also called centigrade. Why do you think this name is used?

Understanding Vocabulary

Create an example that describes someone measuring with precision or with accuracy. Don't use the terms in your descriptions. Exchange examples with a partner. Identify whether precision or accuracy is shown and discuss why with your partner.

Thinking Lab

Unexpected Results

Research scientists regularly encounter laboratory results that are contradictory. Much of their work focuses on figuring out why differences occurred and what the correct answers should be. Unexpected laboratory results need explanation.

Analysis

In your food science laboratory, suppose several students come up with different results after massing a sample of the same liquid. One student might record 102.54 g, while a second masses 10.44 g and a third gets 10.45 g.

Practicing Scientific Processes

1. List possible explanations for why the readings are different.
2. Suppose the first student forgot to tare the balance after placing a container on the balance and before adding the liquid. If the water has already been discarded or used, how could you verify this explanation?

CHAPTER 5

The Scientific Method

Objectives

- Describe in order the steps in the scientific method.
- Explain the role of reasoning skills in forming a hypothesis.
- Identify variables in a food science experiment and explain how they may affect the results.
- Demonstrate completing a data table and report form for a food science experiment.
- Distinguish between a hypothesis and a scientific theory.
- Suggest guidelines for doing a food science research project.

Terms to remember

data
deductive reasoning
dependent variable
experiment
hypothesis
independent variable
inductive reasoning
theory
variable

Suppose you notice a dark stain on the street where the family car is usually parked. Upon closer look, you see a fresh pool of liquid that has a slightly sweet odor. An idea about what has happened forms in your mind and is later confirmed by a trip to the mechanic. Antifreeze, or ethylene glycol, has dripped from a crack in the car radiator, which needs repair before the problem worsens.

Everyday life is full of such puzzles. In trying to solve them, people often use the same process favored by researchers to answer questions and solve problems in science. They observe, suggest explanations, and test their ideas. This system, called the scientific method, is the focus of this chapter.

Forming a Hypothesis

Without curiosity a scientist wouldn't make much progress. Curious people wonder about what they see. They form questions about unexplanable events.

In the same way that you might make a guess about the fluid leaking from a car, scientists' observations also lead to possible explanations for questions and problems. They form **hypotheses** (HY-pah-thuh-seez), *testable predictions that explain certain observations*. A hypothesis is an educated guess. Together with observation, forming a hypothesis is the first step in the scientific method.

A hypothesis also gives direction to an experiment. Asking yourself "What am I trying to prove?" and "What questions do I want to answer?" guides you in structuring an experiment. Stating and writing your hypothesis gives you a goal and a focus.

"Everybody goes out on Saturday night. That means I should have a date."

If deductive reasoning is based on incorrect information, the conclusion can't be correct. What is the flaw in this teen's reasoning?

Reasoning Skills

Everyday experiences prompt you to form hypotheses. To do so, you use both inductive and deductive reasoning.

Inductive reasoning is *drawing a general conclusion from specific facts or experiences.* Suppose a sandwich you wrapped in plastic is dry by lunchtime. You recall the same thing happening to some muffins a few days earlier. From these two particular circumstances, you might use inductive reasoning to conclude that the plastic wrap is faulty.

Deductive reasoning works in the opposite direction. *You reach a conclusion about a specific case based on known facts and general principles.* Such reasoning is often expressed as "if . . . then" statements. Suppose you want a cold drink, but your soft drink can is warm. Would you place it in the freezer for a while? Using deductive reasoning, you might conclude, "*If* water expands when it freezes and my soft drink is mostly water, *then* the can could explode in the freezer." The known fact (if) has lead you to a logical conclusion (then).

Of course, scientific hypotheses can be much more complicated than these examples. They typically come from much observation, experience, and research, as well as previous experimentation.

Experimentation

Experimentation is at the heart of the scientific method. A scientist must design **experiments** that provide *a way to test a hypothesis in order to verify or disprove it.* They must identify and create the conditions that put their hypothesis to the test.

An experiment includes a list of equipment and materials, as well as a procedure that gives step-by-step instructions about how to use these items and record results. As a student, you'll typically conduct experiments designed by others. Eventually, however, you may want to design an experiment of your own.

Many scientists find problems to solve simply by observing the world around them and becoming curious about one particular problem. Through careful experimentation, they search for answers.

Much of a scientist's work involves collecting data. What must be done with that data?

Controlling Variables

To develop a good experiment, you have to control the variables. **Variables** are *factors that can change in an experiment.* By controlling all the variables except one, you get more reliable results.

For example, suppose you want to determine whether glass, plastic, or metal bowls are best for whipping cream. What variables could affect the outcome? Bowl sizes and shapes; the temperature of the whipping cream samples; the type of beating device; the whipping speed—all of these could make a difference.

If a student first tests a metal bowl using whipping cream that's at room temperature and then a glass bowl using cream that's chilled, how reliable will the results be? Was the decisive factor the bowl, the temperature, or a combination of both? By varying only the type of bowl, you can be more confident about your conclusions.

In this experiment, scientists would call the bowl material an **independent variable**. This is *a factor that you change.* A **dependent variable** is *a factor that changes as a result of the independent variable.* What is the dependent variable in the experiment?

Analyzing Data

Explaining what happened in an experiment takes careful analysis of the data (DAY-tuh). **Data** is *the information gathered during an experiment.* The more consistent your data, the more valuable it will be for drawing conclusions.

To determine whether results are consistent, you may need to repeat an experiment. When your class performs an experiment, you can compare posted results. If only you and a partner carry out the experiment, you need several trials to see how results compare.

If the results are consistent, you may draw valid conclusions from them. Inconsistent or confusing results may mean the experiment was poorly planned or conducted. You may need to go "back to the drawing board."

The Data Table

For understanding, scientists arrange data in a clear and logical form. They might use charts, graphs, tables, and explanations. The experiments in this text suggest data tables. Independent variables might be listed along one side of the table and the dependent variables on another. A data table for the whipping cream experiment might look like the one shown in **Figure 5-1**.

Figure 5-1

SAMPLE DATA TABLE

Bowl Material	Results with Brand A Whipping Cream	Results with Brand B Whipping Cream	Results with Brand C Whipping Cream
Glass			
Plastic			
Metal			

Reporting Results

The scientific method isn't complete without reporting the results of experimentation. In the scientific community, results are reported in weekly and monthly journals, where they can be reviewed by other researchers. In turn, these scientists may test the results with experiments of their own.

To be accepted among scientists, an experiment must be reproducible. Other scientists must be able to repeat the experiment. Because scientific questions can be complicated, hypotheses may have to be tested with many experiments to determine whether they are valid. Some hypotheses are tested for years.

The Report Form

To learn as much as possible from an experiment, you need an organized format for reporting laboratory results. **Figure 5-2** shows the form suggested for use with this text. Each report contains several sections. Before class, fill out the purpose and procedures portions to preview what you'll be doing. Finish the report after completing the experiment.

Developing and Revising Theories

Albert Einstein once said, "No amount of experimentation can ever prove me right; a single experiment can prove me wrong." Einstein understood that even his work was not beyond question. Good science is a cycle of observing, proposing hypotheses, and experimentation. A hypothesis may be revised many times as understanding grows.

A hypothesis that is consistently verified may be established as theory. A **theory** is *an explanation based on a body of knowledge gained from many observations and supported by the results of many experiments*. A hypothesis is a possibility; a theory is a probability.

Even long-standing, widely accepted theories are open to ongoing experimentation, however. A theory may be refined, broadened, or disproved as new discoveries are made.

The scientific method, shown in **Figure 5-3** on page 80, offers an organized way to help make sense of a complex world. While studying food science, you'll use this process to become something of a scientist yourself. While you may not develop a theory, you'll have many opportunities to observe, hypothesize, and test ideas.

While conducting experiments in food science, you'll often work with others, both in the lab and in preparing reports. What skills will you need for effective teamwork?

Figure 5-2

Sample Laboratory Report Format

Title of Experiment: "The Effect of Bowl Material on Whipping Cream"

Performed By: Sally Scientist

Partner's Name: Albert Assistant

Date: October 19, 20___

Purpose
State the hypothesis of the experiment or the problem to be solved. You should be able to learn this from reading the experiment.

Procedure
Read the experiment and write a *brief* summary. Describe the general routine of the experiment so that another student could read it and know roughly what you did. You don't need to copy the procedure word for word.

Results
Observations: Include brief, descriptive statements of what you see happen. These should be made throughout the experiment, to note every development, or lack of development, that you observe. Record your observations honestly, even if they don't agree with what the instructions say should occur. Careful observation makes it easier to complete your report.

Data: Include the information you gather during the experiment, arranged efficiently in a table or chart. Each experiment will suggest a format for a sample data table. Label the columns of the table and the individual number values properly.

Calculations: Provide the solutions to any mathematical questions asked in the instructions.

Analysis of Results: Write the answers to the questions included at the end of each experiment.

Conclusion
Your conclusion is an interpretation of the results. It may also be the final figure of any calculations done during the experiment.

Conducting Research

Forming and testing a hypothesis would be slow going if you had to rely on your own experience and observation alone. Instead, the scientific method benefits from research, investigating what others have learned about a subject. While research is valued for its own sake, research and experimentation complement each other. Research may inspire or guide an experiment, and the experimental data reported is added to the body of research.

Choosing a Topic

Every research project starts with a topic. When you're doing research, choose a topic that interests you. You're more apt to enjoy the work and learn from it.

To find a topic in food science, you might start with a walk through a supermarket. Let your observations spark your curiosity. Does this product really need so much packaging? Why is that food kept chilled? What would happen if it were not?

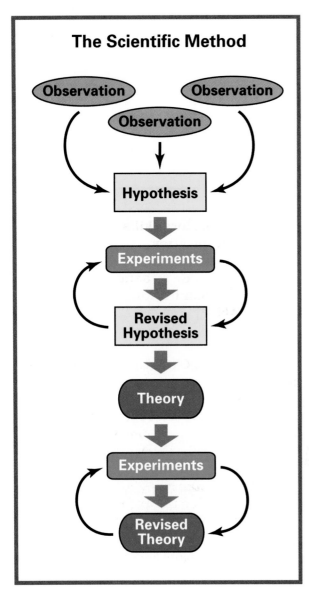

The Scientific Method

Observation · Observation · Observation → Hypothesis → Experiments → Revised Hypothesis → Theory → Experiments → Revised Theory

Figure 5-3
The scientific method is an ongoing process. A hypothesis is continually tested before it becomes a theory. What causes theories to be revised?

You might also review newspapers and scientific journals for articles concerning food science. A skilled Internet user can find numerous web sites suggesting timely topics to investigate.

Narrowing the Focus

Narrowing the topic of research, as shown in **Figure 5-4**, makes research and writing easier. You might be surprised at how specific a topic can be, yet still provide much to explore. You could write an entire book on growing vegetables, for example. The effects of soil on growing carrots is more manageable for a single paper. Really in-depth research might compare the effects of different levels of nitrogen in soil on carrot growth.

If your research leads you to develop an experiment, remember the time factor. The experiment must be completed in the time assigned by your instructor. Also think about resources. Can you obtain the supplies you'll need? Does your food science lab or classroom have the needed space and equipment?

Figure 5-4
When planning a research paper, you may need to do some background reading before narrowing the topic. Why is it a good idea to narrow the topic before you actually start collecting data on note cards?

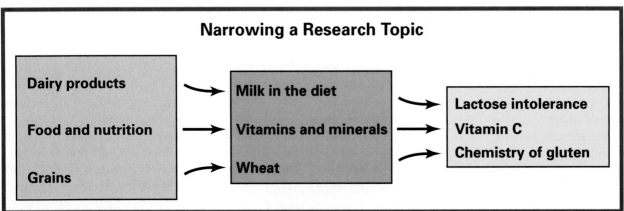

Narrowing a Research Topic

Dairy products · Food and nutrition · Grains → Milk in the diet · Vitamins and minerals · Wheat → Lactose intolerance · Vitamin C · Chemistry of gluten

Careers

Chemical Technician

Are there different jobs in a laboratory?
Ask a . . . Chemical Technician

Like any workplace, a food science laboratory needs people for varied responsibilities," says Isabel Herrera. "In my work as a chemical technician, I assist the scientists. I help prepare for experiments like the one we've been doing to test how much fat we can take out of a cake product without making it crumble. I set up and calibrate the equipment before each experiment. Afterwards, I clean and sterilize the equipment that's been used.

"What I like best is preparing chemical solutions that the scientists need for an experiment. As I take more courses and gain experience, I'll be able to step into more of a leadership role. Being in charge of an experiment of my own is my goal."

Education Needed: Most laboratories prefer that chemical technicians hold a bachelor's degree in chemistry, biochemistry, or chemical engineering. However, many employers will hire someone with an associate's degree, plus expertise with laboratory procedures.

KEY SKILLS
- Laboratory skills
- Organization
- Data analysis
- Attention to detail
- Communication
- Ability to work alone

Life on the Job: Chemical technicians help scientists test the chemical and physical makeup of materials. They follow standard formulas to prepare chemical solutions, which scientists then analyze for various substances. They also help document test results, using written forms and computer-generated graphs and charts.

Your Challenge

Imagine that you're a chemical technician. You test a frozen food sample as directed by a scientist. The results show very high sodium. You try a test of your own and get acceptable sodium limits.

1. How do you determine which set of findings, if either, is correct?
2. If the difference can't be explained, how do you learn what will be done?

Research or Not?

One historical message about nutrition research is that limited information sometimes promotes ideas and practices. A little knowledge can be misleading—and even dangerous.

In the late 1800s, a Dr. James Salisbury proposed that the "natural" human diet should be two-thirds meat since two-thirds of the teeth are "meat teeth," meant for cutting and tearing. He said food from plants should be strictly limited because they "ferment" in the intestines, causing everything from heart disease to insanity.

Some nutrition fads are based on just enough science to sound reasonable. A simple approach may be offered for a serious problem. Evaluating sources carefully can help you identify reliable information.

Applying Your Knowledge

1. What so-called scientific theories about nutrition have you heard?
2. How has technology helped and hindered the scientific community in educating people about sound nutritional theories?

Finding Information

Exploring a new field begins by reviewing already existing literature. To gather information on your chosen topic, you could start with the *Reader's Guide to Periodical Literature* for recent magazine articles. If the library has a computerized database, look for magazine references on your subject. Journals such as *FDA Consumer, Food Technology, Food Engineering, Journal of Food Science,* or *Journal of the American Dietetic Association* may be useful. Most of these journals have a web site that you can check for the latest news on a topic.

Take notes as you review books and articles. Most researchers use note cards, carefully writing a single piece of information and its source on each card. You can arrange these cards to organize your paper before writing it. They will also be handy if you develop an experiment.

Don't overlook knowledgeable people as resources. Teachers can provide insights. If a university is nearby, professors may be able to help you direct your research. Your community might have a food company that makes a product related to your topic. Gathering information from varied sources puts you well on the way toward a worthwhile research project.

Your research becomes more authentic when you include information gathered from authorities. Using a quotation or two from the person interviewed adds interest to your paper.

Properties of Popping Corn

SAFETY FIRST
Review these safety guidelines before you begin this experiment.

If you've ever popped popcorn, you know that some unpopped kernels often remain in the bottom of the popper. In this experiment, you'll try to determine what influences whether corn pops.

A little background will help you get started. Popcorn is bred for its water content; about 14 percent is considered ideal. As popcorn heats, the water inside each kernel expands. Pressure builds, causing the corn to pop. Knowing this, what is your hypothesis for this experiment?

Read through the experiment closely. Notice how the procedure is designed to control the variables.

Equipment and Materials
pin with plastic head   100 popcorn kernels   hot-air popcorn popper   bowl

Procedure
1. On the day before the experiment, help prepare kernels from a newly opened container of popcorn. Puncture some of the kernels of corn with a pin. Your teacher will place other kernels in a food dehydrator overnight to evaporate some of the water. The remainder of the sample will not be altered. This last group is called the *control*.
2. Follow the variation assigned by your teacher.
   a. **Variation 1.** Count out 100 kernels of unaltered popcorn.
   b. **Variation 2.** Count out 100 kernels of popcorn that has been dried.
   c. **Variation 3.** Count out 100 kernels of punctured popcorn.
3. When it is your turn to use the popper, pop your kernels of corn for 2 minutes. Count the number of unpopped kernels that remain.

4. Write your results in your data table and on the chalkboard. In your data table, copy the results from the other variations.

Analyzing Results
1. Which variation had the smallest percentage of unpopped kernels? The largest percentage?
2. How do you explain these results?
3. Does this one experiment prove your hypothesis is correct? Why or why not?
4. How did you control the variables in this experiment?
5. Based on your findings, how would you suggest storing popcorn to preserve its quality?
6. Again based on these results, why do you think cracked kernels are discarded when popcorn is processed for sale?

SAMPLE DATA TABLE

Variation	Number of Unpopped Kernels
Variation 1	
Variation 2	
Variation 3	

Chapter Summary

- The scientific method answers questions about the physical world.
- Reasoning from observation and experience may suggest a hypothesis that explains events.
- Scientists test a hypothesis using the controlled conditions of an experiment.
- Results of experimentation are reported for analysis in the scientific community.
- Data is represented in a format that clearly and logically communicates the information.
- A hypothesis that is frequently demonstrated to be true may become an accepted scientific theory.
- An Internet search, the *Reader's Guide to Periodical Literature,* and experts in the scientific community can help you find information and focus your research.

Using Your Knowledge

1. Describe the value of the scientific method.
2. What is the purpose of a hypothesis in experimentation?
3. Suppose your car stops running and you notice the gas gauge is on "empty." Write one statement showing inductive reasoning and one showing deductive reasoning based on this event.
4. Name two possible variables in an experiment to test the cleaning ability of two laundry detergents.
5. Why is it unwise to change more than one variable in an experimental trial?
6. What is the point in repeating an experiment?
7. How is reporting results of experimentation in a journal different from reporting facts in a data table?
8. Why is filling out an experiment report form helpful for learning?
9. Is it considered appropriate to doubt an accepted theory? Explain.

Real-World Impact

Judging Reliability

Since most scientific studies are only summarized in the news, how can you judge their reliability? Ask these questions: Who funded the research? How was it conducted? Who were the subjects? How large was the study group? A survey co-sponsored by Harvard University looked at news stories on medical research from 1994 to 1998. Fewer than half the stories told about a drug's side effects, and 40 percent didn't give the actual numbers in a study's results. For example, a drug was found to cut the risk of an injury in half, but the risk was only two percent to start with.

Thinking It Through

1. In the study described, only 40 percent of all stories mentioned any ties between a researcher and the group funding the research. Why is it important to know this connection?
2. Results of two different studies may contradict each other. Should you assume that only one set of results is reliable? Explain.

Using Your Knowledge *(continued)*

10. Explain which topic would be more suitable for a research paper: the manufacture of steel; or new uses for computers in the manufacture of steel.

11. What people might be helpful as you undertake a food science research project?

Skill-Building Activities

1. **Science.** Suppose you want to create an experiment on the topics listed here. What hypothesis would you state for each one? a) What happens to uncovered ice cream in a freezer? b) Is aluminum foil or plastic wrap better as freezer wrap? c) Do cookies bake better on greased, metal cookie sheets or on ungreased, teflon cookie sheets?

2. **Science.** In newspapers or magazines, or on the Internet, locate an article describing food science research. (An Internet search might take you to the home page of International Food Technologists, www.ift.org.) Identify the steps in the scientific method mentioned in the article. What variables do you think researchers had to consider? How might they have controlled these variables? What follow-up research or experimentation would you suggest? Could your food science class carry out any of these experiments? Summarize your findings for the class.

3. **Consumer Skills.** Plan a research project to learn which food products are the most economical sources of vitamin C. Briefly explain what processes and resources you would use to choose foods; to ensure a fair and varied sampling; to determine a food's vitamin C content; and to compare nutritional value to price.

Thinking Lab

Developing an Experiment

You've been running experiments and doing research since you were born. How did you learn that bananas taste better than playdough, or that a light switch turns a lamp on and off? Throughout life, people learn about the world by observing, testing, and reasoning.

Analysis

A scientific experiment is considerably more involved than everyday exploration. To answer questions with the exactness required of science, an experiment must be highly structured. Even with simple matters, you must account for numerous details.

Practicing Scientific Processes

Using the scientific method, design an experiment to learn why some brands of paper towels absorb liquids better than others do. Clearly state the purpose of the experiment. Propose a simple hypothesis and develop a procedure to test it. Identify variables and tell how you will limit or control them. Create a data table to report the information you expect to gather.

Understanding Vocabulary

On a piece of paper, write an example that describes each of the "Terms to Remember." Change the order of the terms. Exchange papers with a classmate and match the correct term with each example.

CHAPTER 6

The Sensory Evaluation of Food

Objectives

- Explain how various influences affect food choices.
- Describe sensory characteristics that affect food preferences.
- Plan a setting for successful sensory evaluation.
- Explain the role of sensory evaluation in the food industry.
- Explain the relationship between sensory characteristics and nutrition.

Terms to remember

flavor
garnish
monosodium glutamate
mouthfeel
olfactory
sensory characteristics
sensory evaluation
sensory evaluation panels
taste blind
taste buds
volatile

An old story tells of a restaurant owner and the poor student who lived in the apartment upstairs. One day the owner heard the student tell a friend, "I like living over the restaurant. I can't afford to eat there, but I get all the wonderful smells of the food they serve." The owner took the student before a judge, demanding some sort of payment. Agreeing, the judge had the student drop some coins from the left hand to the right. The owner's payment for the smell of the food was the sight and sound of the student's money.

As the story shows, eating can be enjoyed for more than health reasons. In fact, sometimes other factors have even greater influence.

What Influences Food Choices?

If you fix a meal for a group of people, you may find that even your finest recipe, prepared to perfection, doesn't appeal to everyone. People have different preferences about food, just as they do about music, movies, and clothing. Food preferences can be very personal but usually develop through influences that are common to all people. Some of the major influences are explained here.

Culture and Geography

Cultural heritage may influence food choice. In every region of the world, preferences for certain tastes have been passed down for centuries. These tastes, in turn, were influenced by local climates and geography and by contact with other cultures.

The traditional Greek diet is a good example. The warm, mountainous, and fertile islands of Greece are naturally suited for goats, grapevines, wheat, and olive trees. The seas are abundant with fish. Not surprisingly, goat milk and cheese, wine, bread, and fish are staple foods. Olive oil is the number one staple. Trade with North Africa brought an exchange of seasonings: onion, oregano, garlic, parsley, and cinnamon. Classical Greek culture favored simplicity, which is reflected in its recipes. Many Greek recipes have not changed in 2500 years.

Likewise, foods eaten in Rome today are very similar to those offered to wealthy social classes of ancient times. Butcher shops are filled with cured meats, steaks, and pork chops. The peaches, pears, grapes, beets, lettuce, and cucumbers enjoyed by classical Romans are still part of the modern Italian diet. Bread with cheese remains a popular combination.

In contrast, food preferences in the United States and Canada have an international flavor. They blend eating patterns of many immigrant groups. Some have been adapted to local foods and customs. Pizza, tacos, and chop suey are "Americanized" versions of foods that originated in other lands. The Creole cooking of Louisiana, meanwhile, blends traditions of French-Canadian, African-American, Spanish, Caribbean, and Native American groups that settled around New Orleans.

Emotions and Psychology

Food has a strong psychological aspect. Preferences are often based on emotional associations. If someone has warm memories of family weekends that include a breakfast of pancakes and syrup, that meal might remain a favorite into adulthood. Similarly, when people feel sad or distressed, they often soothe themselves with "comfort foods," from chocolate to chicken soup.

For some people, food choices make a personal statement, like their car and clothing. A gourmet coffee can be as much a status symbol as a luxury automobile.

In a country where people of many cultures live together, different ethnic foods can be enjoyed by everyone. What ethnic cuisines have you tried that are not linked to your own heritage?

Some foods are thought to be comforting during stressful times. Why do you think soup has earned this reputation?

Advertisers are quick to take advantage of this psychological impact to sway food choices. Each element of an ad—words, images, music, and more—is carefully chosen to create a positive impression in the target audience. Consumers need to recognize emotional appeals. Whether you buy a food because you really like it or because you like its advertising campaign can be hard to say. Often it's a combination of both.

Beliefs

Many religious traditions include dietary laws and guidelines in their expression of beliefs. Islam, Judaism, and Christianity all encourage fasting, especially at certain times of the year. Jewish kosher laws concerning diet are very specific about what foods may be eaten and how they must be prepared. Both Jewish and Islamic law prohibit certain foods and specify how animals are to be slaughtered. Buddhists and Seventh-Day Adventists recommend a vegetarian diet.

Personal beliefs can also direct food choices. Many vegetarians are philosophically opposed to killing animals. Someone might buy fresh produce from a farmers market (and carry it home in a recycled bag) to support local food growers and promote simplicity. On the other hand, that same person might plan a lavish meal to celebrate with friends.

Health Concerns

Ideally, health benefits would rank near the top in influences on food choices. Many people do consider nutritional value when deciding what to eat. As the increase in books and magazines dedicated to healthful eating shows, more people are trying to understand and use the vast amount of information available on food and health.

For people with certain physical conditions, choosing or restricting particular foods has a more immediate impact on health. People whose bodies lack a certain enzyme have trouble digesting milk products. (Enzymes are special proteins that help chemical reactions take place in the body. You'll learn more about them in Chapter 19.) People with diabetes choose food based on how it affects their blood sugar level. Some foods trigger serious allergic reactions in certain individuals.

Health concerns change with stages of life, and so do food choices. For breakfast, an active teen male may fix a cheese omelet to supply the calories and nutrients needed for growth. His father, trying to reverse the trend, might pour a bowl of corn flakes with low-fat milk. The teen's grandfather, with even lower energy needs, may choose toast and a fruit spread.

Changes in health can alter eating patterns temporarily. Extra fluids are needed to replace those lost to fever or dehydration, for example.

Interest in organic foods has grown over the years. These foods are grown in compliance with certain standards that prohibit the use of materials that are harmful to human health. Some people are willing to pay the higher cost for these foods.

Some foods may be avoided because they can limit the effectiveness of certain medications, while some medications prevent the body from absorbing certain nutrients. Foods rich in this nutrient may be increased in the diet.

Food Costs

Like nutrition, price plays at least some role in determining what people eat. Depending on your food budget, you may buy every product you want every time you shop; choose a less costly substitute most of the time; or rule out a food altogether. People who rely on government programs to provide groceries are limited in their choice of items.

Technology

Food technologists are like the seafaring explorers of the 1600s, adding new foods to the cultural diet. As proof, try naming all the types of snack foods and crackers on the supermarket shelves. Cheese-flavored, barbecue, low-salt, and fat-free—those varieties and more are the work of food scientists.

Technology has also increased the number of people who have access to some foods. New packaging and preservation methods allow foods to be distributed to a wider area without losing quality. Microwave ovens have become so advanced that even school-age children can fix their own meals, including foods scientifically designed for microwaving.

Sensory Evaluation: A Scientific Approach

Because so many influences are at work, predicting food choices is not an exact science. Sensory evaluation, on the other hand, *is* an exact science. **Sensory evaluation** involves *scientifically testing food, using the human senses of sight, smell, taste, touch, and hearing.* Research in this area has been going on for years.

Launching new or "new and improved" foods is a costly gamble. Before taking that risk,

Preparing home-cooked meals is often less expensive than eating out. You may tend to eat more nutritiously too. What is another advantage?

Fast food offers a quick way to grab a meal. Such restaurants can be found almost every-where, even around the world. What problems can come from eating fast food too often?

food producers make every effort to determine what will appeal to the public.

Those who work in sensory evaluation might take this saying as their motto: "The proof of the pudding is in the eating." Like you, their immediate concern is a food's **sensory characteristics**, *the qualities of a food identified by the senses*—that is, how a food looks, tastes, smells, sounds, and feels when eaten. Like a high-speed computer, the senses interact to give information about these qualities. Food scientists are interested in what each sense contributes to determine the overall reaction to flavor.

The Flavor Factor

When people talk about a food's flavor, they've said a mouthful—in more ways than one. **Flavor** is the *distinctive quality that comes from a food's unique blend of appearance, taste, odor, feel, and sound.* Thus, flavor actually involves all of a food's sensory characteristics. It's a complicated quality, as each sense influences the others.

Appearance

If you take the lesson from the Dr. Seuss book for children, *Green Eggs and Ham,* you know

that looks count for only so much. Concerning foods, appearance can be a sign of quality or ripeness. Blotches of mold warn that a food is inedible. Black streaks on banana skins are clues that the fruit may be fine for banana bread but too ripe and soft for a salad.

Other times, an appealing appearance is based on habit and preconceived notions. This is especially true of color. Purple applesauce is not to be trusted. Green food coloring is often added to chocolate mint ice cream, although natural mint has no such color. It's a matter of expectations.

Combinations of colors are as important as each one individually. Imagine a plate with broiled breast of chicken, mashed potatoes, steamed cauliflower, and canned pears. Now picture grilled breast of chicken, a baked potato, steamed broccoli, and canned peach slices. Which meal would catch your eye in a cafeteria poster ad?

Color can be enhanced with garnishes. A **garnish** is *a decorative arrangement added to food or drink.* A sprig of parsley, leaf of lettuce, or slice of cinnamon apple adds eye appeal, which is important to planning attractive meals.

Size and shape also affect how appetizing food appears. Here, too, variety is pleasing to the eye. Potato wedges and pickle spears are monotonous. Potato wedges and pickle slices are complementary.

Taste and Odor

The next time you get a cold, pay close attention to how food tastes—or doesn't taste. A congested nose can leave you temporarily **taste blind**, *unable to distinguish between the flavors of some foods.* This condition can

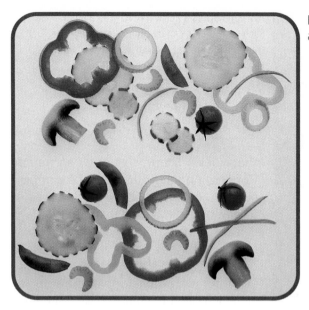

also be caused by some diseases, medications, and medical treatments.

In the case of head colds, taste blindness illustrates the close workings of the senses of taste and smell. They are affected by each other and by the same outside factors, including oxygen and temperature. They respond to the same substances in food. As guides to food safety, both senses are more reliable than sight. Neither one can be fully explained.

The Science of Taste

Researchers assign all food tastes to one of four broad categories: sweet, salty, bitter, and sour. Tastes are detected by the **taste buds**, *sensory organs located on various parts of the tongue.* The cells lining the surface of the taste buds have pores that allow contact between substances on the tongue and the sensory cells below. These pores are tiny. Only substances dissolved in water, which all foods contain, and in saliva in the mouth can gain entry. Substances that carry taste activate the sensory cells.

Some researchers recognize a fifth taste, called *umami* (yoo-MAH-mee). This might be best described as "savory." Umami is associated with the substance **monosodium glutamate** (mahn-uh-SO-dee-um GLOOT-uh-mayt), or MSG. MSG is *a salt that interacts with other*

ingredients to enhance salty and sour tastes. While MSG may be added to a food, it occurs naturally in others. Mushrooms, for instance, are high in glutamic acid, which makes them a natural version of MSG. They impart flavor to many dishes, though they seem to have little taste themselves.

The Science of Odor

Structurally, the sense of smell is more direct than that of taste. The **olfactory** (ohl-FAK-tuh-ree) organs, those *related to the sense of smell,* include a single nerve that ends in sensory cells in the nasal cavity and runs straight to the brain, as explained in **Figure 6-1**. Contradicting this simple setup, olfactory cells can identify countless different aromas, even those that are barely measurable by laboratory equipment.

In contrast to the taste buds, the olfactory organs respond to odors in the form of a gas, as in the steam rising from a bowl of hot chicken soup. Substances dissolved in the gas reach microscopic, hair-like structures called cilia just behind the bridge of the nose. These hairs are bathed in an oily mucus, which further dissolves odor-carrying substances so they can activate the sensitive nerve cells.

The Role of Temperature

A food may taste differently depending on its temperature. **Volatile** (VALL-uh-til) substances, those that are *easily changed into vapor,* are usually released when a food is heated, adding to its odor. Odor, in turn, affects taste. Similarly, a cold food that numbs the tongue seems tasteless until warmed in the mouth, while a food that is normally eaten cold may taste too strong if served warm.

Some types of food are more sensitive to temperature than others. Heat increases the sweet taste of some sugars. Salty tastes, on the other hand, are more intense in cooler foods.

Texture

Qualities related to a food's consistency describe its texture. Sensory evaluators test foods for a variety of textural aspects, such as brittleness and graininess. They note the degrees of qualities as well. On chewiness, a food might rate as hard, tough, moderately chewy, slightly tender, or very tender.

Like taste and odor, texture is not one sensation. Besides consistency, texture results from mouthfeel and sound.

Mouthfeel

Scientists coined the term **mouthfeel** to describe *how a food feels in the mouth.* People who love peanuts but dislike the way peanut butter sticks to the roof of the mouth are reacting against peanut butter's mouthfeel.

As with appearance, certain mouthfeels are associated with quality in certain foods. Premium ice cream feels smooth, not gritty, to the tongue. A tender steak, properly prepared, "melts in the mouth." Pie crusts and crescent rolls should separate into flakes.

Mouthfeel is also affected by temperature. Juices running in a warm steak give a moist, tender feel. That same meat, when cold, may feel tough and chewy.

Sound

A food's texture cannot be separated from its sound. For instance, you don't want to feel *or* hear anything crunch in a mouthful of applesauce. In contrast, crackers that don't crackle also don't appeal to most people, while a soggy cereal is somehow less tasty than the crispy one.

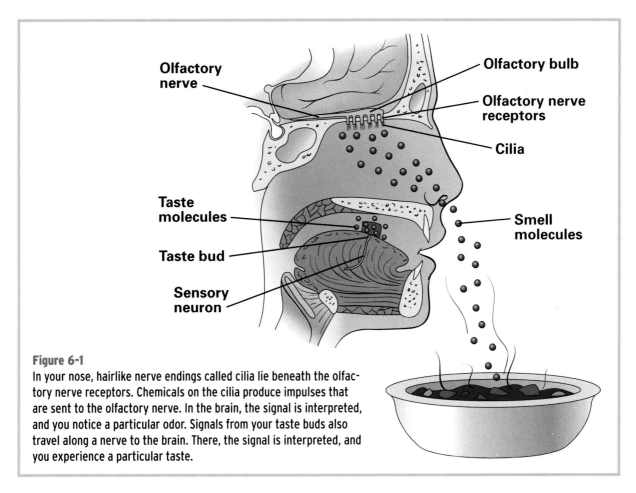

Figure 6-1

In your nose, hairlike nerve endings called cilia lie beneath the olfactory nerve receptors. Chemicals on the cilia produce impulses that are sent to the olfactory nerve. In the brain, the signal is interpreted, and you notice a particular odor. Signals from your taste buds also travel along a nerve to the brain. There, the signal is interpreted, and you experience a particular taste.

Sensory Scientist

What's involved in sensory testing?
Ask a . . . Sensory Scientist

Sensory scientist Adrian Morales knows the value of detail. "I don't actually have a degree in food science," says Adrian. "My master's degree in statistics is what opened the door for me at an international food company. When we do sensory testing of food, the findings must be accurate. That's why I look carefully at how tests are set up, to be sure that they'll tell us what we want to know. Once a sensory test is complete, I analyze the numbers to interpret what they tell. Business decisions about products are based on my findings. That means my attention to detail has high impact on our success."

Education Needed: A bachelor's degree in chemistry, biology, or food science is typical for a sensory scientist. Many firms prefer a master's degree in food science. A good grasp of statistics is essential for analyzing and reporting data. Marketing and general business courses are useful, as sensory scientists often work on product teams.

KEY SKILLS
- Data analysis
- Leadership
- Communication
- Organization
- Teamwork
- Keen senses

Life on the Job: The work of sensory scientists varies with the job. Some design sensory tests and then analyze, interpret, and report results. Some create evaluation procedures and train sensory panels, although a technologist may handle these responsibilities in certain facilities.

Your Challenge

Suppose you're a sensory scientist working with a product development team that aims to come up with new colors for some food products. Staying competitive with attention-getting change is the objective.

1. What food products do you think might sell better with a new color? What impact would sensory testing have?
2. What opinion would you offer?

Sensory Evaluation

Because flavors are so complex, food scientists need effective ways to get information about what people like. They often use **sensory evaluation panels**, *groups of people who evaluate food samples*. These panels fall into three main groups—highly trained experts, laboratory panels, and consumer panels. Their input has a great impact on whether a food makes it from the test kitchen to the shopping cart.

- **Highly trained experts.** These people judge the quality of a product, using standards set by the food industry. They are often involved in testing for very complex foods that require their refined, practiced skills and sensitivity. Coffee, for example, contains over 800 substances that affect flavor. Some sensory evaluation experts work individually.
- **Laboratory panels.** These are usually small groups that work at a company's laboratory. Employees who regularly eat the food might be members. They help develop new products and determine how a change may affect the quality of an existing product.
- **Consumer panels.** Companies use large consumer panels to test foods outside the laboratory—in grocery stores, shopping malls, and at market research firms. Tests may involve several hundred people. Rather

Food products are often evaluated by taste testers. How would an illness, such as a cold, affect a person's ability to do this task?

than using scientific terms and scales, consumers tell how much they like or dislike a product or one of its flavor characteristics.

Uniform Evaluations

To get the most reliable sensory information from people, researchers must make testing experiences as uniform as possible. For instance, all samples that are in a test are served at the same temperature. To design tests that yield valid, scientific results, researchers also take the following steps:

- **Minimize distractions.** Testing takes place in a controlled atmosphere. Lighting and temperature are kept constant. Often individual booths are used to eliminate outside noises and odors. One exception to this practice is the development of a new product, where an effort is made to serve the food as it would normally be eaten.
- **Minimize bias.** Knowing how one sense can affect another, researchers may mask irrelevant characteristics. Color differences may be hidden by using colored lights. Testers guard against the contrast effect, which can occur when a lesser-quality food is offered right after one of higher quality and is rated lower than it would be other-

Some research scientists use sophisticated computer programs to analyze flavor ingredients for food.

Nutrition *Link*

Sensory Appeal

To a five-year-old, eating soup and crackers is just lunch, but "catching" fish-shaped crackers that "swim" in a tomato soup "lake" is fun. Foods with sensory appeal are more apt to be eaten, which can also impact health.

Some elderly adults, for example, are troubled by a declining appetite. Tooth problems and taste blindness can add to eating difficulties. Food textures, colors, and shapes may be all that inspires some elderly people to eat for nourishment.

Depression can dull the appetite. Those with a mental disorder may not see the need for a balanced diet. The physical qualities of well-prepared, nutritious foods can be reminders of the pleasures of eating.

Even those who avoid trying new foods may be encouraged by enticing sensory qualities in food.

Applying Your Knowledge

1. Do you think foods advertised as "fun" are typically healthful? Give examples.
2. Choose a nutritious food with a boring reputation. How would you increase its appeal when serving?

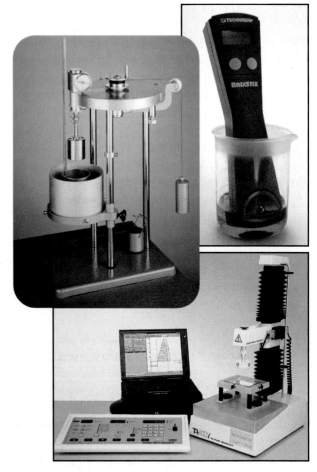

The technical equipment used in sensory testing includes devices that examine food textures (lower photo). The device on the upper left measures viscosity. The refractometer (upper right) measures sugar concentration.

wise. Instead, testers vary the order in which samples are presented, usually limiting tests to three items. Also, people most often select samples labeled "1" or "A" as the best, so items are given random three-digit numbers. As you can see, testers need to understand human psychology in addition to food science.

- **Help evaluators use their senses to the fullest.** Testing is usually done in late morning or midafternoon, when people are most responsive and alert. Food tasters rinse their mouths, usually with water, between samples. Warm water is used to cleanse the mouth of fatty foods. Eating a bland food, such as a saltine cracker, also clears the taste of a previous sample. Evaluators may be given time to notice changes in a characteristic, such as an unpleasant aftertaste.

Objective Evaluation

The senses are amazing evaluators. They can often detect the effects of ingredients, storage time, and other factors related to food quality

EXPERIMENT 6

Odor Recognition

SAFETY FIRST
Review these safety guidelines before you begin this experiment.

Normally, the senses of sight, odor, taste, touch, and sometimes hearing are used in evaluating food. When one sense is isolated, identification of even well-known samples can be difficult. This experiment will test your ability to identify common products by odor alone. Do not taste the samples.

Equipment and Materials

set of 15 test tubes containing samples to be tested

test tube rack

blindfold

Procedure

1. From your teacher obtain 15 samples of odorous material, both food and nonfood, in coded test tubes.
2. Put on the blindfold. Sniff each of the 15 samples as your partner presents them for your evaluation. (He or she should not sniff the samples while presenting them.) Your partner will record your identification of each sample in your data table.
3. Reverse roles, presenting each sample to your partner and recording his or her identifications.

Analyzing Results

1. How many of the 15 substances did you identify correctly? How many did your partner identify?
2. Compare results to other class members. What were the highest and lowest number of correct identifications?
3. Which substances were the hardest and easiest to identify? Why?
4. What does this experiment tell you about the interaction of the senses?

SAMPLE DATA TABLE

Substance Code Number	Your Identification	Actual Identity of Substance

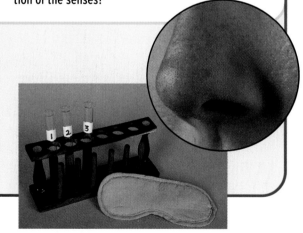

before chemical tests can. That's why people are so important for sensory testing.

Objectivity, however, is another part of scientific experimentation. Compared to sensory evaluation, objective evaluation offers a greater degree of control and consistency. Various laboratory devices can register traits from the tenderness of meat to the thickness of gravy. Using different equipment, scientists can determine a tomato's juiciness, acidity, and even its redness. In combination with sensory testing, objective data enable food scientists to make accurate predictions about a food's potential success in the marketplace.

Chapter Summary

- People's food choices are influenced in many different ways.
- Sensory evaluation is an attempt to scientifically describe and rate foods according to qualities perceived by human senses.
- Food preferences are often based on a food's appearance, taste, odor, and texture.
- Manufacturers use sensory testing, conducted under specific conditions, to evaluate new and existing food products.
- Making nutritious foods more appealing to the senses promotes a healthful diet.

Using Your Knowledge

1. How might the diet of a person living on a rocky seacoast differ from someone living on a rich-soiled prairie?
2. A friend says, "My dad packed a banana in my lunch every day when I was in grade school. I never want to see another banana again." What does this reaction demonstrate?
3. How can health concerns dictate food choices?
4. Using a frozen dinner as an example, explain the impact technology has had on food choices.
5. Why is sensory evaluation used by food producers?
6. What is the difference between flavor and taste?
7. Relate the saying "You can't judge a book by its cover" to foods' appearance.
8. Why might hot cocoa smell more "chocolatey" than chocolate ice cream?

Real-World Impact

Enhancing the Flavor

Why do people spice foods? Social scientists suggest that it's one way to carry a bit of culture and heritage to wherever people travel. Adding native spices to unfamiliar foods makes new surroundings feel more like the home left behind. Food scientists note that such seasonings as cinnamon, cloves, and salt help preserve food and kill bacteria. Early findings from Kansas State University confirm that garlic effectively destroys the E. coli bacteria that causes food poisoning, which thrives in uncooked ground beef.

Thinking It Through

1. A popular dish in Middle Eastern countries is kibbe, which includes ground meat, cinnamon, and garlic. How do you think this recipe was developed?
2. Choose a spice to research, including its origin and uses in different cultures.
3. How can food spices be either an obstacle or an aid to blending into a new culture?

Using Your Knowledge (continued)

9. Why is a food's sound considered an element of its texture?
10. Why might an evaluator be blindfolded when taste testing a food?
11. What kinds of sensory evaluation panels would you use to test a new type of canned pasta? Why?
12. How can companies combine sensory evaluation with objective evaluation to help ensure the quality and success of a food item?

Skill-Building Activities

1. **Consumer Skills.** Find food advertisements in print or electronic media. How does the advertiser try to appeal to your emotions? Bring in or describe the ad to the class, explaining its use of psychology.
2. **Decision Making.** Sample a food that you have never eaten before. Evaluate its sensory characteristics. Also consider any influences that might affect whether you would buy the food. Write your findings in a report for the class.
3. **Critical Thinking.** Write a profile of yourself based on the food choices you make. In other words, what do your food choices reveal about you? To guide your work, list the foods you chose to eat in the last 24 hours. Also consider your favorite foods and what you eat on a typical day.
4. **Social Studies.** Use the Internet or other resources to research one of the following: foods of another country; the dietary laws of a religion (how the laws developed and their significance).
5. **Critical Thinking.** Suppose you burn your tongue on your first bite of pizza. How might this affect your ability to taste the pizza during the rest of the meal, after it has cooled enough to be eaten safely? How well would you enjoy dessert?

Thinking Lab

Food for Thought

Dozens of new varieties of food are introduced to consumers each year. Between new products, reformulated older products, and foods newly introduced to an area, consumers have a wide range of choices to make.

Analysis

Each food product represents a sizable investment of resources for the manufacturer. Profits and jobs rest on an item's success in the marketplace.

Practicing Scientific Methods

1. Choose a favorite food. Describe the sensory characteristics of that food that you think are most important.
2. Design a taste test for comparing several brands of the food selected above. In as much detail as possible, describe: the setting for your test; steps you take to prevent bias; your method of rating or evaluating the food, such as a number scale or list of descriptive terms; and the members of your sensory panel. If possible, conduct the taste test in your class.

Understanding Vocabulary

Create fill-in-the-blank questions that use the "Terms to Remember." Keep your quiz for review at test time.

TechWatch

FROM ENGINE LUBRICANT TO SALAD DRESSING

Who would have guessed that rapeseed oil, once one of the best marine steam engine lubricants, would become the basis for canola oil? But through chemistry, it did. After World War II and the switch from steam engines to diesels, rapeseed oil was no longer needed as an engine lubricant. Though the oil was also a popular cooking oil in Europe and Canada, its erucic acid content was linked with heart disease. Aiming to lower the acid level, two Canadian scientists used gas chromatography equipment to analyze the fatty acid composition of vegetable oils. By 1974 they had replaced the erucic acid with oleic acid, a type of monounsaturated fatty acid. The new oil was named canola, a contraction of "Canadian oil." Today canola oil is valued for its very low level of saturated fat.

UNIT ③ Chemistry Fundamentals

TechWatch Activity
Compare the nutrition labels of several cooking oils available in supermarkets, including canola oil. Put the results in chart form. How do they compare?

Elements, Compounds, and Mixtures

Objectives

- Explain the difference between physical and chemical properties.
- Compare the physical phases of matter.
- Distinguish between pure substances and mixtures.
- Explain the relationship between elements and compounds.
- Compare heterogeneous and homogeneous mixtures.
- Identify chemical symbols and formulas.

Terms to remember

atom
chemical property
compound
element
matter
mixture
molecule
organic compounds
phase
physical property
property
pure substance
solution

Every day, you're a different creation. Your body recreates itself a little at a time, replacing old, worn-out cells with new ones—millions of cells daily. Each cell is made from substances found in the food you eat and the air you breathe. It's true that you are what you eat.

How does this transformation happen? How can broccoli and barbecued chicken become new cells in your body? The chemistry behind this process, which forms the basis for food science, is introduced in this unit. The discussion begins with matter—the material substances that make up the world itself.

What Is Matter?

In science, almost everything is categorized as matter. **Matter** is *anything that has mass and takes up space*. Both the tallest skyscraper and the tiniest computer chip are matter, though vastly different amounts. Matter includes the planets in outer space, the astronauts who may someday reach them, and the air and clouds they sail through to begin their journey. As **Figure 7-1** on page 104 shows, you are matter, as is everything you eat and drink.

Properties of Matter

Properties are one way to identify matter. A **property** is *a feature that helps identify a substance*. It's a defining trait. Sweetness, for example, is one property of sugar.

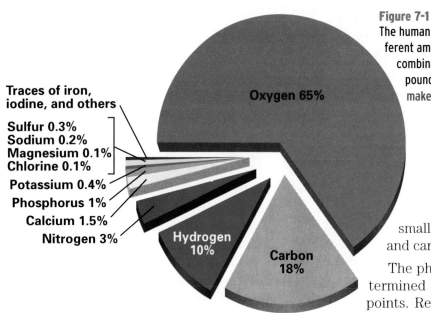

Traces of iron,
iodine, and others

Sulfur 0.3%
Sodium 0.2%
Magnesium 0.1%
Chlorine 0.1%
Potassium 0.4%
Phosphorus 1%
Calcium 1.5%
Nitrogen 3%

Oxygen 65%

Hydrogen 10%

Carbon 18%

Physical Properties

A **physical property** is *a characteristic that can be observed or measured without changing the substance into something else.* Included are such traits as color, melting point, boiling point, odor, and hardness. The physical properties of water include the following: clear; colorless; liquid at room temperature; melts at 0°C; and boils at 100°C.

One physical property you'll notice is **phase,** or *the physical state in which matter can exist.* Matter has the following three basic phases:

- **Solid.** Solids have a definite shape and volume. It's almost impossible to compress a solid to a smaller volume, and it tends to expand only slightly when heated. Aluminum, table sugar, and diamonds are solids at room temperature.
- **Liquid.** Liquids have a definite volume but not a definite shape. So, while liquids are difficult to compress, they will take the shape of their container. Water and rubbing alcohol are liquids.
- **Gas.** Unlike solids and liquids, gases have neither definite volume nor shape. They spread until they are evenly distributed within the largest space that confines them, yet they are easily compressed into the

smallest space. Oxygen, helium, and carbon dioxide are all gases.

The phase a substance takes is determined by its melting and boiling points. Recalling the melting point of water, you can see why the ice in your glass of lemonade changes phase quickly on a hot day.

Chemical Properties

People are more than just a collection of physical traits. They reveal their nature by how they act, or don't act, with others. Like people, other types of matter are described by more than just physical properties. A **chemical property** describes *the ability of a substance to react with other substances.* Whether muffins rise in the oven depends on the reactions of the salt, water, baking powder, and other chemicals in the batter. The chemical properties of substances in poultry give a roast turkey its crispy brown skin.

Classification of Matter

Suppose you want to refer to the object you're now reading. You might describe its physical traits—its size, weight, and color. You might define it by its components: marks of ink that make up words, written on paper that's bound between covers. You could also give it a specific name—a book. More precisely, it's a food science textbook.

Scientists do the same thing. To make the study of matter more manageable and under-

standable, they classify it into categories. They use properties to divide matter into two basic categories, pure substances and mixtures. Examples of both are shown in **Figure 7-2**.

Pure Substances

A **pure substance** is *made of only one kind of material and has definite properties*. It always has the same composition. Its properties

remain definite and constant under a given set of conditions. This means that any pure substance can be distinguished from any other by its physical properties. Two classes of pure substances are elements and compounds.

Elements

At an early age, a child learns basic ideas that are used to build more complicated concepts. Likewise, an **element** is *the simplest form of matter that can exist under normal laboratory conditions*. Elements are the substances from which all other materials are formed. About 110 elements have been identified, each with its own properties. Those of particular interest to food scientists include carbon, hydrogen, oxygen, nitrogen, and sulfur.

Even more elementary than an element is an atom. **Atoms** are *the smallest particles of an element that keep its chemical properties*. The iron in an iron skillet is composed of countless iron atoms, each having all the chemical properties of that element. Since all matter is

Figure 7-2
Classifying is a logical arrangement for a large quantity of items. How does this principle apply to matter?

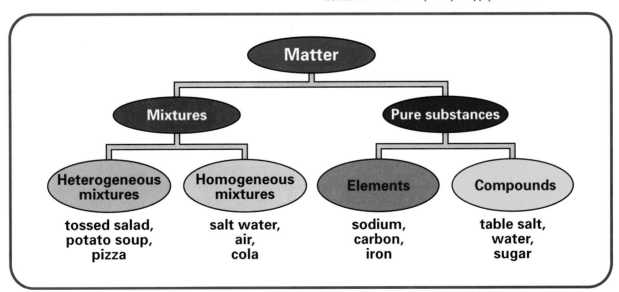

Figure 7-3
Two hydrogen atoms can combine with one oxygen atom to make water. As atoms of different elements come together, what do they create?

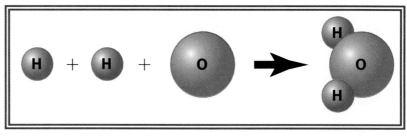

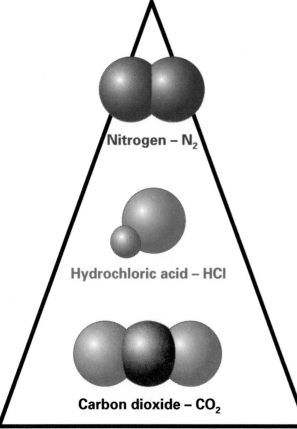

Nitrogen – N$_2$

Hydrochloric acid – HCl

Carbon dioxide – CO$_2$

Figure 7-4
In this and other books, you'll often find molecular structures shown with a space-filling model. Each sphere represents one atom of an element in the structure.

composed of elements, atoms are the building blocks of matter.

Atoms are incredibly tiny, about 0.000000001 (10^{-9}) meter in diameter. To get an idea of their size and number, imagine atoms as peas. A serving of peas on your plate might number a hundred (10^2). To fill your refrigerator would take about a million (10^6) peas. A billion peas (10^9) might fill your house. A trillion (10^{12}) peas would fill every house in a small town. A quadrillion (10^{15}) peas would fill every building in a city the size of Seattle.

Now imagine a blizzard over North Dakota—a blizzard of peas, covering the state in a blanket 1.2 m deep. You've imagined a quintillion (10^{18}) peas. Such a blanket over the whole planet would contain roughly a sextillion (10^{21}) peas.

Finally, imagine 250,000 planets, each the size of Earth, 1.2 m deep in peas. Now you have a septillion (10^{27}) peas—and that, give or take, is the number of atoms in the human body.

Compounds

A second category of pure substance is compounds. A **compound** is *a substance made of two or more different elements chemically joined together*. The properties of a compound are completely different from those of the elements that form it.

Just as an element is made of atoms, some compounds are made of molecules. A **molecule** is *the smallest unit of a molecular compound*. Each molecule contains atoms of all the compound's elements, in the same number and proportion. **Figure 7-3** shows how atoms combine to make one molecule of the compound water. Other molecular structures are shown in **Figure 7-4**.

The chemical processes that form compounds can be undone, and the individual elements separated again. Table sugar, for instance, is composed of carbon, hydrogen, and oxygen. Slow heating causes the sugar to undergo chemical changes, creating new substances. Eventually, the sugar will completely decompose into carbon and water vapor; the water vapor contains the hydrogen and oxygen from the original compound.

Most of the compounds you will study in food science are called **organic compounds**. This means they *contain the element carbon*.

Mixtures

Combining pure substances may create a mixture. A **mixture** is *a combination of two or more substances in which each substance keeps at least some of its original properties*. Components in a mixture are physically blended but not chemically attached to each other. Properties of a mixture vary according to the substances it contains, as well as the proportion of those substances. An oil-and-vinegar salad dressing is a food that is a mixture.

Mixtures can be further classified as either heterogeneous or homogeneous.

Heterogeneous Mixtures

In heterogeneous mixtures (he-tuh-ruh-JEE-nee-us), individual substances are dissimilar and can be recognized by sight. These mixtures are uneven in makeup and properties; often the materials are randomly distributed. If you've spiced a recipe with a seasoning blend, you created a heterogeneous mixture. One dash of seasoning might contain mostly grains of salt, with some larger bits of crumbled parsley, and fewer but larger flakes of dried pepper. Another dash might include more pepper than parsley. Even if you don't know spices, you can distinguish one ingredient from the others.

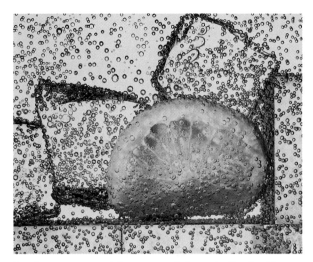

In a homogeneous mixture, one substance is dissolved in another. What substance is dissolved in this mixture?

Homogeneous Mixtures

In contrast, a *homogeneous* mixture (ho-muh-JEE-nee-us) is the same in every part of a given sample. Its components cannot be identified by sight. You can't see the salt in a glass of salt water, for example, but you can taste the salt if you take a sip. Another name for a homogeneous mixture is a **solution**. A solution is *a homogeneous mixture in which one substance is dissolved in another*.

Solutions may consist of any combination of solids, liquids, or gases dissolved in each other. Air is a solution of gases. Carbonated colas are solutions of the gas carbon dioxide dissolved in a liquid. Like chemical properties, solutions play an important role in food science.

Chemical Symbols and Formulas

By now you may be thinking that chemists have a different name for everything. They certainly have different symbols for every element. These symbols provide a shorthand method for representing elements and compounds and how they react with each other. Without these symbols and the formulas formed from them, recording chemical information could give you eye strain and writer's cramp.

In a heterogeneous mixture, you can see and distinguish the parts that make up the whole. Why is this seasoning mix heterogeneous?

Symbols

A chemist's abbreviations of the names of elements are called symbols. Each symbol contains one or two letters. Most often these are the first letter or two of the element's name.

If only one letter is used as a symbol for an element, it's always capitalized. When the symbol contains two letters, only the first is capitalized. I is the symbol for the element iodine, while helium is represented as He. Some symbols are derived from an element's older Latin name. From the Latin *aurum* comes the symbol for gold, Au. Iron is symbolized by Fe, from the Latin *ferrum*. Note that chemical symbols are not followed by a period.

The names and symbols of the elements you are most likely to encounter in food science are as follows: hydrogen (H); carbon (C); nitrogen (N); oxygen (O); sodium (Na); aluminum (Al); phosphorus (P); and sulfur (S).

Formulas

A chemical formula represents a compound. Just as a child might be named after both parents, a chemical formula includes the symbols of the elements that create the compound. If a molecule of the compound contains more than one atom of an element, the number of atoms is noted by a subscript, a small number written next to and slightly below the elemental symbol. The subscripts in a compound's formula tell you the proportions of the elements to each other in the compound. The formula for a specific compound is always the same, because the compound's composition never changes.

Suppose you are writing about two gases you exhale: carbon dioxide and water vapor. Carbon dioxide is composed of carbon (C) and oxygen (O). One part carbon combines with two parts oxygen to create the chemical formula CO_2. Water vapor is made of two parts hydrogen (H) to one part oxygen (O); its formula is H_2O. Listed below are the names and formulas of some compounds frequently found in food. You'll encounter some of these again in your study of food science.

- Sodium chloride (NaCl)
- Sodium bicarbonate ($NaHCO_3$)
- Caffeine ($C_8H_{10}N_4O_2$)
- Artificial sweetener Aspartame® ($C_{14}H_{18}N_2O_5$)

Propane is the gas used in a gas grill. The formula for propane can be shown in several ways. You'll see all of these used in this text.

Propane - C_3H_8

$$H-\overset{\overset{\displaystyle H}{|}}{\underset{\underset{\displaystyle H}{|}}{C}}-\overset{\overset{\displaystyle H}{|}}{\underset{\underset{\displaystyle H}{|}}{C}}-\overset{\overset{\displaystyle H}{|}}{\underset{\underset{\displaystyle H}{|}}{C}}-H$$

FOOD SCIENCE
Careers

Food Technologist

How do companies decide what foods to produce? Ask a . . . Food Technologist

KEY SKILLS
- Organization
- Analytical skills
- Problem solving
- Leadership
- Communication
- Creativity

F ood technologist Althea Tinsley enjoys seeing the effects of her work in the real world. "Not everyone has a career that impacts practically everyone. Since we all have to eat, my work with food makes a difference. I have to keep up with the trends to know what products will work. Just as an example, where would busy people be these days without instant foods? My company manufactures a number of products that can be prepared quickly and easily. Instant potatoes taste more like homemade today because we've continually experimented to improve their quality. I've been involved in that and many other advancements that I think consumers really appreciate."

prise and public programs. Companies use technologists to evaluate the sensory and nutritional qualities of their own products, and the competition's. Technologists also work on teams to improve quality control in the manufacturing process. Others, especially in government work, compile data on such topics as food composition and how processing methods affect nutrient value.

Education Needed: A food technologist needs a bachelor's degree in food science, food technology, or biology. Knowledge of chemistry is especially useful. An advanced degree may lead to leadership roles in related fields, including production and market research.

Life on the Job: Food technologists work in both private enter-

Your Challenge

Suppose you're a food technologist working with a team studying product ideas for the future and focusing on social trends.

1. How might these situations affect a family's food choices: all adults in family employed; three generations in one household; teens with part-time jobs?
2. What other trends in society might your team consider? Why?

Separating Mixtures

A mixture is composed of two or more substances that have not reacted with each other. The substances keep at least some of their original properties and can be separated by physical means. In this experiment you'll mix known masses of sucrose (table sugar) and cornstarch and then add water to this mixture. One component of the mixture will dissolve in the water while the other won't.

By filtering the mixture, you'll separate the component that doesn't dissolve (insoluble) from the other component. Drying the residue that is trapped on the filter paper as well as evaporating away the water that passes through the paper will allow you to recover both of the original solids.

Equipment and Materials

balance	weighing or waxed paper	filter paper	wash bottle
2, 150-mL beakers	water	funnel	hot plate or burner
sucrose	100-mL graduated cylinder	ring stand	beaker tongs
cornstarch	stirring rod	iron ring	

Procedure

1. Using weighing or waxed paper, mass between 1.0 and 1.5 g of sucrose and record the amount in your data table.
2. Repeat Step 1 using a second piece of weighing paper and cornstarch.
3. Add both the sucrose and the cornstarch to a clean 150-mL beaker.
4. Using your 100-mL graduated cylinder, measure 30 mL of tap water and add to the beaker. Stir the mixture with a stirring rod to dissolve as much solid as possible.
5. Mass a piece of filter paper and record in your data table. Fold the filter paper in half, then in quarters. Open the folded paper to form a cone, with one thickness of paper on one side and three thicknesses on the other. Write your initials on the edge of the paper.

6. Place the paper in a funnel. Place the funnel in an iron ring that has been clamped to a ring stand. Moisten the filter paper with a small amount of water from a wash bottle.
7. Mass a second clean, dry, 150-mL beaker. Place this beaker under the funnel to catch the liquid that passes through the filter paper. (This liquid is called the filtrate.)
8. Decant (pour off) the liquid from the solid, leaving as much of the solid as possible in the beaker. Do this by pouring the liquid down a stirring rod into the funnel. Be careful to keep the liquid in the funnel below the top edge of the filter paper at all times. After nearly all the liquid has passed through the filter paper, use a stream of water from the wash bottle to wash all of the solid out of the beaker onto the filter paper.

9. After all the liquid has passed through the filter paper, wash the solid on the filter paper with more water from the wash bottle.

10. Carefully remove your filter paper from the funnel and place it on a tray designated by your teacher.

11. Heat the beaker containing the filtrate over medium heat until a solid begins to appear in the bottom of the beaker.

12. Using beaker tongs, remove the beaker from the heat and place it on the lab table. Note the appearance of the contents of the beaker.

13. Place the beaker on a piece of paper with your initials on it on a second tray designated by your teacher.

14. The next day, mass the beaker and its contents, as well as the filter paper and the residue on it. Record these figures in your data table.

15. Subtract the mass of the beaker from the mass of beaker and contents. Subtract the mass of the filter paper from the mass of paper and residue. The remainder is the mass of sucrose and cornstarch recovered. Record these figures in your data table.

Analyzing Results

1. Which component of the original mixture dissolved in the water? Which one did not?

2. Were the sucrose and cornstarch completely separated by this procedure? Explain.

3. Did the sucrose have the same mass before and after the filtration process? If not, why might it be different?

4. Do you think you could use this same procedure to separate sodium chloride (table salt) and sucrose? Why or why not?

SAMPLE DATA TABLE

Substance	Mass
Sucrose	
Cornstarch	
Filter paper	
Beaker	
Filter paper and residue	
Residue	
Beaker and contents	
Contents	

Chapter Summary

- Everything that has mass and takes up space is matter.
- A substance can be identified by its physical and chemical properties.
- Physical properties include a substance's phase, or state.
- Matter can be identified as either a pure substance or a mixture.
- Atoms are the basis of all pure substances and mixtures.
- Chemists use symbols and formulas to represent elements and compounds.

Using Your Knowledge

1. Why is music not considered matter?
2. Using physical properties, how would you describe a gallon container of ice cream?
3. If you pump air into a 100-mL balloon, how far will that air spread?
4. Why is it important to know the chemical properties of two substances before mixing them?
5. Which is more complex: an element or a compound? Why?
6. Compare atoms with molecules.
7. Classify the following foods as homogeneous or heterogeneous mixtures: a) tomato juice; b) apple juice; c) coleslaw; d) pudding; e) chicken soup.
8. Why is sweetened tea considered a solution?
9. How are symbols related to formulas?
10. What do the subscripts in the formula $C_6H_{12}O_6$ tell you?
11. What elements would you expect to find in the compounds represented as: a) NH_3; b) Al_2O_3; c) H_2SO_3; d) $C_8H_{10}N_4O_2$?

Real-World Impact

The Power of Copper

Copper was one of the first elements recognized by early civilizations. One copper pendant has been dated to 8700 B.C. Copper is infinitely recyclable. The copper in the penny you have in your pocket may have been part of a bracelet worn by a pharaoh of ancient Egypt. This quality is not lost on the copper industry, which pays almost as much for recycled copper as for newly mined ore.

Copper is also an essential nutrient for the human body and widely found in foods. In addition, copper released from new copper plumbing may help purify the water.

Thinking It Through

1. What foods are rich in copper?
2. What other elements can you name that are also nutrients?

Skill-Building Activities

1. **Science.** Using a chemistry textbook or other resources, find the common name for each of the following compounds, which are often found in foods: a) $C_6H_8O_8$; b) $K_2C_4H_4O_6$; c) $C_{12}H_{22}O_{11}$.
2. **Teamwork.** Design a symbol bingo game. Create cards for four players, each containing nine symbols of elements frequently used in food science. What will you use or create as markers?
3. **Teamwork.** With a partner, create a list of homogeneous and heterogeneous mixtures. Compare lists with those made by others in the class. Which mixtures are easier to identify? Why?
4. **Science.** The model on page 102 shows part of a DNA molecule. Look for information in other references or on the Internet to find out why this molecule is so complex. Report your findings.
5. **Communication.** Create a rhyme, song, or other memory-jogging device to help classmates recall information in this chapter.

Understanding Vocabulary

Turn the "Terms to Remember" into true-or-false statements. Take turns quizzing a partner, correcting any false statements.

Thinking Lab

The Chemistry of Food

Whenever you eat or drink, you consume chemicals that your body needs for energy, growth, and repair. Thinking about food in a chemical way can be surprising.

Analysis

Foods are mixtures of many different compounds. If you eat scrambled eggs, you're consuming a mixture of water, ovalbumin, coalbumin, ovomucoid, mucia, globulins, amino acids, lipovitellin, livetin, cholesterol, lecithin, lipids, fatty acids, butyric acid, acetic acid, lutein, zeaxanthine, and vitamin A, with a little sodium chloride (salt) sprinkled in. Doesn't that sound appetizing?

Organizing Information

1. Which of the compounds in scrambled eggs are already familiar to you? Why?
2. Many of these compounds will be discussed in this course. Scan the table of contents in this text to see where you will likely learn more about many of these compounds.
3. Choose one food and find out what its chemical components are.

CHAPTER 8

Chemical Reactions and Physical Changes

Objectives

- Compare chemical reactions to physical changes.
- Compare the parts of an atom.
- Calculate the mass of one mole of an element or compound.
- Explain how ionic and covalent bonds are formed.
- Identify the parts of chemical equations.
- Distinguish between reversible and irreversible reactions and changes.
- Describe how chemical and physical changes in food affect nutrition.

Terms to remember

atomic mass
atomic number
chemical bond
chemical equation
chemical reaction
covalent bond
covalent compound
electron
ion
ionic bond
ionic compound
mass number
mole
nucleus
neutron
periodic table
physical change
product
proton
reactant

Candles lit on a birthday cake are a festive sight—and a dazzling display of chemistry. Have you ever wondered what happens when match meets wick? This chapter offers an illuminating discussion on the physical and chemical processes that are part of burning candles, eating food, and just about everything else.

The Cycle of Change

The candles flickering on a birthday cake are a good example of two basic scientific concepts, physical change and chemical reaction.

Catching fire, a candle's wick undergoes a **chemical reaction**, *the change of substances into other substances.* The string in the wick decomposes into the elements from which it was formed, which combine with oxygen in the air. As this chemical reaction continues, the heat from the fire begins to melt the wax. The wax undergoes a **physical change**, *an alteration of a substance that does not change its chemical composition.* In this case, the substance—wax—changes its phase from solid to liquid. The liquid wax moves up the wick toward the flame and also catches fire. Catching fire, the wax then undergoes a chemical reaction.

115

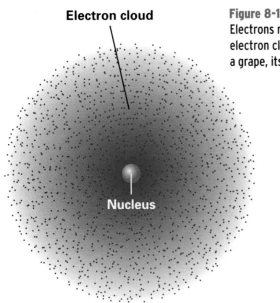

Electron cloud

Nucleus

Figure 8-1
Electrons move around the nucleus in an area called the electron cloud. If the nucleus of an atom were the size of a grape, its electrons would be about 1.6 km away.

protons contained in an atom determines its identity. That is, atoms of the same element have the same number of protons. *The number of protons in an atom is the* **atomic number** of the element.

- **Neutrons.** A **neutron** is *an uncharged atomic particle.* Atoms of one element can have different numbers of neutrons. Protons and neutrons are both found in the *dense core of an atom* called the **nucleus** (NOO-klee-us), the plural of which is *nuclei* (NOO-klee-eye). The protons give the nucleus a positive charge.

- **Electrons.** An **electron** is *a negatively charged particle that moves around the nucleus*, as shown in **Figure 8-1**. Electrons have almost no mass compared to protons and neutrons, which are extremely light themselves. However, it is the electrons that determine whether a chemical reaction takes place.

Chemical reactions produce one of the following three results: either elements join to form compounds; or compounds break apart into elements; or compounds form other compounds. How do these changes occur? Why is the process important to life? The second question will be answered over the course of upcoming chapters. Answering the first question takes an understanding of the atom—the building block of all matter. All chemical reactions involve the rearrangement of atoms.

Atoms: The Building Blocks of Matter

As you've learned, an atom is the smallest particle of an element that retains its chemical properties. As you know, atoms are extremely tiny. One gold atom is so small that it takes billions of gold atoms to make a speck of gold large enough to be seen under a microscope.

Structure of Atoms

Tiny as they are, atoms are made of even smaller parts called subatomic particles. The three most important particles are as follows:

- **Protons.** A **proton** is *a particle that has a positive electrical charge*. The number of

Figure 8-2

Properties of Subatomic Particles
Protons   • Found inside the atom's nucleus.   • Have an electrical charge of +1.   • Have a mass of 1.*
Neutrons   • Found inside the atom's nucleus.   • Have no electrical charge (neutral).   • Have a mass of just slightly more than 1.**
Electrons   • Found outside the atom's nucleus.   • Have an electrical charge of -1.   • Have a mass of 1/1836.**
*The mass of a proton is set at 1.   **Compared to the mass of a proton.

Figure 8-3
Electrons are distributed over a range of energy levels. How many electrons are at each level in this sulfur atom? Which electrons have the most energy?

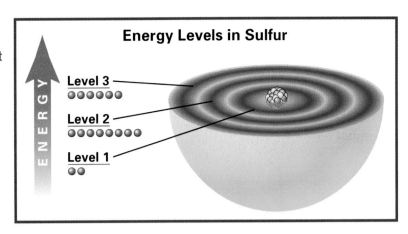

Energy Levels in Sulfur

Level 3
Level 2
Level 1

While the atoms of one element are different from the atoms of any other, subatomic particles are not. A proton in a carbon atom is identical to a proton in an oxygen atom. The same is true for electrons and neutrons. **Figure 8-2** compares properties of subatomic particles.

Energy Levels of Electrons

Opposite electrical charges attract. Positively charged protons in the nucleus hold the negatively charged electrons within the atomic structure. An atom stays neutral as long as it has an equal number of protons and electrons.

Electrons move about the nucleus in regions of space called energy levels, as shown in **Figure 8-3**. The first energy level, which is closest to the nucleus, is lowest in energy. More energetic electrons occupy levels increasingly farther from the nucleus. An electron in the third level, then, has more energy than one in the second. This pattern continues through all of an atom's energy levels.

Each energy level can hold only a certain number of electrons—two in the first, for instance, 18 in the third. Most atoms are chemically stable, or unlikely to react with others, when they have eight electrons in their outermost energy level. Hydrogen and helium are exceptions. With just one energy level, their atoms achieve stability with only two electrons.

Atomic Mass

By combining *the total protons and neutrons in an atom's nucleus,* you can determine that atom's **mass number**. Not every atom has the same mass number, however. Recall that while every atom of an element has the same number of protons, the number of neutrons may vary. This is shown in **Figure 8-4**. Thus, mass numbers may also vary. For example, every atom of chlorine has 17 protons. If one chlorine atom has 18 neutrons, its mass number would be 35. A different chlorine atom with 16 neutrons would have a mass number of 33.

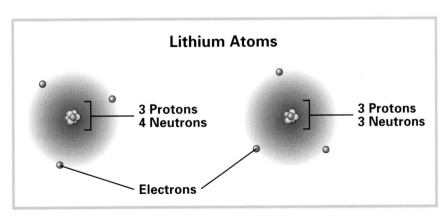

Lithium Atoms

3 Protons
4 Neutrons

3 Protons
3 Neutrons

Electrons

Figure 8-4
An element can have atoms that are not the same. These two lithium atoms contain different numbers of neutrons. How can you tell that these are lithium? What is the mass number of each of these atoms?

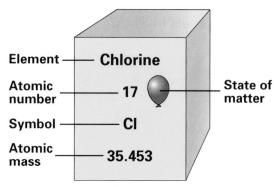

Figure 8-5
Each block in the periodic table gives you information about an element. What is the relationship between the atomic number and atomic mass?

Mass numbers are used to determine the atomic mass of an element. When the mass numbers of many chlorine atoms were actually averaged, the number 35.453 resulted. This decimal number is the atomic mass of chlorine. The **atomic mass** of any element, then, is *the average mass of a sample of atoms of that element found in nature.*

Hydrogen, with the simplest atomic structure (one proton and no neutrons), has an atomic mass of 1.0. Carbon's atomic mass is 12.0, and sulfur's is 32.1.

Atomic mass and other facts about each element are found in a **periodic table**. This is *a chart that arranges elements by atomic number into rows and columns according to similarities in their properties.* Elements with similar properties are placed in the same columns. A periodic table is printed inside the back cover of this textbook. One block of the table is shown in **Figure 8-5**.

The Mole

Suppose a recipe for oatmeal cookies tells you to "add 1000 oats." Can you imagine counting oats one by one? Items that you use in large amounts are more conveniently measured in large units than counted individually.

Chemists face the same situation with atoms and molecules. Working with some-

thing that tiny and numerous requires a very large unit of measure. For this purpose, scientists use the mole. A **mole** is *the unit of measure used to count atoms or molecules. It is a quantity equal to 6.02×10^{23}.* Written out, that's 602 followed by 21 zeros. Just as a dozen is 12 of anything, whether eggs or doughnuts, a mole of anything is 6.02×10^{23} items. A mole of carbon is the same number of atoms as a mole of hydrogen. The same is true of molecules in compounds. As you can see in **Figure 8-6**, a mole is a very large number.

As with any quantity, however, moles may be different in other properties. A mole of feathers and a mole of pennies are the same number of items. In mass and volume, however, they are vastly different.

Moles and Mass

When scientists work with a certain amount of an element or a compound, they often need to know its mass. Thanks to earlier discoveries, they have a few rules that simplify this task.

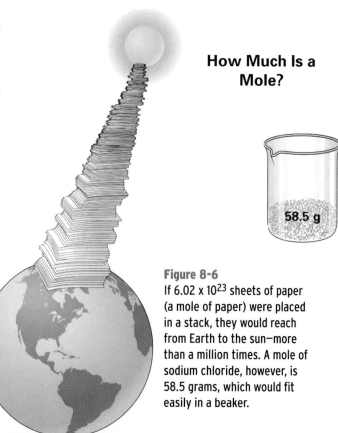

How Much Is a Mole?

58.5 g

Figure 8-6
If 6.02×10^{23} sheets of paper (a mole of paper) were placed in a stack, they would reach from Earth to the sun—more than a million times. A mole of sodium chloride, however, is 58.5 grams, which would fit easily in a beaker.

How do food retailers keep up with regulations?
Ask a . . . Consumer Safety Officer

I have seen many food retailers who were frustrated by trying to understand FDA codes," says consumer safety officer Tonya Halliburton. "That's why communication is such a huge part of my work. I'm a liaison between the FDA and retail food providers, so I teach people about government regulations and make sure they comply. The best part of my work is setting up training sessions. Every time we put on a program in some part of my region, I meet new people. Food service establishments send their employees to our programs, and many managers come to get updates and help with their problems. A consumer safety officer takes public safety seriously. I want to instill that same feeling in the people I meet."

Education Needed: The United States government generally requires consumer safety officers to have a bachelor's degree in biological sciences, chemistry, food science, or nutrition. Coursework in public health and law, especially in areas related to food safety, is also recommended.

KEY SKILLS
- Public speaking
- Leadership
- Teaching skills
- Knowledge of food safety
- Organization
- Relationship skills

Life on the Job: Consumer safety officers are the FDA's "field officers," providing technical assistance to local and state food safety agencies. They interpret and help those groups carry out regulations. They often develop and conduct training sessions for government and industry food inspectors, so that regulations are understood and uniformly implemented and enforced.

Your Challenge

Suppose a fast-food restaurant chain is sending its restaurant managers to an FDA food safety training program. As a consumer safety officer, you are designing the program.

1. What are some topics that your program will cover?
2. How will you impress the importance of safe food handling on attendees?

Each of these—water, salt, sugar cubes, and gas in the balloon—contains one mole of atoms or molecules. Their masses, however, are different.

One of the rules set by science is this: the mass of one mole of an element is called its *molar mass* and is equal to that element's atomic mass in grams. In the periodic table, for example, you see that the atomic mass of sodium (Na) is 23.0. A mole of sodium atoms, then, has a mass of 23.0 g. The atomic mass of chlorine (Cl) is 35.5, so a mole of chlorine atoms has a mass of 35.5 g.

To find the mass in grams of one mole of a compound, you must add the atomic masses of the elements in the compound. For example, a mole of sodium chloride (NaCl) has a mass of 58.5 g. The molar mass of NaCl equals the combined molar masses of sodium (23 g) and chlorine (35.5 g).

What if a compound contains more than one mole of an element? The formula for carbon dioxide, CO_2, for instance, shows two moles of oxygen. Here, the atomic mass of that element must be multiplied by its subscript. The molar mass of carbon dioxide equals the molar mass of carbon (12.0 g) plus twice the molar mass of oxygen (2×16.0 g), for a sum of 44.0 g.

Counting moles of atoms and figuring their mass may sound trivial, but it's essential to studying chemical reactions. By knowing the ratio of atoms and molecules involved in a reaction, chemists can figure the mass of elements needed to duplicate it. Just as a ratio of butter to sugar and eggs is a recipe for a cake, a ratio of certain elements is a recipe for a certain compound. From such "recipes," fuels and medicines are made.

How Compounds Form

Compounds form when two or more elements join chemically. Their atoms give in to attractive forces strong enough to bind them. *The forces that hold atoms together* are known as **chemical bonds**. Chemical bonds are of two types. Each one produces a particular kind of compound.

Ionic Bonding

Atoms are most stable when their outer energy level contains eight electrons. An atom that doesn't have this number is apt to interact with another atom to satisfy this need. It may transfer the electrons in its outer level to another atom, or accept electrons from an atom. **Figure 8-7** shows how electrons transfer.

For instance, a sodium atom has one electron in its outer level, but eight in the level just before it. A chlorine atom has seven electrons in its outermost energy level. When these atoms collide, sodium's single outer electron is transferred to the chlorine's outer level. In solving their "problem," the elements create the compound sodium chloride (NaCl)—table salt.

With an equal number of protons and electrons, an atom is neutral. After electrons transfer, however, a "donor" atom has more positive protons in its nucleus than negative electrons moving about it. It thus becomes positively charged. Likewise, the atom that accepts the extra electrons now has more negative electrons than positive protons. It takes a negative charge. These *charged particles* are called **ions** (EYE-ahns).

With opposite electrical charges, these positive and negative ions are attracted and held together by an **ionic bond**, *the bond formed*

Transferring Electrons

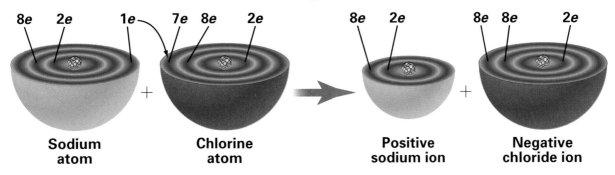

8e 2e 1e 7e 8e 2e → 8e 2e 8e 8e 2e

Sodium atom **Chlorine atom** **Positive sodium ion** **Negative chloride ion**

Figure 8-7
The transfer of an electron from a sodium atom to a chlorine atom forms sodium and chloride ions. Why did the sodium atom give up an electron?

by the transfer of electrons between atoms. This occurs only between atoms of different elements.

Atoms of metallic elements most easily achieve a stable energy level by losing electrons from their outer level. They tend to form positive ions. Nonmetals generally achieve stability by gaining electrons. *Compounds that result when metals and nonmetals bond ionically with each other* are called **ionic compounds.** They are also called salts.

An ionic compound's properties are completely different from those of its "parent" elements. Again, sodium chloride, shown in **Figure 8-8**, is a good example. Pure sodium is a highly reactive metal. It is kept under oil to prevent it from reacting with oxygen and igniting. Chlorine is a poisonous gas best known for purifying drinking or pool water because it kills bacteria. The transfer of a single electron between the two, however, allows them to form a white solid that's safe to eat.

Covalent Bonding

Ionic bonding occurs only between metals and nonmetals. Can you see why? Nonmetals need to gain electrons to be stable. When two nonmetals come together, all the atoms are "needy"; none has an electron to spare. Instead, both atoms need to gain one or more electrons.

With electron transfer impossible, two nonmetals acquire the electrons needed for stability by sharing them. This creates a remarkably strong bond, chemically speaking. A **covalent bond** (ko-VAY-lunt) results when *atoms share electrons with each other.* As you might guess, *compounds formed by the sharing of electrons* are called **covalent compounds,** and sometimes molecular compounds.

In covalent compounds, neither atom gains or loses electrons; both stay neutral. This distinguishes covalent compounds from ionic compounds.

Sometimes more than two atoms form covalent bonds. Hydrogen atoms, for example, have only one electron. In nature they occur in pairs,

Figure 8-8
When sodium and chloride react chemically to form ions, they create cube-shaped salt crystals.

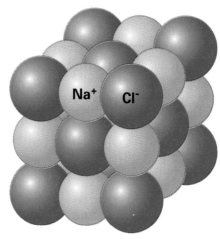

Na⁺ Cl⁻

Sharing Electrons

H   H   O   H₂O

Figure 8-9
When hydrogen and oxygen collide, water forms. Since neither atom is strong enough to take an electron from the other, they share. Look at the outer level on the right. You'll see that the oxygen atom needed two hydrogen atoms to achieve the eight electrons that make it stable. **How many electrons does each hydrogen atom now claim?**

sharing electrons for stability. Oxygen, meanwhile, has six electrons in its outer level. These two elements can become stable if each of two hydrogen atoms shares an electron with an oxygen atom. Now each hydrogen atom has two electrons, the oxygen atom has eight outer electrons, and the earth has one of its most precious resources. That molecule, represented as H_2O, is water. **Figure 8-9** shows this sharing of electrons.

Chemical Equations

If a formula is a word in chemistry's language, then a **chemical equation** is a sentence. This is *a written description of a chemical reaction, using symbols and formulas.*

The first symbols or formulas in a chemical equation are the **reactants** (ree-AK-tunts), *the elements or compounds present at the start of a reaction.* A plus sign separates the reactants. An arrow, called a yield arrow, represents change. The symbols or formulas to the right of the yield arrow, also separated by a plus sign, are the **products**. These are *the elements or compounds formed during the reaction.*

When more than one molecule is present in an equation, this is shown by the *coefficient.* The coefficient is a numeral written before the symbol or formula.

During a chemical reaction, matter is neither created nor destroyed, but is said to be *conserved.* This principle, the law of conservation of mass, is a basic one in chemistry. It means that both sides of a chemical equation must always be in balance, as shown in **Figure 8-10**. Although atoms rearrange during the change from reactants to products, their number and kind stay the same.

The following equation describes what happens when glucose, a sugar, reacts chemically with oxygen.

A Balanced Equation

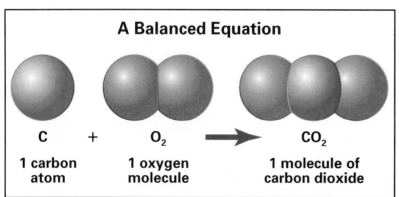

C   +   O₂   CO₂

1 carbon   1 oxygen   1 molecule of
atom   molecule   carbon dioxide

Figure 8-10
This equation is balanced because the same number of atoms exist on each side. If a piece of coal contains 10 billion atoms of carbon, how many oxygen molecules will it react with when it burns? How many molecules of carbon dioxide will be formed?

reactants   yield   products
arrow

$$C_6H_{12}O_6 + 6\ O_2 \rightarrow 6\ CO_2 + 6\ H_2O$$

glucose oxygen carbon water
dioxide

What information does the equation give besides the reactants and products? The coefficient shows the ratio of oxygen to glucose molecules in the reaction—six molecules of oxygen to one molecule of glucose—and of carbon dioxide to water in the products. These coefficients also show the ratio of moles and molecules. You can compare the number of atoms present on each side of the equation by multiplying the coefficients by the subscripts. You'll find they are the same. This means the moles of atoms are also the same.

Reversible Chemical Reactions

The chemical change described above is reversible. The products can be converted again into the reactants. In this case, you exhale carbon dioxide and water—the equation's products—which plants use to make glucose and oxygen, the equation's reactants. In this course you'll find other reversible reactions that occur in foods and in the body.

Many chemical changes are not reversible, as the burning candle described at the start of the chapter illustrates. The ash, smoke, and gases produced by the burning wick cannot be recombined into fibers.

Physical Changes

In contrast to chemical reactions, physical changes alter only properties, such as shape, size, and phase. The formation of a mixture, when substances combine without reacting, is an example of a physical change. A substance dissolved in water to form a solution also undergoes physical change. The substance disappears, yet keeps at least some of its original properties. As with chemical reactions, mass is conserved during physical changes.

Phase Changes

One type of physical change occurs when matter changes phase. Freezing is the change from a liquid to a solid. Melting changes the solid to a liquid. Another phase change occurs when a liquid changes to a gas, called vaporization (vay-pur-ih-ZAY-shun). Condensation (kahn-den-SAY-shun), in turn, changes a gas to a liquid.

Throughout its phase changes, a compound remains the same substance, with the same chemical makeup and formula. Water, ice, and steam all go by one chemical formula, H_2O.

Reversible Physical Changes

As with chemical changes, some physical changes are reversible; others are not. All phase changes are reversible, given the needed changes in environment. Physical changes that produce solutions are also reversible. Salt crystals lose their structure in a saltwater solution, and the water changes taste. However, the salt can be reclaimed, intact and unchanged, by heating the solution. If you heat the solution in a closed system, the water vapor can be captured as well. Both pure substances regain their previous condition.

Preparing food often involves irreversible physical changes. You can't, afterall, unmash a potato.

A phase change occurs when conditions cause matter to switch from one phase to another. What phase change will occur here?

Physical Changes and Chemical Reactions

SAFETY FIRST
Review these safety guidelines before you begin this experiment.

When matter undergoes a physical change, only size, shape, temperature, or physical state of the substance alters. Examples of physical changes are freezing or melting, boiling or condensing, dissolving, grinding, tearing, and breaking into smaller pieces. Physical changes produce no new substances. On the other hand, one or more new substances always result during chemical reactions. In this experiment you will cause several substances to undergo change. By observing the products, you'll determine whether a physical change or a chemical reaction occurred.

Equipment and Materials

iron filings (Fe)	magnet	ring stand	safety goggles
sodium bicarbonate ($NaHCO_3$)	6 test tubes	iron ring	hotplate or burner
sodium chloride (NaCl)	test tube rack	funnel	crucible tongs or forceps
sucrose ($C_{12}H_{22}O_{11}$)	water	filter paper	beaker tongs
sand (SiO_2)	3, 150-mL beakers	plastic wash bottle	vinegar
magnifying glass	stirring rod	watch glass	

Procedure Part A

1. Place small samples of each of the five solids (Fe, $NaHCO_3$, NaCl, $C_{12}H_{22}O_{11}$, SiO_2) on a piece of paper. Examine each with a magnifying glass. Record your observations in data table A.

2. Test the effect of a magnet on each substance by lifting the paper and passing the magnet under it.

3. Test solubility by placing half of each sample in a clean, dry test tube and adding 5 mL of water. Tap the side of the tube with your finger to mix the solid and water completely.

Procedure Part B

4. Mix the remaining iron filings with the sand on a single piece of paper. Test with the magnet as before.

5. Transfer the mixture of iron filings and sand, along with the remaining sodium chloride, to a clean 150-mL beaker. Add 30 mL of water and stir. Record your observations in data table B.

6. Attach an iron ring to a ring stand. Place a funnel in the iron ring. Fold a piece of filter paper in half, and then in half again to form a cone. Open the cone so that three layers of paper are on one side, and one layer is on the other. Place the paper in the funnel. Moisten it slightly with water from a wash bottle to hold it in place. Place a second 150-mL beaker under the funnel to catch the filtrate (the liquid passing through the filter paper).

7. Decant (pour off) the liquid from the solid by pouring it down a stirring rod into the funnel. Be careful to keep the liquid below the top edge of the filter paper at all times. After nearly all the liquid has passed through the filter paper, use a stream of water from the wash bottle to wash all of the solid out of the beaker onto the filter paper.

8. After all the liquid has passed through the filter paper, wash the solid on the filter paper with more water from the wash bottle.

9. Examine the solid remaining on the filter paper. Record your observations in data table B. Set the filter paper aside for later examination.

10. Pour a small amount of the filtrate onto a watch glass. Put on safety goggles. Heat the watch glass over medium heat on the hotplate or burner until the

liquid has completely evaporated. Remove the watch glass from the heat, using crucible tongs or forceps. Examine the solid remaining with a magnifying glass and describe it in your data table.

11. Place the remaining sucrose in a clean, dry 150-mL beaker. Place it over medium heat, noting any changes in appearance. Every two or three minutes remove the beaker from the heat, using beaker tongs, and check for odors by fanning the fumes toward your nose. Place the beaker over high heat for 1 to 2 minutes longer. Remove it from the heat and allow it to cool to room temperature.

12. Carefully unfold the filter paper reserved in Step 9. Spread out the residue on the paper with your stirring rod. Pass the magnet under the paper and record what you observe in your data table.

13. After the beaker used to heat the sugar has cooled to room temperature, examine the solid remaining

in the beaker. Add a few milliliters of water to test its solubility.

14. Add the remaining sodium bicarbonate to a clean, dry test tube and add 3 mL of vinegar. Feel the bottom of the test tube. Record your observations.

15. Follow your teacher's instructions for proper disposal of materials.

Analyzing Results

1. Did the properties of any of the substances change when they were mixed on the paper with other substances?

2. Was dissolving the substances in water a physical or chemical change? How do you know?

3. Did heating produce a physical or chemical change in the filtrate? In the sugar?

4. Was dissolving the sodium bicarbonate in the vinegar a physical or a chemical change? How do you know?

SAMPLE DATA TABLE A

Substance	Formula	Color	Effect of Magnet	Solubility in Water
Iron filings				
Sodium bicarbonate				
Sodium chloride				
Sucrose				
Sand				

SAMPLE DATA TABLE B

Substance	Observations
Iron and sand mixture	
Iron, sand, and salt mixed with water	
Residue (solid) on filter paper	
Solid remaining after evaporation of filtrate	
Sugar heated	
Sodium	

Chapter Summary

- Physical changes in a substance affect such properties as size, shape, and phase.
- Chemical reactions produce a change in a substance's basic identity.
- Atoms are the smallest particles of an element that retain the chemical properties of the element.
- Atoms are composed of subatomic particles: protons, neutrons, and electrons.
- Chemists figure the mass of elements and compounds by using moles.
- Elements combine to form compounds through reactions of electrons in their outer levels.
- Chemical bonds between atoms can be either ionic or covalent, depending on whether electrons are transferred or shared.
- Chemical reactions are expressed by chemical equations.
- Some chemical reactions and physical changes are reversible.

Using Your Knowledge

1. How are chemical reactions and physical changes different?
2. How can you determine an element's atomic number?
3. Can a chemist identify an atom by its number of neutrons?
4. Why is the number of electrons in an atom's outer level significant to chemical reactions?
5. What is the difference between atomic mass and mass number?
6. What is a mole? Why is it practical when dealing with chemicals?
7. How does an element's molar mass compare to its atomic mass?
8. Explain the process of ionic bonding.
9. Why do nonmetals form covalent bonds?
10. When do chemists write chemical equations?

Real-World Impact

Creating Compounds

Food scientists create some compounds to give foods a good taste and aroma. They can combine both artificial and naturally occurring substances to create tastes and smells as varied as pineapple and peppermint, almond and wintergreen. One compound gives sour apple candy its tartness. Another sweetens low-calorie foods. Natural substances extracted from foods are more complex, providing richer tastes and aromas. Flavorings made "from scratch" in a laboratory may be noticeably artificial.

Thinking It Through
1. What, if anything, is the difference between the claim "no added flavors" and "no artificial flavors"?
2. Why do you think artificial flavors are sometimes used instead of natural?

Using Your Knowledge *(continued)*

11. Identify the following examples as physical changes or chemical reactions: a) souring milk; b) slicing a cake; c) adding sugar to coffee; d) baking a cake; e) burning a piece of paper; f) melting ice.
12. Are any of the processes in question 11 reversible? If so, how would you reverse them?

Skill-Building Activities

1. **Research.** Scientists are still discovering even smaller subatomic particles. Find information on these discoveries and summarize information about them to your class.
2. **Communication.** Choose a concept discussed in this chapter. Think of a way to explain it to someone who isn't familiar with the topic. What comparisons or imagery could you use to make the ideas clear? Refine your explanation and try it on a friend or family member. Ask for feedback to check that you clarified the concept. Report on the success of your "lesson" to the class.
3. **Critical Thinking.** Develop an analogy between a chemical formula and a recipe. If a soup recipe called for 4 potatoes, 2 onions, 8 carrots, 4 stalks of celery, and 1.5 L of water in order to serve 4 people, how would you alter this to serve 8 people? 40 people? How does this compare to preparing 2 moles of a compound? 10 moles?
4. **Mathematics.** Using the periodic table, determine the molar mass of each of the following compounds: a) NaCl; b) H_2O; c) H_2SO_4; d) $C_6H_{12}O_6$.
5. **Science.** What is missing in the following equation? Why is it needed?
$$2 H_2 + O_2 \rightarrow H_2O$$

6. **Mathematics.** How many grams of water are produced in the following reaction when one mole of C_2H_5OH reacts with three moles of oxygen?
$$C_2H_5OH + 3 O_2 \rightarrow 3 H_2O + 2 CO_2$$

Understanding Vocabulary

On a sheet of paper, reorganize the "Terms to Remember" by placing related terms into different groups, such as subatomic particles or parts of a chemical equation. Create as many different groupings as you need.

Thinking Lab

Laboratory Language

Like other studies, chemistry has developed a specialized language to refer to equipment, processes, and products.

Analysis

As you might expect in science, substances tend to be named using a fairly logical system. Prefixes such as *mono-* and suffixes such as *-ate* are used consistently to describe specific qualities of a substance.

Organizing Information

1. Using a chemistry text or other resource, identify common prefixes and suffixes used in names of chemical compounds. Learn their meanings and locate examples of each.
2. Present your findings in a table or chart.

CHAPTER ⑨ Water

When the poet Samuel Coleridge wrote, "Water, water, everywhere," he was referring to the sea, but the words are also descriptive of the earth as a whole. With oceans, lakes, and rivers covering nearly three-quarters of its surface, Earth is indeed "the blue planet." In addition, underground beds called aquifers store huge amounts of water, which on occasion percolate to the surface as vapor in geysers and volcanoes. You may marvel at the stark beauty of icecaps and glaciers in the polar regions, especially if you struggle with humid summer days at home. Without water, life as you know it could not have developed and would soon cease to exist.

It's easy to appreciate the value of water. To understand how it serves its essential functions, though, you need to know it chemically. This chapter takes a closer look at water—what it is and what it does.

The Water Molecule

You may recall the water molecule (H_2O) as an example of covalent bonding. Two hydrogen atoms share electrons with one oxygen atom, creating chemical stability for all. Different elements may form different types of covalent bonds, however. The properties of oxygen and hydrogen affect the nature of the bond.

Polar Covalent Bonds

One important property that affects the makeup of a water molecule is called *electronegativity.* This is a measure of how strongly an atom attracts electrons in a chemical bond. Oxygen is much more electronegative than hydrogen. As oxygen exerts a stronger pull on shared electrons, the electrons are drawn closer to the oxygen atom. The oxygen end of the molecule is somewhat negative, while the hydrogen end is somewhat positive, as shown in **Figure 9-1** on page 130. This effect makes water a **polar molecule**, *a molecule with a clear division of opposite electrical charges.* ("Polar" means "opposite.") The bonds in such molecules are termed polar covalent bonds, and the substances that result are called polar covalent compounds.

Objectives

- Relate water's composition and structure to its properties.
- Compare bonds in water.
- Explain the functions of heat of fusion and heat of vaporization.
- Explain the effect of air pressure changes on boiling point.
- Explain sublimation and surface tension.
- Explain the functions of water in food preparation.
- Describe hard and soft water.
- Describe how the body uses water.

Terms to remember

boiling point
bound water
colloidal dispersion
density
emulsifier
emulsion
free water
hard water
heat of fusion
heat of vaporization
hydrogen bond
immiscible
latent heat
medium
melting point
polar molecule
solute
solvent
sublimation
surface tension

The Water Molecule

Negatively charged end

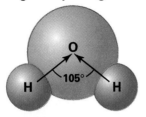

Positively charged end

Figure 9-1

The oxygen end of a water molecule is slightly negative, while the hydrogen end is slightly positive. What name is given to this type of molecule? Why do the hydrogen atoms form a 105° angle?

The arrangement of electrons also affects a molecule's shape. Since all electrons have a negative charge, they repel each other. In covalent compounds, however, the unshared electrons repel the shared electrons more forcefully than the shared electrons repel each other. In a water molecule, the two shared electrons are driven closer together, and their atoms with them. An angle drawn between the nuclei of the hydrogen atoms and the nucleus of the oxygen atom would measure about 105°, which is somewhat less than the 109.5° angles in a perfect tetrahedron.

Hydrogen Bonds

The polar nature of water molecules leads to their attracting each other far more strongly than would be expected. As you can see in **Figure 9-2**, the negatively charged oxygen atoms attract the positively charged hydrogen atoms in other molecules. This **hydrogen bond** is *an attractive force between any molecules in which hydrogen is covalently bound to a highly electronegative element.*

Hydrogen bonds between molecules are much weaker than the covalent bonds that hold a single molecule together. They are much stronger, however, than any force that exists between nonpolar molecules. Because water is a highly polar molecule, its hydrogen bonds are unusually strong.

Properties of Water

The properties that make water valuable to everything that lives come from its molecular makeup. In fact, water is the **medium**, *a substance through which something is transmitted or carried,* for nearly every chemical reaction that sustains life.

Like sodium chloride, water has surprising qualities, considering its "roots." Hydrogen is a highly flammable gas, reacting with oxygen in the air in order to burn. How is it, then, that water can put out fires? Once hydrogen and oxygen have combined to form water, they no longer have a need to react. Thus, they can be used to cool burning substances as they cut off the supply of oxygen that keeps a fire burning. Again, how does water remain liquid at room temperature, while oxygen and other substances with greater molecular masses have turned to gas? Hydrogen bonding explains this behavior.

Figure 9-2

As these molecules come together to form water, hydrogen bonds make the connections. Inside the molecules, covalent bonds hold the hydrogen and oxygen together.

Bonding in Water

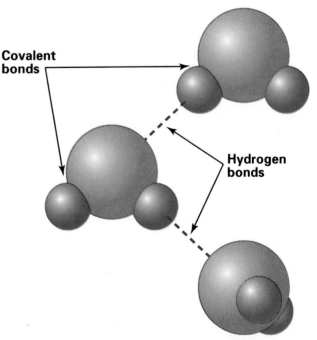

Covalent bonds

Hydrogen bonds

Because freezing and melting occur at the same temperature, both phases of water, solid and liquid, can be present when the temperature is 0°C.

Density of Water at Various Temperatures

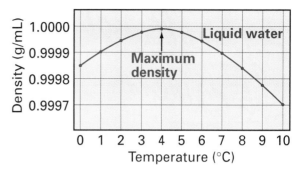

Figure 9-3

Water achieves its minimum volume, and therefore its maximum density, at 3.98°C.

Freezing and Melting

If you've ever returned from ice skating to enjoy a steaming cup of hot cocoa, you've experienced water in its three phases—solid ice, liquid in the cocoa, and gaseous steam.

Like other substances, water changes phase at certain temperatures. The **melting point** is *the temperature at which a substance changes from a solid to a liquid.* Melting occurs as heat energy enables molecules to break out of their solid structure. For pure substances, melting and freezing occur at the same temperature. What you call it depends on which process is occurring. Water's melting point is much higher than you might expect of such a small, "lightweight" molecule. The strong hydrogen bonds between molecules take considerable energy to break.

These same bonds cause water's unusual behavior when it freezes into a solid. As a liquid cools, its atoms or molecules generally draw closer. They become more orderly and usually more compact. One property of a solid, remember, is that it can be compressed only slightly. The crystal structure of solids usually increases **density**, *a substance's mass per unit of vol-*

ume. The frozen form of most substances is denser than the liquid form.

Water acts in this way, to a point, as **Figure 9-3** shows. Due to hydrogen bonds, the most orderly arrangement of water molecules contains a good deal of open space between them—more space, in fact, than in their "jumbled," liquid state. It's similar to lining up equal-sized, cardboard boxes that have the flaps stuck open. Water has a more open structure at temperatures below 4°C than it has in warmer conditions. As water turns to ice at 0°C, the molecules arrange into a solid that is 10 percent less dense than its liquid. This is why ice floats in cold drinks, and icebergs drift in oceans.

As water freezes, hydrogen bonding causes the molecules of liquid water to separate slightly from one another. The open arrangement can result in a six-pointed snowflake like this one.

Careers
Agricultural Engineer

Can agriculture promote a balanced ecology?
Ask an ... Agricultural Engineer

Neil Robinson takes a long-range view of his career. "Ever since I was in high school, I've worried about what's happening to the environment. Working in a field that relates to these issues is just natural for me. As an agricultural engineer, I deal with farming problems. Right now I'm helping a farmers' group devise a means of sustainable irrigation. They've been diverting river water for their crops, but they're worried about possible long-term effects on other communities, including the prairie. Someday I'd like to tackle ag problems in another part of the world. This is a life-long commitment for me."

Education Needed: Agricultural engineers need a bachelor's degree in agricultural, environmental, or civil engineering. Work in chemistry, water and soil science, and surveying offers insight into the complex natural conditions that affect agricultural production. An understanding of environmental issues, such as pollution and soil erosion, is essential.

KEY SKILLS
- Problem solving
- Relationship skills
- Communication
- Desire to work outdoors
- Organization
- Interest in environmental issues

Life on the Job: Agricultural engineers apply engineering principles to raising crops, striving for solutions that serve the farmer, society, and the environment. They may work for government agencies, makers of drainage and irrigation materials, or as consultants. They may specialize in such areas as machinery design or soil and water management.

Your Challenge

Imagine you're an agricultural engineer for a soil and water conservation agency. After several years of floods, farmers who live along the river are pressing for action to prevent more damage to their crops.

1. Who is responsible for a solution to this problem? Why?
2. What means to restrict or redirect the river's course do you consider?

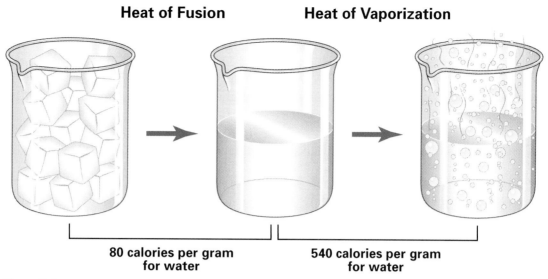

Heat of Fusion

Heat of Vaporization

80 calories per gram
for water

540 calories per gram
for water

Figure 9-4
The heat of fusion and the heat of vaporization are measures of the energy needed to cause phase changes.

Heat of Fusion

Why do you put ice, and not cold water, around foods in a cooler? Obviously, ice is colder, but how does that translate into keeping foods colder for a longer time? The answer has to do with **latent heat**, *the heat required to create a phase change without a change in temperature.* Think of latent heat as the "cost" of changing phases.

One example of latent heat is the **heat of fusion**. See **Figure 9-4**. This is *the amount of heat needed to change 1.0 g of a substance from solid to liquid phase. When the situation is reversed, and the liquid phase becomes a solid, this heat energy is released.*

For water, the heat of fusion is 80 calories per gram (80 cal/g). This means that it takes 80 calories of heat energy to change 1.0 gram of water from solid to liquid. As ice melts, it absorbs heat energy without a rise in temperature. Instead, water molecules use the energy to break out of the solid ice structure.

When water freezes, heat energy releases from the ice. Eighty calories of heat energy are released for every 1.0 g of liquid water that freezes. Eighty calories isn't much. You use more energy than this in an hour simply breathing, blinking, and tapping your toe.

Heat of fusion is the reason foods are put on ice to keep cold. Each gram of ice absorbs 80 calories of heat energy as it melts. Cold water, by comparison, absorbs just one calorie per gram for every degree C as it warms to the temperature of its surroundings.

Boiling and Condensing

A liquid's **boiling point** is *the temperature at which its vapor pressure equals the air pressure above the liquid.* At normal atmospheric pressure, water boils at 100°C. Like its melting point, water's boiling point is very high given its makeup and structure. Again, this is due to hydrogen bonds. It takes much more energy to break the hydrogen bonds between the molecules and allow the water to turn to steam than to separate most molecules the size and mass of water molecules.

Latent heat is involved in this phase change also, as the **heat of vaporization**. This is *the amount of heat needed to change 1.0 g of a substance from the liquid phase to the gas phase.* The heat of vaporization of water is 540 calories per gram (540 cal/g). This means it takes 540 calories to change 1.0 gram of water from liquid to steam. Like heat of fusion, heat of vaporization doesn't increase a substance's temperature. During any phase change, the temperature of the substance stays constant.

Notice, too, that boiling water over a higher heat simply causes it to boil more rapidly and evaporate more quickly. Adding heat faster doesn't make boiling water any hotter; it only wastes energy.

Just as the heat absorbed during melting is released during freezing, the heat absorbed when a liquid turns to gas is released when the gas condenses to liquid. This is why steam at 100°C is even more dangerous than boiling water at 100°C. When steam contacts your skin, it condenses because your body temperature (about 37°C) is well below the boiling point of water. As it condenses, it releases its heat of vaporization. This added heat, which water at 100°C does not possess, causes the initial damage. As the 100°C water cools on your skin, you absorb more heat, which intensifies the burn.

Air Pressure and Boiling Point

The definition of boiling point tells you that air exerts a downward pressure. This force results from the movement of atoms and molecules, which also causes some of the effects you'll read about in later chapters.

Air pressure over liquid water controls how rapidly water molecules can escape into the air. Vaporization can only occur when molecules gain enough energy to exert a force equal to the force of the air pressure over the liquid. Molecules gain the energy they need to exert this force by being heated.

Pure water boils at 100°C at standard atmospheric pressure, which is the pressure of the air at sea level on a clear day. At lower air pressure, water molecules meet less resistance in forming gas bubbles and escaping from the surface. Thus, lowering atmospheric pressure lowers the boiling point. **Figure 9-5** illustrates this principle.

This has practical implications for food preparation, as the "high altitude" directions on some food packages indicate. Atmospheric pressure drops as altitude increases. As a result, the boiling point of water also drops—1°C for every 293-m rise in altitude. Foods prepared in water take longer to cook because water boils at lower temperatures, as shown in

Figure 9-5
What happens to the boiling point of water when atmospheric pressure lowers? Why?

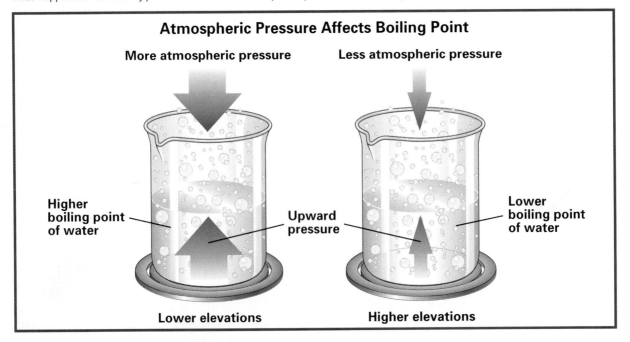

Atmospheric Pressure Affects Boiling Point

More atmospheric pressure

Less atmospheric pressure

Higher boiling point of water

Upward pressure

Lower boiling point of water

Lower elevations

Higher elevations

Figure 9-6

Boiling Point at Various Elevations	
Elevation Feet (meters)	Temperature of Boiling Water
Sea level	100°C
1,000 (305)	98.9°C
2,000 (610)	97.8°C
3,000 (915)	96.7°C
4,000 (1220)	95.6°C
5,000 (1525)	94.4°C
6,000 (1830)	93.3°C
7,000 (2135)	92.2°C
8,000 (2440)	91.1°C
9,000 (2745)	90.0°C
10,000 (3050)	88.9°C

Figure 9-6. The proportions of recipe ingredients, especially liquids, may need to be altered.

Daily, less drastic changes in air pressure can affect food preparation. If weather forecasters predict a "low pressure system" moving in, be prepared for stormy weather and lower boiling points. You might plan a different day to make candy or jelly, where success depends on the boiling point of the syrup.

Seeing the effect of air pressure on boiling point, you can understand how a pressure cooker works. As water in the cooker boils, an airtight seal traps the steam. The pressure of the steam is added to the atmospheric pressure inside the cooker. The increased pressure raises the boiling point of the water, causing foods to cook more quickly.

Sublimation

Another kind of phase change bypasses the liquid stage. **Sublimation** (sub-luh-MAY-shun) is *a change from the solid phase directly into the gas phase*.

Relatively few substances sublime. The compound carbon dioxide is one that does. At normal atmospheric pressure, the vapor pressure of its solid state is greater than that of its liquid state, so it changes from a solid to a gas without becoming a liquid. Since it doesn't melt, it's never wet; and since it absorbs heat so efficiently, it's popular for refrigeration. Thus, solid carbon dioxide is better known as dry ice.

Sublimation is induced in a method of commercial food preservation called freeze-drying. You're probably most familiar with the sublimation that sometimes occurs in food that has been kept too long in the freezer. This dry, grayish, off-tasting state is aptly known as freezer burn. It's caused when ice in the food sublimes.

Surface Tension

Have you ever filled a water glass so full that the water's surface seemed to stand just above the rim? Have you ever seen an insect rest or skip across the surface of a pond? These feats are not optical illusions. They are displays of a property of water known as **surface tension**. Surface tension is *an inward force or pull that tends to minimize the surface area of a liquid*. It's the same reason why rain falls in

Sublimation occurs when dry ice goes directly from the solid phase to the gas phase. Is this a common occurrence?

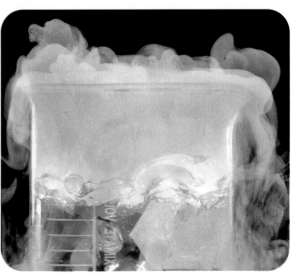

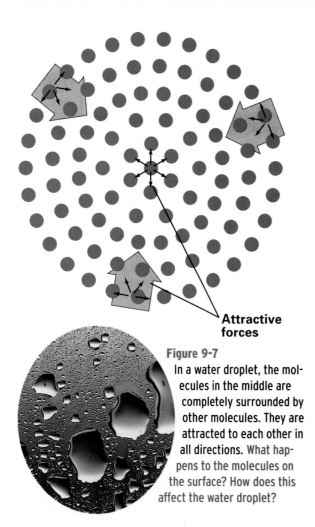

Attractive forces

Figure 9-7
In a water droplet, the molecules in the middle are completely surrounded by other molecules. They are attracted to each other in all directions. What happens to the molecules on the surface? How does this affect the water droplet?

spherical droplets, as though shrinking from contact with the air. A sphere is the most compact geometric shape any material can assume.

The surface tension of water, too, directly results from the strength of the hydrogen bonds between water molecules. Within a body of water, molecules experience the force of hydrogen bonds equally in all directions, as shown in **Figure 9-7**. Molecules on the surface experience a pull from below and to the sides, but not from above. As a result, they are drawn together and downward, behaving almost as a skin. It's a thin skin, however, and those water-skipping bugs would sink if they broke through the fragile outer surface.

Lowering surface tension is one way that detergents clean clothes. Detergents interfere with the formation of hydrogen bonds. This reduces the attraction of water molecules for

each other, and thus surface tension. Water penetrates fabrics more easily.

Water and Foods

Anyone who has heated a can of condensed soup or bitten into a juicy pear knows that water is a main ingredient in many foods. Experienced cooks value water as an important medium in food preparation.

Water Content in Food

You might not know it from appearances, but some foods are mostly water. Fruits, vegetables, and meats are 70 to 90 percent water.

Some water, fittingly called **free water**, *readily separates from foods that are sliced, diced, or dried*. A slice of bread gets crisp in the toaster as the heat evaporates the free water at the bread's surface.

Water that cannot be easily separated in food is called **bound water**. The various chemical groups in a food's molecules hold bound water tightly. It cannot react with other substances as free water does. It freezes only at very low temperatures and doesn't evaporate. Adding dry gelatin to water is a good example of what happens when water becomes bound. Within minutes the water becomes unable to flow.

Free water readily separates from fruit. What other food examples illustrate this point?

When you dissolve one substance in another, which is the solute and which is the solvent?

Water in Heat Transfer

Imagine heating a potato in a saucepan on a range top. The bottom of the potato would scorch before the rest of the vegetable cooks. If you add water to the pan, it acts as a medium for heat transfer. It first absorbs heat through the bottom of the pan, then bathes the potato evenly in heat. You'll learn more about the forces behind heat transfer in an upcoming chapter.

Water as a Solvent

In a solution, *the substance that dissolves another substance* is called the **solvent** (SAHL-vunt). *The substance that is dissolved* is the **solute** (SAHL-yoot). A substance that can be dissolved in another substance is said to be soluble. Thus, a solute is soluble in a solvent.

As a polar covalent compound, water can dissolve many substances. It is in fact called "the universal solvent."

As a polar molecule, water can dissolve ionic compounds. Each charged end of the water molecule attracts ions of the opposite charge, pulling them apart and surrounding them with other water molecules. When sodium chloride (table salt) is mixed in water, for instance, the negative oxygen ends of the water molecules attract the positive sodium ions, while the positive hydrogen ends attract negative chloride ions. Examine **Figure 9-8** to see what happens.

Water can also dissolve many covalent, or molecular, compounds. Here again, water's at-

tractive forces are able to pull the molecules apart. Sugar, for example, dissolves in the free water of strawberries to form a sweet syrup. Water dissolves the flavor molecules in tea leaves and coffee beans, creating popular beverages.

Water's excellence as a solvent can be a problem, for it's also effective with some vitamins and minerals present in foods. These nutrients may dissolve in cooking water and be lost during food preparation.

Hard Water

The term **hard water** may sound like a contradiction. It refers to *water that contains calcium or magnesium ions*. These metallic elements often dissolve in water as it passes through the ground. Water that doesn't contain these metal ions is considered soft.

While most water contains some dissolved substances, the metal ions in hard water can affect the quality of food prepared with it. Hard water interferes with water's tenderizing effect.

Figure 9-8

A polar water molecule has a negative end and positive end. When sodium chloride dissolves in water, one end of the water molecule attracts sodium ions and the other end attracts chloride ions. Why?

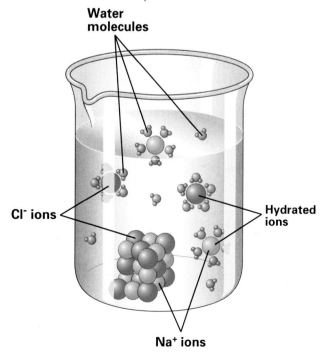

Water molecules

Cl⁻ ions

Hydrated ions

Na⁺ ions

The final rinse in a car wash uses distilled water. What is the purpose?

For example, dried beans, peas, and lentils must cook longer in hard water. Such beverages as iced tea may turn cloudy.

Hard water doesn't dissolve soap as effectively as soft water. Laundry washed in hard water tends to have a grayish film. Dishes washed in hard water in a dishwasher often have water spots.

Soft Water and Sodium

To solve their hard water problems, some households use a water softening device that works by ion exchange. The water softener exchanges each calcium or magnesium ion with two sodium ions.

You may have heard that sodium has been linked to high blood pressure, which in turn contributes to heart attack and stroke. Are people risking their health for the sake of cleaner clothes?

The amount of sodium added to soften water is not a problem for most people. Even in especially hard water, sodium is added at a rate of 160 mg per liter. The recommended limit for sodium is 2400 mg daily. You could drink one liter of sodium-softened water a day and still get less than ten percent of that amount. For most people, table salt and processed foods add more sodium to the diet.

People can learn the sodium levels of their drinking water from their local health department, since some amount of this mineral occurs naturally in most water. Those with concerns may buy distilled (demineralized) water or choose a water purification system that removes both sodium and hard water ions.

Applying Your Knowledge

1. Why doesn't sodium-softened water taste noticeably salty?
2. To soften moderately hard water takes 80 mg of sodium per liter. If you drank the suggested eight, 250-mL glasses of this water each day, what percentage of the recommended sodium limit would you consume?

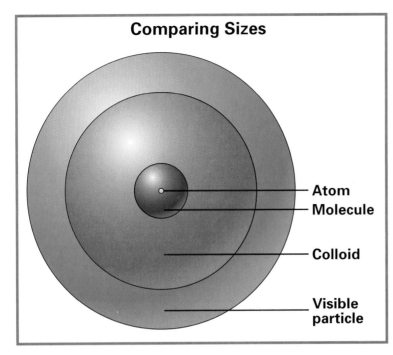

Comparing Sizes

- Atom
- Molecule
- Colloid
- Visible particle

Colloid particles are generally clumps that are ten to 100 times larger than typical ions or molecules dissolved in solutions. They are still not large enough to be seen.

For these reasons, people usually want to remove the undesirable ions from hard water. The method they use depends on whether water is temporarily or permanently hard.

Water that is temporarily hard contains bicarbonate ions. Metal ions bond to bicarbonate ions, and the resulting compound dissolves in the water. Such water can be softened by boiling. Boiling changes bicarbonates to insoluble carbonates. These compounds, formed between the metal ions and the carbonate ions, don't dissolve in water. Instead they settle to the bottom of the container. You may have seen this effect on a teakettle that has been used for a long time to boil water.

On the other hand, if the metal ions are present as compounds called sulfates, the water is considered permanently hard. Such water must pass through an ion exchange filter or undergo chemical treatment to remove the calcium or magnesium ions. In areas with hard water, many consumers use appliances called water softeners to remove the metal ions.

Water As a Dispersing Medium

Water can be a medium for a mixture called a **colloidal dispersion** (kuh-LOYD-ul dis-PUR-zhun). This is *a homogeneous mixture that is not a true solution.* Compared with true solutions, colloidal dispersions contain relatively large solute particles. These particles are called colloids (KAHL-oyds).

Although larger than atoms, ions, and most molecules, colloids are still too small to be seen with a microscope. They're also not large enough to settle out of the colloidal dispersion. Instead they remain scattered throughout, or dispersed in, the solvent.

Proteins as a group form colloidal dispersions. Recipes for many molded salads, for example, tell you to make a colloidal dispersion with boiling water and gelatin, which is a protein. Colloidal dispersions are described in more detail in Chapter 20.

One common colloidal dispersion that includes water is an **emulsion** (ih-MUL-shun). An emulsion is *a mixture of two liquids containing droplets that don't normally blend with each other.* The droplets remain separate because each liquid's molecules are attracted more strongly to each other than to molecules of the other liquid. Liquids that *don't blend or mix* are called **immiscible** (ih-MISS-uh-bul).

Emulsions frequently consist of a liquid composed of polar molecules, such as water, and a liquid of nonpolar molecules. An oil-and-vinegar blend is a common example. While these two molecules naturally avoid each other, they can be permanently blended with the addition of an **emulsifier** (ih-MUL-suh-fy-ur). An emulsifier is *a substance composed of large molecules that are polar at one end and nonpolar at the other.* The polar end is attracted to water molecules; the other end is attracted to the mole-

A vinegar-and-oil blend is a common example of an emulsion. What quality makes it an emulsion?

A healthy body maintains a remarkable balance between water intake and output, an average of about 2.6 liters of water each day. A person typically consumes about 700 to 1000 mL of water a day in food and 1200 to 1500 mL in liquids. Chemical reactions that take place within the body contribute another 200 to 300 mL.

On the outgoing side, about 150 to 200 mL of water are lost each day in solid waste, and 1000 to 2000 mL as urine. You sweat off about 350 mL, and exhale roughly 350 mL as water vapor.

Because it's so basic, water is often overlooked as a component of good health. The next few chapters may give you new appreciation for this simple but vital compound.

cules of oil. Simply put, the emulsifier glues the two types of molecules together.

Water and the Body

When a person steps on a scale, about 50 to 65 percent of the weight recorded is water. The exact percentage depends on the amount of fat, bone, and muscle in the body. Men, with their higher proportion of muscle to fat, usually have a slightly higher percentage of water in their bodies than women.

In general, the body obtains water from two sources: food and liquid consumed, and internal chemical reactions. To help ensure an adequate intake, the body stimulates feelings of thirst. This alerts you that your personal water supply is running low. Drinking the six to eight glasses of water a day recommended by nutritionists avoids triggering this "warning signal."

Water helps your body maintain a constant temperature. Normally, if your body temperature begins to rise, you perspire. The perspiration then evaporates off your skin. Energy is required for evaporation to occur, and you are the nearest source of energy. The water removes energy from your body as it evaporates, which helps maintain a constant body temperature.

Water is the medium for many chemical reactions within the body. How many glasses should you drink daily?

The Solvent Properties of Water

SAFETY FIRST

Review these safety guidelines before you begin this experiment.

Water is a very effective solvent for a variety of solutes. In this experiment you will compare the solvent properties of water to those of two other household liquids. **Note:** Alcohol is a flammable liquid. Keep it away from flames.

Equipment and Materials

10-mL graduated cylinder	test tube rack	sodium chloride	rubbing alcohol
water	balance	sucrose	vegetable oil
8 test tubes	stirring rod		

Procedure

1. Using the 10-mL graduated cylinder, measure and add 10 mL of water to each of four clean test tubes in a test tube rack.
2. Mass 2.0 g of sodium chloride and add to the first test tube. Stir with a clean stirring rod and record your observations in your data table.
3. Repeat Step 2 with the second test tube and 2.0 g of sucrose.
4. Measure 5.0 mL of rubbing alcohol with the 10-mL graduated cylinder. Add this to the water in the third test tube. Stir with a clean stirring rod. Record your observations in your data table.
5. Repeat Step 4 with the fourth test tube, using 5.0 mL of vegetable oil.
6. Using the 10-mL graduated cylinder, measure and add 10 mL of alcohol to each of two clean test tubes.
7. Add 10 mL of vegetable oil to each of the remaining two clean test tubes.
8. Add 2.0 g of sodium chloride to one test tube of alcohol, and 2.0 g of sucrose to the other. Record your observations in your data table.
9. Repeat Step 8 with the two test tubes of vegetable oil.

Analyzing Results

1. Which liquid dissolved the two solids completely?
2. Which liquid or liquids dissolved in water?
3. Given that water is a polar molecule, alcohol is a slightly polar molecule, and vegetable oil is nonpolar, explain your answer to the second question.
4. Again remembering their relative polarities, explain each liquid's ability or inability to dissolve sodium chloride, an ionic compound, and sucrose, a polar molecular substance.

SAMPLE DATA TABLE

Mixture	Behavior	Mixture	Behavior
Sodium chloride in water		Sodium chloride in alcohol	
Sucrose in water		Sucrose in alcohol	
Alcohol in water		Sodium chloride in salad oil	
Salad oil in water		Sucrose in salad oil	

Chapter Summary

- A water molecule's composition and electron arrangement create a division of electrical charges.
- As a polar molecule, water forms hydrogen bonds.
- The level of atmospheric pressure directly affects the boiling point of water, which can affect food preparation.
- Food contains both free and bound water.
- Hard water contains metal ions, which can cause problems in food preparation.
- A healthy body balances water intake with water output.

Using Your Knowledge

1. How does water's composition of oxygen and hydrogen create a polar molecule?
2. How does water form hydrogen bonds?
3. How can you explain water's unusually high melting point?
4. Why is liquid water denser than ice?
5. Why does temperature remain constant when a substance melts?
6. Why doesn't pasta cook more quickly at a high boil than at a low boil?
7. Explain why water boils sooner at higher altitudes, yet foods take longer to cook.
8. What is the chemical explanation for sublimation?
9. Explain the difference between free water and bound water.
10. Why is water an effective solvent of ionic compounds?
11. Why is water said to be a dispersing medium in cooking?
12. Should you wait until you feel thirsty to drink water? Why or why not?

Real-World Impact

World Water Day

Since 1993, the United Nations has observed March 22 as World Water Day. The purpose is to focus attention on the global water supply. Various UN agencies hold conferences and seminars for scientists and government policy makers. They publicize the serious lack of clean water, which contributes to some 25 million deaths each year, especially in developing nations. They support community education and conservation programs.

Thinking It Through

1. The theme of World Water Day 1999 was "We All Live Downstream." What does that statement mean? Why was it chosen?
2. As students of food science, what can your class do to promote World Water Day in your community?

Skill-Building Activities

1. **Consumer Skills.** Do a blindfold taste test comparing bottled water and tap water. Let the tap water sit for at least 24 hours to allow the chlorine to escape. Be sure all samples of water are at the same temperature. What differences, if any, do you notice? Which do you prefer? Why do you think bottled water is so popular?

2. **Health.** Using the web site of a reputable medical association, investigate the use of diuretics. What is their function? What are their legitimate and "unofficial" uses? Summarize the information for your class.

3. **Citizenship.** Research issues concerning the water supply at the web site www.uswater-news.com/archives/. Choose a topic and view several related articles. How does the issue affect the individual and society? What problems are posed? What solutions are suggested? Which course do you recommend? Report your findings to the class.

4. **Communication.** Why does a drop of water form a sphere? Create a poster that illustrates your answer.

5. **Consumer.** Take a class survey on the most popular beverages and the amounts consumed daily. Discuss the nutritive value of each choice.

Understanding Vocabulary

Work with a partner. Give the definition of one "Term to Remember." Then have your partner identify the term. Reverse roles and continue until you have defined all of the terms.

Thinking Lab

Pretreating Water

For water that is safe for cooking, drinking, and bathing, most people rely on public water companies. These utilities filter and chemically treat water at sewage treatment plants to remove impurities.

Analysis

Some communities are experimenting with methods of "pretreating" water before it enters the sewer system. Home builders layer sand and crushed rock along driveways to trap engine oil drippings from cars. Marshes are created to filter out other pollutants.

Thinking Critically

1. What might be the advantages of such types of pretreatment?
2. How can you practice water conservation while preparing foods?

CHAPTER 10 Acids and Bases

Objectives

- Relate the process of ionization to the formation of acids and bases.
- Explain qualities of acids and bases.
- Compare the acidity of substances, using the pH scale and pH indicators.
- Use molarity and titration to determine the concentration of an acid.
- Contrast the concepts of strength and concentration in acids and bases.
- Compare general qualities of acids and bases in foods.
- Explain the importance of pH to physical health.

Terms to remember

acid
base
buffer
concentration
equivalence point
indicator
ionization
molarity
neutral
neutralization
pH scale
titration

For many people the term "acid" brings to mind the caustic fluids used in a chemistry laboratory. While that's true, acids are also found in foods, such as lime juice and vinegar. Many of the foods you eat gain flavor from one acid or another. Acids are also part of everyday life in other ways—in chlorine bleach and in car batteries, for example.

While you might not recognize bases as easily as acids, they are also common. Taking an antacid for an upset stomach and improving garden soil with lime are both examples of how bases are used. Although few foods are basic, blood is, and so are many household cleansers. Baking soda and lye are bases.

So just what makes a substance an acid or a base? How are the acids in foods different from those that are dangerous to consume or touch? You'll soon see what those differences are.

Ionization of Water

An understanding of acids and bases begins with water. Scientists have found that a very small portion of molecules in water (about two in one billion) form tiny charged particles that you know as ions. The *process of forming ions in water solution* is called **ionization** (eye-uh-nuh-ZAY-shun).

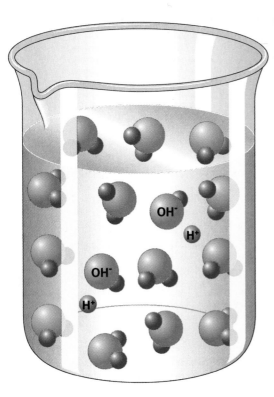

When water ionizes, both hydrogen (H^+) and hydroxide (OH^-) ions are produced. See **Figure 10-1**. The equation for this process is shown below.

$$H_2O \leftrightarrow H^+ + OH^-$$
Water   **Hydrogen**   **Hydroxide**
              ion   **ion**

In the equation, notice the plus and minus sign beside each ion. These signs are known as superscripts, small letters, numbers, or symbols placed just above and to the side of another.

Hydrogen ions are hydrogen atoms that have lost their electron and become positively charged. Thus, the hydrogen ion carries a plus sign by its symbol.

Hydroxide ions are composed of one oxygen atom bonded to one hydrogen atom. They have a negative charge. Thus, the symbol carries a negative sign.

The two-directional yield arrow in the equation shows that the ions can recombine to form water. All water contains some hydrogen ions and some hydroxide ions. *Since pure water*

has an equal number of hydrogen and hydroxide ions, it is said to be **neutral**.

Defining Acids and Bases

Most water on Earth, however, isn't pure. It contains many dissolved substances. When some substances dissolve in water, they release hydrogen ions, while others release hydroxide ions. These substances are acids and bases, and they change the balance of ions in a water solution.

Acids contain hydrogen. *When* **acids** *dissolve in water, their molecules break apart and release hydrogen ions into the solution.* **Bases**, which are the chemical opposites of acids, *release hydroxide (hy-DRAHK-side) ions in water solution.* Simple bases are composed of metal ions and hydroxide ions, which break apart when they dissolve in water. A few bases, such as ammonia, are covalent compounds that produce hydroxide ions by reacting with water. **Figure 10-2** shows an example of a base in an ionization equation.

Some compounds that dissolve in water, of course, are neither acids nor bases. Sodium chloride, for example, splits into positive sodium ions and negative chloride ions, which doesn't change the ratio of hydrogen to hydroxide ions in the water solution.

Strength of Acids and Bases

Not all acids and bases produce ions in solution to the same degree. *Strong* acids and bases are said to ionize completely when they dissolve in water. This means that when acid molecules are added to a solution, nearly all of the acid molecules break apart and release hydrogen ions into the solution. The same is true of bases, except they release hydroxide ions.

Hydrogen chloride (HCl), for instance, separates into H^+ and Cl^- in water solution. This

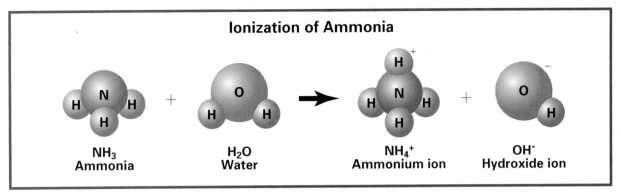

Ionization of Ammonia

NH$_3$
Ammonia

H$_2$O
Water

NH$_4^+$
Ammonium ion

OH$^-$
Hydroxide ion

Figure 10-2
A reaction between the gas ammonia (NH$_3$) and water produces ammonium ions containing one nitrogen atom and four hydrogen atoms, plus hydroxide ions. What does the superscript on the H mean? Is ammonia an acid or a base? How can you tell?

reaction occurs in nearly every hydrogen chloride molecule. The resulting solution, hydrochloric acid, is not only an acid, but also a strong acid, as shown in **Figure 10-3**.

On the other hand, when acetic acid (HC$_2$H$_3$O$_2$) dissolves in water, only about 0.5 percent of its molecules (five in 1000) ionize; the remaining molecules are neutral. Therefore, this acid is considered to be *weak*. The acetic acid found in the vinegar added to a potato salad is a weak acid.

Neutralization of Acids and Bases

What happens when you mix an acid with a base? As the proportion of hydrogen ions and hydroxide ions changes, the solution becomes either less acidic or less basic. Suppose, for example, you start adding a little base to a concentrated acid solution. You are adding hydroxide ions, but there are still more hydrogen ions present. The solution is still acidic, but

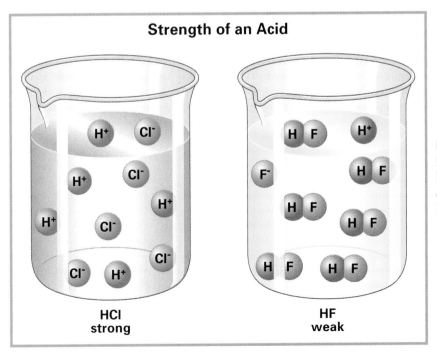

Strength of an Acid

HCl
strong

HF
weak

Figure 10-3
HCl is a strong acid. As you can see, it ionizes completely. HF is a weak acid in which only a small fraction of HF molecules ionize in water.

less so than before the base was added. If you add enough base to the solution, a neutral condition can eventually be reached.

The process just described, which brings a solution to a neutral state, is called neutralization. **Neutralization** is *a chemical reaction in which hydrogen ions from an acid react with hydroxide ions from a base to produce water molecules.* When equal numbers of hydrogen and hydroxide ions are present to react with each other, neutralization occurs and the acidic and basic properties of the solution disappear. The equation for neutralization is shown below.

$$H^+ \quad + \quad OH^- \quad \rightarrow \quad H_2O$$

Hydrogen ion   **Hydroxide ion**   **Water**

General Properties of Acids and Bases

Acids and bases have distinct properties. The intensity of these properties varies according to the quantity of hydrogen or hydroxide ions released in solution. The general properties described below, then, will not be exactly the same in every substance that is either an acid or base.

- **Acids.** *Acidus* is the Latin word meaning sour, which describes how acids taste. The citric acid in a lemon or grapefruit is an example of acidic tartness. Acids also change the color of some foods, and even certain flowers. Hydrangeas may produce pink flowers when planted in a neutral soil. Blooms

can be made or kept blue when soil is treated with an acidic compound.

- **Bases.** Bases, if edible, taste bitter. Magnesium hydroxide, better known as milk of magnesia, is a good example. Bases feel slippery. The bar of soap that slips away from you in the shower is a base. Some commercial glass cleaners contain ammonia, making windows feel slick after you wash them.

The pH Scale

Acids and bases play important roles in science. Their functions also have many applications in foods and nutrition. The impact of acidity on buttermilk biscuits, for example, can mean failure or success.

Since scientists and others need to know how acidic or basic substances are, a scale was devised as a way to express this. The **pH scale** is *a mathematical scale in which the concentration of hydrogen ions in a solution is expressed as a number from 0 to 14.* This scale indicates the relative number of hydrogen and hydroxide ions in a solution.

On the pH scale, a value of 7.0 indicates that a solution is neutral. When a solution is neutral, the number of hydrogen ions equals the number of hydroxide ions, so acidic and basic properties disappear.

These products are examples of bases. Where would they be on the pH scale?

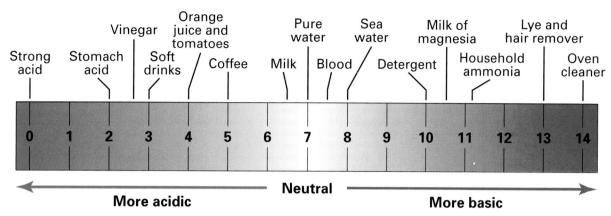

The pH Scale

Strong acid — Stomach acid — Vinegar — Soft drinks — Orange juice and tomatoes — Coffee — Milk — Pure water — Blood — Sea water — Detergent — Milk of magnesia — Household ammonia — Lye and hair remover — Oven cleaner

0 1 2 3 4 5 6 7 8 9 10 11 12 13 14

More acidic ← Neutral → More basic

Figure 10-4

The pH scale is used to measure how acidic or basic a solution is. Which end of this scale shows items with more hydrogen ions?

Due to the mathematics used in the pH scale, greater acidity results in values lower than 7.0. Thus, a solution that has a pH below 7.0 contains more free hydrogen ions than free hydroxide ions. The lower the pH, the more acidic the solution is. Values above 7.0 indicate an increasingly greater ratio of hydroxide to hydrogen ions in the solution. The most basic solutions have the highest pH values.

Using **Figure 10-4**, you can compare the acidity of various common solutions. Note that strong bases can be as hazardous as strong acids. Lye, or sodium hydroxide (NaOH), is a powerful caustic (a burning or dissolving agent) used in drain openers. If you've ever accidentally inhaled ammonia fumes, you know the searing, stinging effect.

Measuring pH

The pH of a solution can be determined simply and precisely with a pH meter. This hand-held instrument gives an electronic read-out of pH from a probe dipped in the solution.

An alternative is to use paper that has been soaked with an **indicator**, *a substance that changes color depending on the pH*. The dry indicator paper is dipped in the solution, and its color then compared with a scale that matches colors with pH values. Indicators give less exact results than meters. Most respond only within a limited pH range.

Indicators for laboratory use are commercially prepared. Certain foods having a blue or red shade get their hue from pigments called anthocyanins (an-thoh-SIGH-uh-nins). These colorings are very sensitive to pH changes. In acid solutions, anthocyanins turn red. They turn light violet or colorless in neutral solutions, and basic solutions leave them blue. Litmus, a popular dye for indicator papers, works in this way.

You can test the action of indicators yourself. To see the effect, add lemon juice, which is an acid, to juice from blackberries, concord grapes, or shredded red cabbage. Next stir in baking soda, a base. Observe the change after each addition. What reactions take place?

Expressing Concentration

Very early in the morning on a particular street, you might see only a few cars. During rush hour, you might see many cars on the same street. This image illustrates the concept of concentration.

In science, **concentration** is *the measure of the amount of a substance in a given unit*

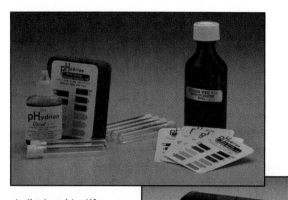

Indicators identify approximate pH levels. What colors are associated with acids and bases?

With a pH meter, a scientist can measure the exact pH of a solution.

of volume. The concentration of an acid, for example, is the number of acid molecules in a given volume of water. You'll recall that chemists report the quantity of atomic particles in *moles*, rather than a specific number of particles. Likewise, it's easier to express the concentration of a solution as the number of moles of material dissolved in it.

Calculating Molarity

One measure chemists often use to express the concentration of solutions is molarity (moe-LAR-uh-tee). **Molarity** is *the number of moles of solute per liter of solution*. To calculate molarity, divide the number of moles of solute in solution by the volume of the solution in liters, as shown here.

$$\text{Molarity} = \frac{\text{Number of moles of solute}}{\text{Volume of solution in liters}}$$

A solution with a molarity of 1.0 has 1.0 mole of solute dissolved per liter of solution. This is called a one-molar solution and is written 1.0 M. Suppose a one-liter solution contains 175.5 g of NaCl. Use the equation to figure its molarity. You learned in Chapter 8 that 1.0 mole of NaCl is 58.5 g. To find the number of moles of solute, divide 175.5 by 58.5 to get 3 moles per liter. The molarity of this solution, then, is 3.0 M.

This information enables you to prepare solutions of specific concentrations. A 1.0 M solution of NaCl in water would be prepared with 58.5 g of NaCl per every liter of solution.

Figure 10-5 shows you how to prepare a 2.0 M solution of sodium chloride. A 2.0 M solution has 2.0 moles of solute in each liter of solution. Therefore, a 2.0 M solution of NaCl would be prepared with 117 g of NaCl per liter of solution (2 × 58.5 = 117). How many grams of NaCl would you need per liter of solution to make a 3.5 M solution?

Titration

A common method used in the laboratory to determine the concentration of an acid or base is a **titration** (tie-TRAY-shun). The main equipment needed to do a titration is shown in **Figure 10-6**.

Suppose you need to find the concentration of an acid solution. To carry out a titration, you first add an indicator to a measured amount of the solution. Then you very slowly add a measured amount of basic solution of known concentration. The indictor changes color when the two solutions neutralize each other. This *point at which neutralization occurs* is

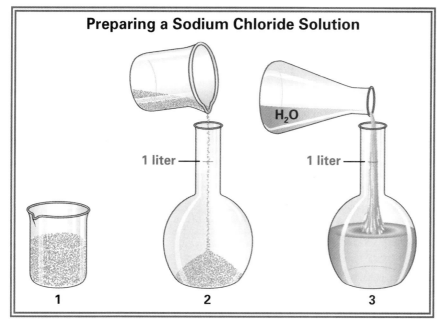

Preparing a Sodium Chloride Solution

1

2

3

Figure 10-5
To prepare a 2.0 M solution of NaCl, follow these steps. First, mass 117 g of NaCl. Then transfer it to a 1-liter flask. Finally, add water, filling the flask to the 1-liter mark. How was the figure of 117 g calculated?

called the **equivalence point**, or end point. By knowing the concentration of the base and the volumes of acid and base needed to reach the equivalence point, chemists can follow-up this titration with calculations that determine the concentration of the acid solution.

Concentration Versus Strength

In discussing acids and bases, the concepts of concentration and strength are often confused. It's worth clarifying the difference here.

The terms concentrated and dilute describe a solute-to-solvent ratio; they compare the number of moles of a solute dissolved in a given amount of solvent. Concentrated solutions contain more moles of solute per liter than do dilute solutions. A 6.0 M solution of hydrochloric acid is more concentrated than a 1.0 M solution of that same acid. Similarly, a 0.001 M solution of sodium hydroxide is more dilute than a 0.1 M solution.

The terms concentrated and dilute can describe any solution. The qualities of strength and weakness, however, apply specifically to acids and bases. As described earlier, these qualities are related to a compound's tendency to ionize. For example, a 4.0 M solution of hydrochloric acid is more dilute than a 6.0 M solu-

Figure 10-6
A ring stand and clamp, a buret, and an Erlenmeyer flask are basic equipment needed for titration. What must be added in order to complete the titration?

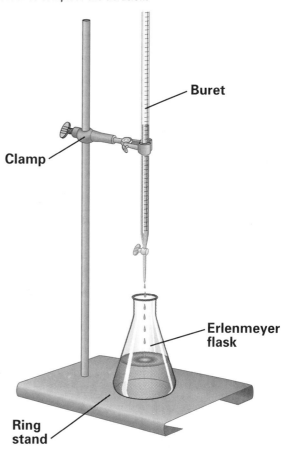

Buret

Clamp

Erlenmeyer flask

Ring stand

tion of acetic acid. However, hydrochloric acid is the stronger acid. Thus, it is possible to have a dilute solution of a strong acid or base.

Acids and Bases in Foods

Most bases associated with food are strong, but have a relatively low solubility. Thus, concentrated solutions are impossible to prepare. Working with foods that contain bases poses no health hazard.

In contrast, most food acids are weak. Since only concentrated solutions of strong acids present any risk of chemical burns to skin or other tissue, food acids are not a health problem. Unless you squirt juice in your eye while cutting a lemon, these acids are harmless.

Almost no food ingredient has a pH of exactly 7. Even distilled water can be slightly acidic if it absorbs carbon dioxide from the air.

In later chapters, you'll learn the importance of hydrogen ion concentration to foods, their preparation, and storage. Pickles keep longer than fresh cucumbers, for example, because their vinegar bath lowers their pH, which helps prevent spoilage. Similarly, a high pH in eggs is a sign of age and thus a loss of quality. The pH balance is important to making baked goods rise.

Acids and Bases in the Body

Like a cake baking in the oven, your body needs the proper pH to carry out the chemical reactions that sustain your life. A balanced pH is most notable in two basic body fluids: blood and digestive juices.

Blood pH

Blood pH is an exception to the rule, "you are what you eat." The pH of foods in your diet has no effect on blood chemistry. This is due to the body's remarkable way of regulating blood pH. A system of **buffers**, *substances that help maintain the balance of hydrogen and hydroxide ions in a solution,* holds the pH of blood steady at a slightly basic 7.4. In a healthy person, this figure never varies more than one-tenth of a pH unit up or down.

One helpful tool that a physician has is a blood test, which provides information about the acid-base relationships in blood. Such tests commonly contribute to diagnosing certain illnesses.

Carbon dioxide (CO_2) plays a major role in this system. As a waste product of cell activity, carbon dioxide gas is carried in the bloodstream to the lungs to be exhaled. The gas dissolves in blood to form carbonic acid (H_2CO_3), which keeps the blood from becoming too basic. At the same time, some blood molecules contain bicarbonate or phosphate ions. These buffering ions act as bases to neutralize any excess carbonic acid.

A number of conditions can upset this system of checks and balances, sometimes with serious results. A blood pH of 7.6 or higher is called alkalosis (al-kuh-LOW-sis). If pH is 7.2 or lower, the imbalance is called acidosis (as-ih-DOE-sis). Untreated, both disorders can dull the breathing reflex, leading to fatigue, confusion, coma, and possibly death.

Digestive Juices

The digestive system, in contrast, relies on a variety of pH levels to break down foods and nutrients. Digesting protein in the stomach requires a highly acidic environment. Gastric acid, which is mostly hydrochloric acid, has a pH around 2. Its release in the stomach is regu-

Nutrition *Link*

A Matter of Taste

Sour, bitter, salty, sweet—these sensations, you recall, are four tastes perceived by the human tongue. You now recognize sour as a property of acids, and bitter as a property of bases. What is the connection between a food's taste and its hydrogen ion content?

- *Sour* is related to a food's concentration of hydrogen ions, or with the food acid's potential for ionization. Stronger acids or more concentrated solutions of weak acids, then, make for stronger sour tastes.

- Taste buds that register *bitter* are less specialized than some taste buds. They sense taste from a variety of molecular makeups, so foods other than bases can taste bitter. Because bases in foods have low solubility, they take longer to dissolve in saliva. Thus, bitter flavors take longer to perceive and tend to linger.

- *Salts* are one product of acid-base neutralization. (Water is the other.) Thus, while acids and bases are not directly involved in salty taste, they do help create substances that are.

- *Sweet* is a complex perception. Acids, bases, and various chemical groups (some of which you'll read about later) bond to form molecules of sweet substances.

Acidic and basic ingredients also influence your perception of tastes. Acids that are present in levels too low to taste can make table salt (sodium chloride) taste saltier.

Applying Your Knowledge

1. What might be the connection between strong acids and bases, and people's natural dislike for strongly sour and bitter tastes?
2. Based on your studies in this course so far, do you think taste accurately indicates the presence of acids and bases in foods? Why or why not?

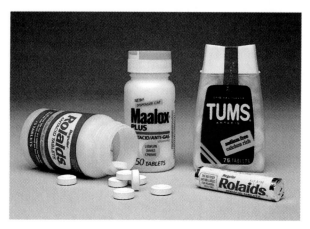

lated to maintain an overall stomach pH between 1.5 and 1.7.

As food moves from the stomach to the small intestine, a neutralizing effect is needed. Basic body fluids are secreted to reduce the acidity of the digested material.

Americans purchase more than $250 million worth of antacids every year. They are taken to help neutralize stomach acids that are produced to help digest food. Overeating is one common cause of the problem.

FOOD SCIENCE
Careers

Product Evaluation Scientist

How does a food maker know what will sell?   Ask a . . . Product Evaluation Scientist

"The secret to creating successful new food products," says Julia Fontes, "is to do your homework." As a product evaluation scientist, Julia explores marketing ideas for new food items. "We look at surveys and trends to find out what people want and test prototypes for the right qualities. Salt is a basic example. How much salt would you put in a product like potato chips? Many people want less salt today, yet too little salt seems bland. The 'wrong' amount can hurt sales, but figuring out what's 'right' can be difficult. Alternative potato chip flavors have offered one solution. Our decisions are based on statistical analyses, not personal preferences. We've learned that as individuals, we don't always agree with the public."

Education Needed: Product evaluation scientists need a degree in food science, chemistry, or biology. To design and interpret consumer tests, they need to know both sensory evaluation and statistics. An evaluation scientist with a master's de-

KEY SKILLS
- Teamwork
- Communication
- Data analysis
- Organization
- Creativity
- Leadership

gree in business could head a marketing division.

Life on the Job: Evaluation scientists work in research and development to plan and interpret consumer testing. They usually plan and direct consumer market research and sensory panels, translating consumer opinion into specific recommendations for new or improved food items.

Your Challenge

As a product evaluation scientist, you learn about a new food trend in another country. You wonder whether the item would be popular here at home.

1. As a scientist, what questions do you have about marketing the food?
2. What other experts and resources could help you in trying to predict success?

Neutralization

SAFETY FIRST
Review these safety guidelines before you begin this experiment.

In this experiment, you'll compare the acetic acid content of several brands of vinegar by titrating equal volumes of the vinegar with the base sodium hydroxide (NaOH). You'll determine how much base is needed to neutralize the acid in the vinegar. The more acid present, the more base you will need to neutralize it.

Equipment and Materials
2, 50-mL burets

1.0 M sodium hydroxide solution

sample of vinegar

250-mL Erlenmeyer flask

phenolphthalein solution

Procedure
1. Put on safety goggles and wear them throughout this experiment. NaOH can harm the eyes.
2. Clamp a clean, 50-mL buret to a ring stand. Fill the buret with 1.0 M NaOH solution.
3. Fill a second buret with the brand of vinegar assigned by your teacher.
4. Add 20 mL of vinegar from the buret to a clean 250-mL Erlenmeyer flask. Add three drops of the indicator phenolphthalein to the vinegar. Place the flask under the tip of the buret containing the NaOH.
5. Slowly add the NaOH to the vinegar. A pink color appears where the base first contacts the acid. Gently swirl the flask until the color disappears. Add the base drop by drop, swirling after each drop, until the base turns the solution to a pale pink that does not disappear.
6. In your data table, record the volume of NaOH used.
7. Wash out the flask and repeat the titration. If you used more than 25 mL of base, be sure to refill the buret before beginning the second titration.
8. Again record the volume of base required. Average the two amounts.
9. Report your figures on the chalkboard.
10. In your data table, record the figures for each of the brands of vinegar tested by the other experiment groups.

Analyzing Results
1. What observation can you make about the amount of acetic acid in the various brands of vinegar? How do you explain this?
2. What was the function of the indicator in this experiment?
3. Would you use the same vinegar for pickling, where very low pH is important, as you would in a salad? Why or why not?

SAMPLE DATA TABLE

Brand of Vinegar		Volume of Base	
	Trial 1	Trial 2	Average

Chapter Summary

- Acids dissolve in water to form hydrogen ions; bases form hydroxide ions.
- Different compositions give acids and bases different properties.
- The relationship of hydrogen and hydroxide ions in a solution is measured on a pH scale.
- The pH of a solution can be determined by using indicators.
- The concentration of moles of a solute in solution is expressed as molarity.
- The concentration of acids and bases can be determined through titration.
- Concentration refers to the amount of a substance in solution; strength refers to a compound's rate of ionization.
- Proper pH is vital to the chemical reactions that sustain life.

Using Your Knowledge

1. What happens when water ionizes?
2. What happens when acids and bases ionize in water?
3. What is the difference between strong and weak when referring to acids and bases?
4. Compare the general properties of acids and bases.
5. What do the numbers on the pH scale mean?
6. What do you know about the concentration of a calcium carbonate solution described as 2.4 M?
7. Why is the equivalence point significant in a titration?
8. What is meant by the statement, "Strength is determined by ionization, not by solubility"?
9. Can a solution be both weak and concentrated? Explain.
10. How dangerous are the acids in foods?
11. How does the body regulate blood pH?
12. How does the pH environment of the stomach compare with that of blood?

Real-World Impact

Acid Rain

The burning of coal, oil, and oil products produces the gases nitrogen dioxide, sulfur dioxide, and sulfur trioxide. In reacting with moisture in the air, these compounds form acids called acid rain. Acid rain is blamed for corroding statues and buildings and for killing marine life in lakes and streams.

Thinking It Through

1. How do you think acid rain affects the pH of a lake? Compare this to a similar condition affecting the pH of blood.
2. Some lake beds are high in limestone, a rock that reacts with carbon dioxide and water to form a basic solution. Do you think this makes them more prone or less so to the damaging effects of acid rain?

Skill-Building Activities

1. **Critical Thinking.** Why do you think a chemical reaction between an acid and a base that produces water is called neutralization?
2. **Mathematics.** Suppose you need a 2.0 M sodium hydroxide (NaOH) solution. How many grams of sodium hydroxide would you dissolve in each liter of water?
3. **Mathematics.** The pH scale increases and decreases by powers of ten. A solution with a pH of 2.0 is ten times more acidic than one that registers 3.0; it is 100 times more acidic (10 × 10) than a solution with a pH of 4.0. How much more acidic is a pH 2.0 acid solution than an acid solution that registers 5.0 on the pH scale?
4. **Consumer Skills.** Read the labels of different antacid products. Which ingredients do you think are bases? What do you think the other ingredients do?
5. **Science.** Use a chemistry textbook to research other definitions of acids and bases. How do these compare to the descriptions in this chapter?

Understanding Vocabulary

Write a one-page description of a person working in a laboratory with acids and bases. Include each of the "Terms to Remember" in your description.

Thinking Lab

pH-Balanced Shampoo

Before 1940 the cleaning agent in shampoo was soap. It tended to leave hair filmy and brittle, especially if the water was at all hard.

Analysis

Today's shampoos are frequently advertised as "pH balanced." They are designed to be compatible with the chemical makeup of your hair, and range in pH from about 5 to 8.

Thinking Critically

1. What is the advantage of using a pH-balanced shampoo?
2. Before air pollution became widespread, many people claimed that rainwater was superior for washing hair. Why might this have been true?
3. Why do you think hard water makes soap shampoos less effective?

Objectives

- Compare units of heat measure.
- Describe the relationship between molecular motion and temperature.
- Compare processes of heat transfer.
- Explain what affects rates of chemical reaction in food.
- Analyze the relationship between food intake and body weight.

Terms to remember

absolute zero
anorexia nervosa
bulimia
calorie
conduction
convection
energy
heat
joule
kilocalorie
microwaves
obesity
radiation
specific heat

Two thousand years ago the Greeks called "action" and "operation" *energeia*. In English today, the same concept is called **energy**, *the ability to do work*. After two millennia, the meaning of the word is essentially the same, although understanding of it has changed greatly. The same force that powered the world of the ancient Greeks is still active today—in every reaction in the universe. In fact, scientists say, the world contains all the energy it ever had, and all it ever will. The planet is a giant energy recycler.

While energy is a vague concept, its impact is easily seen and felt. As this chapter explains, successfully transferring energy from one source to another is a foundation for cooking, eating, and health.

Energy on Earth

The sun is the source of physical energy and life on Earth. Solar energy, flowing in waves, supplies both light and heat. A plant, perhaps a corn plant, absorbs this energy to fuel its life activities, including growing ears of corn. This energy is stored, then transferred to anything that eats the corn—a crow, a cow, or you. Finally, energy is converted in the body to mechanical energy, or energy of motion. This energy is used to flap a crow's wing, swish a cow's tail, or turn a page in your food science text.

Energy is also involved in phase changes, such as the boiling of water, and in chemical changes, such as the burning of a campfire.

Thermal or heat energy is the form of greatest interest in food science. **Heat** is *energy transferred from an object of higher temperature to one of lower temperature.* Doesn't this also describe how food is cooked? Transferring heat energy to food is part of a chain of physical and chemical events that produces interesting and satisfying results.

Nutrition Facts

Serving Size 1/2 cup (114 g)
Servings Per Container 4

Amount Per Serving

Calories 90   **Calories from Fat** 30

The calories you find listed on a nutrition label are not the same measurement that a scientist uses. How are they different?

Measuring Heat Energy

When is a calorie not a calorie? What is a calorie in the first place? You may have seen the word on food packages and dietary guides. This use of "calorie," however, is different from the chemist's.

In chemistry, a **calorie** (cal) is a unit of heat measurement. One calorie is *the amount of energy needed to raise the temperature of 1.0 g of water 1.0°C*. When most people say "calories," however, they are actually talking about kilocalories. A **kilocalorie** (kcal or kcalorie) equals *1000 calories*. Kilocalories are also referred to as Calories. Discussions in this text will distinguish between calories and kilocalories.

Another unit of heat energy used by chemists, and most of the world, is a metric measure called the **joule** (JOOL). One joule (J) is equal to *0.239 calorie*. Conversely, one calorie is equal to 4.18 joules, and one kilocalorie is equal to 4.18 kilojoules (kJ).

Heat Energy and Matter

Some struggling cooks complain that they "can't even boil water without burning it." Why does water boil but not burn? For that matter, why does water boil? To understand these and other ways that heat energy affects matter, you need to understand properties of matter that affect heat transfer.

Molecular Motion

Whether a substance is a gas, liquid, or solid, its molecules are always in motion. As shown in **Figure 11-1**, gas molecules travel at very high speeds in a straight line until they collide with each other or with some physical barrier. Then they bounce off in another direction. The empty space between molecules far exceeds the size of the molecules themselves.

Molecular motion in liquids is more restricted because the molecules are closer together—touching one another, in fact. These molecules can tumble past each other, allowing the liquid to flow or move out of the way easily when an object is dropped into it. While liquids have no definite shape, they do have a definite volume, because the forces of attraction between the molecules are strong enough to hold the liquid together.

Molecular motion in a solid is most restricted of all, because its particles are locked in place. Since molecules can only vibrate and rotate, solids maintain a definite volume and a definite shape.

Despite these differences, molecules in all three states of matter respond the same way when heat energy is added: they speed up. Their increased activity can be measured as an increase in temperature. Thus, temperature is a measure of molecular motion.

In contrast, lowering the temperature slows molecular motion. Scientists theorize that at −273°C, or **absolute zero**, *all molecular motion stops and no heat energy remains*. Since no one has actually achieved absolute zero in a laboratory, a total absence of molecular motion has never been observed.

Specific Heat

Every substance has a **specific heat**, *the amount of heat needed to raise the temperature of one gram of a substance 1°C*. Water, for instance, has a specific heat of 1.0 calorie per gram per degree Celsius, written "1.0 cal/g°C." It takes 1.0 calorie to raise the temperature of 1.0 g of water 1.0°C. Iron heats much more quickly; only 0.11 of a calorie raises

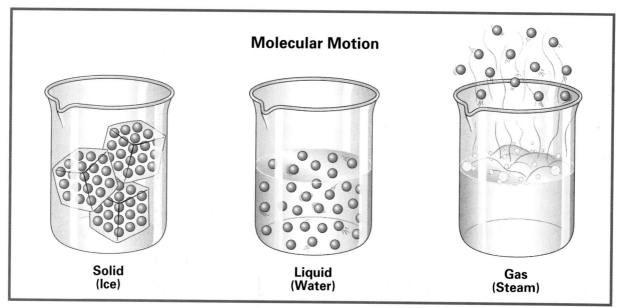

Molecular Motion

Solid
(Ice)

Liquid
(Water)

Gas
(Steam)

Figure 11-1
Molecular motion increases when a substance such as water changes from a solid, to a liquid, to a gas. Do the molecules in a solid move at all? Do liquids and gases have volume?

1.0 g of iron 1°C. Its specific heat is 0.11 cal/g°C. Glass heats more quickly than water but less quickly than iron. It takes an added 0.20 of a calorie to raise 1.0 g of glass 1°C. How would you describe the specific heat of glass?

Methods of Heat Transfer

Heat transfers by conduction, convection, and radiation. Cooking typically involves two or more of these methods at the same time.

Conduction

In **conduction** (kun-DUK-shun), **Figure 11-2**, *heat energy is passed by the collision of molecules.* When two substances make contact, the molecules of the warmer substance transmit some of their energy to the molecules of the cooler one. The warmer substance cools as its molecules lose energy and

slow down. The energy transfer continues until both substances are the same temperature. Only if energy is continuously supplied will the warmer substance maintain its temperature.

As an example of cooking by conduction, imagine making pancakes. Batter is poured into a pan that is made hot by energy released from a heating element. The pan transfers the heat by conduction to the batter, which begins to cook. You cook the top of the pancake by flipping it over, allowing the pan to conduct heat to that side. The pancake absorbs energy, yet the pan remains hot due to the constant heat

Figure 11-2
Heat energy transfers directly from the heating element to the food during conduction. Why does a cook turn a hamburger over in the skillet?

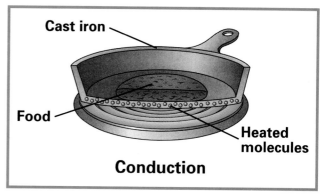

Cast iron

Food

Heated molecules

Conduction

source beneath it. If you leave the heating element on and control the temperature, you can make a whole stack of light and fluffy pancakes.

Convection

Heat also transfers by circulatory movement in a liquid or gas. Called **convection** (kun-VEK-shun), this method is shown in **Figure 11-3**. When a liquid (or gas) is heated, that portion on the bottom nearest the heat grows hot first. As they absorb heat energy, these molecules speed up and move farther apart, causing the liquid to become less dense. The less dense portion of the liquid rises, allowing the cooler, more dense liquid to sink to the bottom. There, the cooler molecules are heated until they rise. This circular flow, called a *convection current,* continues until all the liquid is evenly heated.

Food on a grill cooks with radiant heat. How does this heat transfer method differ from convection and conduction?

Radiation

Heat energy can also be transferred by **radiation**—*heat transmitted as infrared rays.* Infrared refers to a range of electromagnetic waves of particular wavelengths.

The sun is the ultimate source of radiant energy on earth. Both infrared radiation and visible light move through space at the same speed: 3.0×10^8 meters per second, or 186,000 miles per second. Standing in the sun, you feel the warmth of energy coming from 150 million kilometers away. Bread in a toaster and a steak on the grill experience heat in this same way, although at a closer distance. You, the bread, and the steak all absorb radiant heat without touching the heat source.

Usually, foods cook by several methods at one time. When you bake cookies, for example, air in the oven is heated as it absorbs energy and circulates in the confined space. That's convection. The air transfers heat by radiation to the cookies, which also absorb heat conducted through the baking sheet. How many types of heat transfer are involved in boiling potatoes?

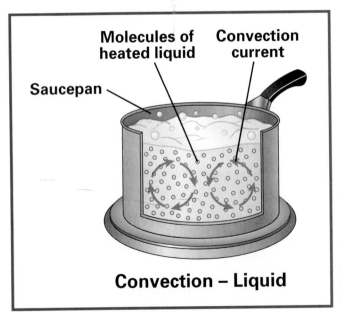

Saucepan

Molecules of heated liquid

Convection current

Convection – Liquid

Figure 11-3
Convection heating occurs in a liquid or a gas. How are the vegetables in this pan cooked?

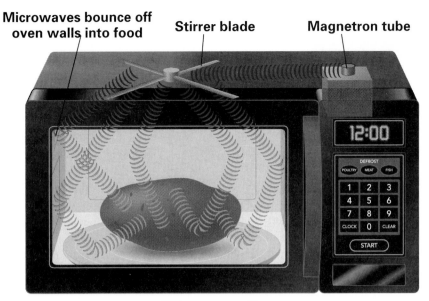

Microwaves bounce off oven walls into food **Stirrer blade** **Magnetron tube**

Microwave Oven

Microwaved foods don't always heat evenly. Why do you think authorities recommend that baby bottles not be warmed in a microwave oven? What can you do to heat foods evenly in a microwave oven?

Microwaves

Like infrared radiation, **microwaves** are *electromagnetic waves*. They are lower in energy than infrared waves, however. In fact, they have little effect on most substances, but have their greatest effect on polar molecules, such as water. As they absorb the energy from microwaves, water molecules in food start to vibrate. This energy is then conducted to neighboring molecules, which heats the food.

The fact that microwaves are absorbed by food to a depth of 5 to 7.5 cm means they heat food faster than traditional methods. Heat from traditional sources is absorbed only at the surface, then spreads slowly by conduction. Thus, microwaves usually cook foods in less time.

Induction

Induction cooking makes use of an induction coil that is placed just below the cooking surface. The cooking surface is made of a smooth, ceramic material. An alternating electric current in the coil produces an alternating magnetic field. This field causes iron cooking utensils placed on the cooking surface to become hot, while the surface itself remains cool.

The hot cooking utensil transmits heat to the food it contains.

Rates of Reaction

Successful food preparation depends on more than applying heat. Otherwise, you could bake a cake and scramble eggs with the same method. Obviously, other factors are at work. The rate of the chemical reactions that cook food also affect the process—and the final product. The following factors determine whether a cooked food is actually fit to eat:

- How much energy the food absorbs.
- How quickly energy is absorbed.
- The rate at which energy is transmitted.

For example, if a piece of meat is left in a pan over medium heat for two minutes, it may be warm but still be completely raw. If the same size piece of meat is placed in an extremely hot pan for two minutes, it may char or scorch on the outside, yet still be raw on the inside.

Temperature

The temperature you use to cook food is your main control over the rate at which chemical

The Microwave Alternative

Compared to other methods of heat transfer, microwave cooking can save nutrients. With little added water, fewer nutrients dissolve away. Shorter cooking times protect heat-sensitive nutrients. Also, less need for fats and oils helps avoid the related health risks.

On the down side, "hot spots" may develop when certain nutrients—usually water, sugars, and fats—attract microwaves that become concentrated. Biting into these can be painful. "Cold spots" that develop in other areas of a food may allow disease-causing organisms to grow. The stuffing in poultry is a special risk. Stirring, rotating, or otherwise distributing heat as food cooks promotes safe microwave cooking.

Applying Your Knowledge

1. Why do you think microwave cooking takes little added water?
2. What are some similarities between using a microwave and a conventional oven?

changes take place. Temperature affects the reaction rate in two ways.

First, higher temperatures raise the speed of molecular motion. The more quickly molecules move, the more often they collide. This increases the overall rate of reaction. Also, higher temperatures increase the impact of molecular collisions. To cause a reaction, molecules must collide with enough energy to break the bonds between atoms. Adding heat increases the chances that this will occur.

As a general rule, the rate of a reaction doubles for every 10°C increase in temperature.

Lowering the temperature reverses the process by slowing the rate. Later in this course, you'll see how a food's composition helps determine what rate of reaction is best for cooking—that is, should a high or low temperature be used?

Surface Area

Contrary to an old saying, a watched pot *does* eventually boil. However, if you put the same volume of water in a shallow pan, it boils more quickly. The pan has more surface area, exposing more water molecules to the heat source and speeding the physical change of phase.

Likewise, greater surface area increases the rate of chemical change in cooking foods. As the air or liquid surrounding food is heated, energy transfers to food molecules by conduction. This process begins at the surface of the food and continues to the center until the food is cooked. Thus, a food with more surface area exposed to heat will cook more quickly as long as the foods have similar volume. The closer the center of the food is to the surface, the more quickly heat can reach it.

When cooking food, you regulate the heat. Molecules in the food move more rapidly. What happens to the molecules to cause the chemical reactions of cooking?

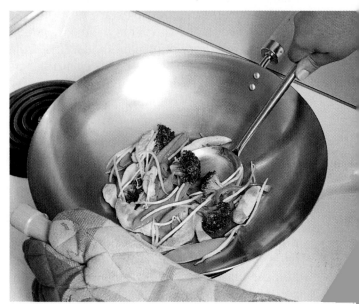

When cooking a hamburger patty, you turn it so that heat can penetrate the surface on both sides. Which of these patties will cook faster? Why?

This principle holds for microwave cooking as well. Although microwaves penetrate beyond the food surface, they, too, are limited by the amount of surface area available. Conduction also occurs in microwave cookery, so the depth the energy must travel to reach the center has an effect.

Heating Media

As you've gathered from your reading, heat is carried by different media. The four common media for transferring heat in cooking are air, water, steam, and fat.

From your studies of water, you know that certain media are capable of transmitting more energy than others. Steam is hotter than the hottest liquid water at ordinary pressure, for example, and fat can be heated to much higher temperatures than liquid water. Since the rate of chemical reactions increases with temperature, the medium used to transfer heat directly affects cooking times.

Roasting, baking, broiling, and grilling are cooking methods that rely primarily on air as the heating medium. Boiling, braising, stewing, and simmering use water. Often two or more media are combined. A turkey in a plastic oven bag, for example, is cooked by both air and steam.

Energy Needs and Health

As energy travels the universe, one important destination is your body. The kilocalories in the foods you eat fuel all of your body's activities. Even asleep, your body releases energy from the nutrients in your diet to maintain its basic functions.

As with cooking, heat energy supplied by kilocalories may come from a number of sources, and each source has a specific as well as general effect on your health. In the next unit you'll see how different foods provide different classes of nutrients, and how these nutrients work together to store energy within the body and unlock its energy reserves. You'll learn how people can have enough to eat—and then some—and still not have enough energy.

Energy and Body Weight

Probably the most obvious relationship between energy intake and health is that of weight control. Concerns about consuming and storing excess energy (that is, overeating and getting fat) are common. Extra weight, especially extra fat, increases the workload on the heart, lungs, and joints. In time, this raises the risk for numerous health problems, including high blood pressure, heart disease, and diabetes. Extreme overweight is termed **obesity**,

Foods are usually cooked with more than one heat transfer medium. Which ones are indicated here?

FOOD SCIENCE
Careers

Exercise Scientist

What role does motivation play in exercise?
Ask an . . . Exercise Scientist

According to exercise scientist, Pete Ramirez, "Motivation is continually on my mind. I own a fitness center, so I don't want to lose clients because they lose the motivation to continue. Some people like a difficult workout, but others want to move at a slower pace, so I have to figure out personalities. Then I use that information to design individual programs and also to choose my approach with each person. I like to see the changes in people—how they gain energy and feel more confident as they develop muscle tone. If I can keep them going enough to see progress, the motivation to continue is usually there."

Education Needed: Exercise scientists typically have a bachelor's degree in exercise science or athletic training. Courses in anatomy or biology are useful for understanding the mechanics of the human body. A background in nutrition helps in understanding how diet affects energy levels and contributes to health and fitness.

KEY SKILLS
- Physical fitness
- Desire to work with people
- Communication
- Leadership
- Motivational skills
- Organization

Life on the Job: Exercise scientists work in a variety of settings, from hospitals, to wellness centers, to professional sports. They develop exercise prescriptions and activity programs designed to help individual clients gain muscle, lose weight, or increase strength and energy. They may work with dietitians to create diet and exercise plans for people coping with such illnesses as diabetes and heart disease.

Your Challenge

Imagine you're an exercise scientist. A doctor has referred a patient to you. The client, who is overweight and not very active, is recovering from heart surgery.

1. What short- and long-term goals do you set for this client?
2. What psychological obstacles might you need to overcome in the client? How do you do this?

Energy transfers from the sun, to the food supply, and then to you. Your energy level depends on how well the foods you choose to eat supply you with what you need.

ties. The more active you are, the more kilocalories you need. The opposite is also true.

When you figure a financial budget, any money that you earn and don't spend might be kept as savings. The dietary version of a savings account is body fat. Typically, every 3500 kilocalories of energy "income" that your body doesn't need is stored as about 500 grams, half a kilogram (about one pound) of body fat.

To lose half a kilogram of excess body fat, then, you need to eat 3500 kilocalories less than your body needs. By gradually decreasing kilocalorie consumption while increasing physical activity, you lose weight slowly. Eating 500 kilocalories a day less than your body needs, for instance, results in a half-kilogram loss of body fat in a week. Weight lost at this rate is more likely to be kept off because it reflects a reasonable diet that can be maintained over time. Increasing daily activity makes the process faster and more efficient.

having excessive fat and a body weight that is 20 percent or more above a healthful range. Obesity is additionally linked to some types of cancer.

Many people worry about consuming too many kilocalories. Fewer people recognize the risk of getting too many kilocalories from nonnutritious sources. Your body weight may be a healthful one, but if your energy supply comes from foods that don't provide needed nutrients, you may feel listless and lack energy. You may be more prone to infection and other illnesses. Even your concentration can be affected.

Like increases in height, some weight gain is healthy and expected during the teen years. Few people would want to have the same height or weight on their eighteenth birthday as on their twelfth. Rates of growth vary widely among teens. Tables listing healthful weight ranges are available from physicians, dietitians, and medical references.

The Energy Balance

Maintaining a healthful weight is like living on a balanced budget: intake should equal output. You should consume roughly the same number of kilocalories as you use in your daily activi-

Maintaining or losing weight is easier when you burn kilocalories through exercise. When you are active, you can take in more kilocalories without putting on weight.

A brownie might add around 300 kilocalories to your daily intake. A 12-ounce soft drink adds about 150 kilocalories. By eliminating that many kilocalories each day, how much weight could you take off in a week?

To gain a half-kilogram a week, you would need to eat 500 kilocalories a day over your body's needs. Staying physically active is a good idea; it helps ensure that weight gained is the healthful sort, muscle rather than fat.

Whatever your weight goal is, remember that nutrition counts at least as much as kilocalories. Losing weight on a diet of saltine crackers is no better for you than gaining weight on a diet of ice cream. A chart on caloric values can help you find an energy balance. The next unit, particularly the next chapter, explains the chemistry, and the value to your health, of achieving nutritional balance as well.

Eating Disorders

Mirrors at a carnival funhouse purposely distort your reflection. You may look half as tall and twice as wide as you really are. Part of the fun is knowing that these images are not real. Some young people, however, actually see themselves in an unreal way. Women between ages twelve and twenty-five are especially prone to distorted perceptions of their own size.

People with **anorexia nervosa** (an-uh-REX-ee-uh ner-VO-suh) have *an eating disorder characterized by self-starvation, resulting in the intake of fewer kilocalories than needed for good health.* Anorectics believe that they are never thin enough. In reality they are dangerously underweight, depriving the body of vital nutrients and upsetting its delicate balance of chemicals. Some people literally diet and exercise themselves to death.

Ironically, while they often resemble "walking skeletons" themselves, people with anorexia typically like to prepare food for others. Eating for them is a matter of control. Control over what they and other people eat equals control over life.

Similar issues underlie the eating disorder called **bulimia** (byoo-LIM-ee-uh). Bulimia is marked by *cycles of gorging on large amounts of food followed by self-induced vomiting and use of laxatives.* People with bulimia overeat to cope with stress, loneliness, or depression, then purge themselves from anger or self-disgust.

Bulimics typically maintain a stable weight but do themselves other damage. Repeated vomiting bathes the teeth with hydrochloric acid from the stomach, eroding protective tooth enamel. Loss of hydrochloric acid, in turn, causes a loss of the mineral potassium, which can lead to heart failure in extreme cases.

While the symptoms of eating disorders are physical, their causes are emotional. People may need professional help to understand and manage the psychological roots of their condition.

When untreated, eating disorders are very difficult to overcome. Programs are available for those who have such problems. Professional counselors can offer helpful treatment and advice.

Effect of Surface Area on Cooking Rate

SAFETY FIRST

Review these safety guidelines before you begin this experiment.

In this experiment, you will use different-sized pieces of potato to study the effect of surface area on cooking rate.

Equipment and Materials

100-mL graduated cylinder	potato	wooden toothpicks	2.5-cm square, cut from an index card
400-mL beaker	paring knife		

Procedure

1. Measure 200 mL of water using the 100-mL graduated cylinder. Pour the water into a 400-mL beaker. Place the beaker over medium-high heat on a range burner.

2. While the water is heating, obtain a potato from your teacher. From the potato, cut a cube measuring 2.5 cm on each side. Use the index card square as a guide for cutting to size.

3. Follow one of the following variations as assigned by your teacher:
   a. **Variation 1.** Use the single cube of potato.
   b. **Variation 2.** Cut the cube into eight equal-size cubes by cutting the original cube in half in all three directions.
   c. **Variation 3.** Cut the cube into 27 equal-size cubes by cutting the original cube in thirds in all three directions.

4. Place the potato cube or cubes in the water after it has come to a boil.

5. At 1-minute intervals, test the potato with a wooden toothpick. Based on how easily the toothpick punctures the potato, determine its degree of doneness, for example, uncooked, slightly cooked, cooked. Record this information in your data table.

6. When the toothpick punctures the potato easily, note the time in your data table. Turn off the burner and remove the beaker from the range.

7. On the chalkboard, record the number of the variation you followed and the time needed to cook the potato.

Analyzing Results

1. Was there a difference in the total volume of potato present in each beaker?
2. Was there any relationship between the size of the potato cubes and their cooking time? If so, describe it.
3. Explain the difference in cooking time.

SAMPLE DATA TABLE

Variation Number _____

Time in Minutes	Degree of Doneness

Chapter Summary

- Energy is either released or absorbed in all phase changes and chemical reactions.
- Temperature is a measure of molecular motion in a substance.
- Heat is transferred through conduction, convection, and radiation.
- Heating a substance increases the rate of reaction, while cooling the substance decreases it.
- The greater the surface area of a food, the faster it will cook.
- The most common media for heat transfer in cooking are air, water, steam, and fat.
- The energy needed by the body is obtained from the energy in food as measured in kilocalories.
- The number of kilocalories consumed affects weight gain and loss.

Using Your Knowledge

1. What is the relationship between calories, kilocalories, and joules?
2. How does a liquid's molecular structure affect its shape and volume?
3. What happens on a molecular level as an object cools?
4. What does it mean to say that wood has a specific heat of 0.42 cal/g°C?
5. How does frying an egg illustrate conduction and boiling potatoes illustrate convection?
6. How can microwaves cook foods more quickly than other types of heat transfer?
7. Why does one liter of cake batter cook faster as 12 cupcakes than as a single, square cake?
8. How does the medium used for heating affect the rates of chemical reactions in food?
9. Does maintaining a healthful weight guarantee that you're eating right? Why or why not?
10. How do kilocalories relate to weight loss and weight gain?
11. Compare anorexia nervosa and bulimia.

Real-World Impact

Concerns About Weight

While more Americans today seem to be health- and nutrition-conscious, some evidence suggests that many have returned to, or never changed, their poor eating habits. Food trend analysts note that products marketed as low-fat, low-sodium, and otherwise healthful have slowed in sales. Meanwhile, the Mayo Clinic Foundation reports that about $70 billion a year is spent to treat obesity and related problems in the United States, which contribute to some 300,000 deaths. Each year, the number of American adults who are overweight increases by almost one percent.

Thinking It Through

1. Studies show that Americans spend billions of dollars a year on weight-loss products. How can these figures be correct, given the findings described above?
2. A diet plan claims that you will lose weight simply by avoiding the fattening foods it identifies. How do you respond?

Skill-Building Activities

1. **Management.** Bring several of your favorite recipes to class. Identify the methods and media of heat transfer used in each. Which media are associated with each method?
2. **Critical Thinking.** Three diners are enjoying tea with sugar. The first diner drops a sugar cube in a glass of iced tea; the second puts a sugar cube in hot tea; the third pours granulated sugar in hot tea. Assuming each diner used the same amount of sugar, explain whose will dissolve most quickly.
3. **Communication.** In groups or as a class, debate this statement: "Medical advances designed to reduce the health problems of a poor diet, such as supplements and cholesterol-lowering drugs, actually encourage poor eating habits."
4. **Mathematics.** Derrick wants to lose some weight at a rate of one-half kilogram of body fat per week. He needs to know how this translates to cutting foods from his diet. Using a chart with the caloric values of foods, suggest foods or food combinations that might be eliminated each day to achieve such a goal.
5. **Research.** With a team of students, plan a panel discussion that presents information on eating disorders. After researching the topic, present your findings to the class.

Understanding Vocabulary

For each "Term to Remember," list three facts that you have learned about it from this chapter.

Thinking Lab

Rates of Chemical Reactions in Cooking

Cooking food quickly generally preserves nutrients by limiting the food's exposure to heat and water.

Analysis
You can alter the rate of chemical reactions that cook foods by your choice of temperature, surface area, and heating media.

Thinking Critically
1. If you increase temperature to speed cooking, should you also change the food's surface area? Why or why not?
2. Rice and pasta both cook by slow cooking in very hot water. Which would you guess retains more nutrients? Why?
3. Steam is hotter than boiling water. Why is steaming considered a more "nutrient-friendly" way to cook food?

FUNCTIONAL FOODS

As early as 1000 B.C., Chinese documents linked diet to health. Today's "functional foods," including nutraceuticals, designer foods, and pharmafoods, carry on that link. In 1994 the National Academy of Sciences' Institute of Medicine defined functional foods as "any modified food or food ingredient that may provide a health benefit beyond the traditional nutrients it contains." Functional foods are intended to treat or prevent symptoms and diseases, not simply eliminate deficiencies. Research shows that soy foods, for example, may reduce osteoporosis and cardiovascular diseases. Functional foods are often sold as dietary supplements. Health claims can be made for supplements without prior FDA approval as long as research supports the claims.

UNIT ④ The Science of Nutrition

● **TechWatch Activity**
Use the Internet and other resources to find information on functional foods. Create an informational pamphlet on a nutraceutical food, designer food, or pharmafood.

CHAPTER 12 Nutrition Basics

Objectives

- Relate earlier scientific findings to today's understanding of nutrition.
- Explain the role of respiration and oxidation in nutrition.
- Identify and briefly describe essential nutrients.
- Explain how different nutritional guidelines are formulated and used.
- Choose healthful foods according to the Dietary Guidelines for Americans.
- Demonstrate how to use food labels to compare nutrients in foods.
- Plan healthful menus using the Food Guide Pyramid.
- Relate the understanding of nutrition to physical well-being.

Terms to remember

Daily Values (DVs)
Dietary Reference Intakes (DRIs)
enzymes
essential nutrients
nutrient dense
nutrients
nutrition
oxidation
Recommended Dietary Allowances (RDAs)
respiration
scurvy

Suppose you're late for school in the morning. Do you eat a good breakfast? Later you drop some coins in a vending machine. Do you choose low-fat pretzels or high-fat potato chips? The food decisions you make affect your overall health. Just as the family car runs better and longer when it's well maintained, the same is true of your body.

Some teens don't see how the food they eat connects to their health. What they're not seeing is that many health problems develop over time. Poor eating habits learned early in life increase the risk of health problems later.

The effects of poor nutrition aren't always obvious. One teen concentrates well when she studies and takes tests. Proper nutrition helps make that possible. Another teen is irritable and lacks the energy to enjoy a full day's activity. He also seems to catch infections easily. Poor nutrition can create these effects. The subtle impacts of nutritional deficiency are many. You can get used to feeling and looking a certain way and not even realize the possibilities for improvement.

To choose foods that keep that magnificent machine called the human body in good working order, you need to know something about **nutrition**, *the science of how the body uses food*. This chapter looks at healthful food choices that fuel the chemical processes of nutrition.

Nutrition History Highlights

The instinct to consume food for survival is as old as humanity. Some ancient people went so far as to credit food with having magical powers. Around 400 B.C., Hippocrates, "the father of modern medicine," taught that people could learn what was good for them by reasoning from observable facts. His approach began to sweep away some of the myth and mystery surrounding health and nutrition.

When you understand nutrition principles, you can make food decisions that help you stay healthy. What impact might good health have on your appearance?

Hippocrates' studies led him to believe that foods contain a life-giving substance. He was onto something, but it would be another 2,000 years—the mid-1700s, in fact—before the scientific method proved him right.

An Early Discovery

One event in the 1700s illustrates how early nutrition discoveries were made. At the time, British sailors often suffered from **scurvy**. *This circulatory disease, now known to result from vitamin C deficiency, causes bleeding gums, weakened blood vessels, and extreme fatigue.* It was frequently fatal.

In the 1740s, a British physician set out to find a cure for scurvy. He chose a group of twelve British sailors who had the illness. These men were fed a variety of foods, including limes, lemons, cider, vinegar, seawater, nutmeg, and oil of vitriol (a metal compound). The illness disappeared only in those sailors who ate lemons or limes. As a result of this early experiment in nutrition, lemons and limes became required provisions for British sailors. That practice gave the sailors a nickname that is sometimes applied to all British citizens today: limeys.

Lavoisier: Oxidation and Respiration

Some 30 years later, chemist Antoine Lavoisier (luv-WAH-zee-ay) was literally playing with fire in his laboratory in Paris as he explored what happens when things burn. He discovered that burning required a certain gas, which he named oxygen. Scientists have since used the term **oxidation** to describe *those chemical reactions in which elements combine with oxygen.*

Compounds in the wick of a candle chemically combine with oxygen in order to release heat energy in the form of a flame. What happens if the oxygen is removed?

When you inhale, you take in air, a mixture of nitrogen and oxygen gases with small amounts of argon, water vapor, and carbon dioxide. The mixture of gases that you exhale contains more than 100 times the amount of carbon dioxide that was in the air you inhaled. The quantity of oxygen is reduced by five percent. What process was at work?

eating, etc.), Lavoisier measured his oxygen consumption. With the results, Lavoisier confirmed his notions about respiration. He was able to link the bodily processes of motion and digestion to its use of oxygen.

Lavoisier believed that what the body loses during respiration is replenished by nourishment from food. Thus, the study of nutrition was launched. Because of his work, Lavoisier is known as "the father of nutrition."

Following Lavoisier's experiments, many scientists conducted more calorimetric studies, which measure the heat produced by various chemical reactions. By the 1880s they could show that cells stay alive by using oxygen to break down substances, sometimes combining them into new ones. This basic concept is central to understanding nutrition.

Today research into nutrition continues at universities and in laboratories throughout the world. Nutrition is an exciting field; discoveries continue to reveal how diet directly affects your health. So far, scientists have identified about 50 different compounds that prevent deficiency diseases, promote normal growth, and support reproduction in human beings.

Lavoisier suspected that living organisms also use oxygen to create heat energy, similar to the way a candle does, and he was right. Living organisms are made of carbon compounds. *As organisms take in oxygen from the air, the carbon and hydrogen in their cells oxidize. Carbon dioxide and water vapor form during the reaction, releasing heat energy. The heat maintains body temperature and provides energy for the organism's activities.* This process is known as **respiration**. Although you might think of respiration as breathing, you can see it's much more than that.

To prove his theories, Lavoisier conducted many experiments on guinea pigs. He even enlisted the help of his assistant Seguin, who dressed himself in a special airtight suit and breathed through a tube. While Seguin was in different situations (inactive, before and after

Lavoisier's work gave him recognition as the first modern chemist. Although highly regarded in his time, some were suspicious of certain political links. He was arrested, tried, and sent to the guillotine shortly after the French Revolution.

Elements of Nutrition

Concepts of cell respiration and oxidation may seem far removed from everyday life. The saying, "You are what you eat," may be easier to apply. What you eat are **nutrients**, *substances that are found in food and needed by the body to function, grow, repair itself, and produce energy.* Nutrients are the fuel for cell activity.

A fish filet, a cup of tomato soup, and a brownie all contain nutrients, but different types and in different amounts. What's more, while some can be stored in the body for long periods of time, others must be replenished almost daily. A chronic shortage of an essential nutrient in the diet usually causes some type of deficiency disease. You read what a lack of vitamin C did to British sailors. That's why food choices matter. You need to give your body the best fuel you can to keep it working well for a lifetime.

Essential Nutrients

By the definition above, all nutrients are essential; however, experts limit the term **essential nutrients** to mean *those that the body cannot*

Figure 12-1

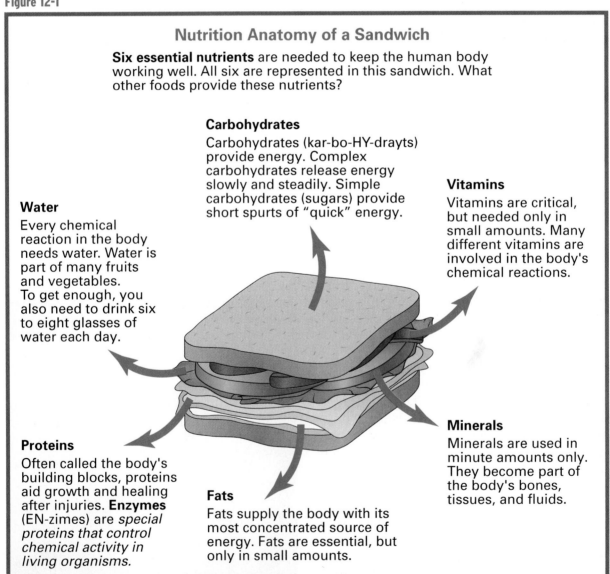

Nutrition Anatomy of a Sandwich

Six essential nutrients are needed to keep the human body working well. All six are represented in this sandwich. What other foods provide these nutrients?

Carbohydrates
Carbohydrates (kar-bo-HY-drayts) provide energy. Complex carbohydrates release energy slowly and steadily. Simple carbohydrates (sugars) provide short spurts of "quick" energy.

Vitamins
Vitamins are critical, but needed only in small amounts. Many different vitamins are involved in the body's chemical reactions.

Water
Every chemical reaction in the body needs water. Water is part of many fruits and vegetables. To get enough, you also need to drink six to eight glasses of water each day.

Minerals
Minerals are used in minute amounts only. They become part of the body's bones, tissues, and fluids.

Proteins
Often called the body's building blocks, proteins aid growth and healing after injuries. **Enzymes** (EN-zimes) are *special proteins that control chemical activity in living organisms.*

Fats
Fats supply the body with its most concentrated source of energy. Fats are essential, but only in small amounts.

Exercise takes energy. Which of the essential nutrients are most useful when you're physically active?

make itself but that are needed to build and maintain body tissue. It is essential that these be taken in from the diet. Interestingly, the nutrients considered essential for humans are different from those of other animals. Humans and trout don't make their own vitamin C, but horses and monkeys do.

The six classes of essential nutrients that are needed by humans are shown in **Figure 12-1**. Like characters in an epic novel, these nutrients work together in fascinating relationships with one goal: to keep you alive. Their individual stories will unfold in later chapters of this unit.

Nutrients and Energy

The six essential nutrients might be divided into two complementary groups: energy producers and energy releasers.

The energy-producing nutrients are carbohydrates, fats, and proteins. These are organic compounds because they contain carbon atoms. The presence of carbon indicates that these nutrients come from living organisms. It is also what gives them their energy-producing capacity. Carbon burns easily and creates ready heat. As Lavoisier proved, the body's use of nutrients is basically a burning, heat-producing process.

All that energy would stay locked inside the nutrients, however, without a trigger to release it. The other essential nutrients—vitamins, minerals, and water—are energy releasers. They help create the conditions in which food molecules are taken apart, freeing their various components for the body's use.

Nutritional Guidelines

A soccer player with a sprained ankle isn't much use to the team. Likewise, a worker with a health problem caused by poor nutrition can't contribute as much to the nation's productivity.

Think, too, of what the future holds for a country where many children can't learn because they are poorly fed.

Prompted by this sense of national interest, governments throughout the world research and promote healthful eating habits among their citizens. In the United States, results of government research are as close as the nearest label on a food package. You can also find information in pamphlets from the health department and on the web site of a university's nutrition department. Some of the information is in the form of tables about nutrients in the diet.

Recommended Dietary Allowances

To help people know how much of each nutrient they need for good nutrition, the U.S. government maintains a table of the **Recommended Dietary Allowances (RDAs)**. These numbers give *th*

Nutrition information is readily available for you. What resources would be most reliable?

adequate amount of a specific nutrient needed by most healthy people. RDAs were first established over fifty years ago. They currently cover protein, ten vitamins, and seven minerals.

The RDAs supply a detailed description of the nutrient needs of different population groups. Eighteen groups are identified, including infants, children of a certain age range and gender, and pregnant women. These categories are used because not everyone has the same nutritional needs. Medical conditions may also affect individual nutrient needs.

As the name says, RDAs are recommended, not required, amounts. They include a 30-percent margin of safety. They're about 30 percent higher than the actual average requirement. The higher figure allows for personal variations in needs. It also reflects differences in how much of a nutrient found in a food is actually absorbed and used by the body.

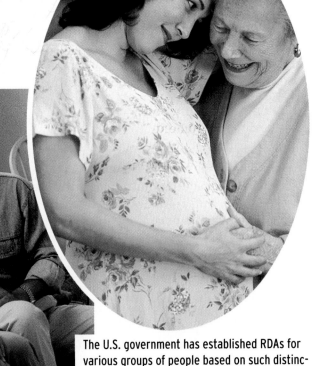

The U.S. government has established RDAs for various groups of people based on such distinctions as age, gender, pregnancy, and lactation.

Why aren't cereals fortified with vitamin D?
Ask a ... Nutritional Formulator

Most people eat cereal with milk, which already has vitamin D added," explains Mary Ann Estes, a nutritional formulator for a cereal producer, "so vitamin D isn't really needed in our products. This is just one example of what I have to think about. The nutritional makeup of our cereal products has to be balanced with cost, taste, and practicality. Otherwise, the healthiest food in the world won't sell.

"Obviously I can't keep track of everything myself. My contribution is technical data: what nutrients are in a food, what should be added, and what's the best way. My decisions are influenced by what I learn from other experts in marketing and product development."

KEY SKILLS
- Organization
- Leadership
- Communication
- Teamwork
- Analytical skills
- Problem solving

Life on the Job: Nutritional formulators usually work for manufacturers of processed foods and nutritional supplements. In developing formulations that meet product quality and cost efficiency standards, they control ingredient use and inventory. Duties also include communicating with corporate heads and manufacturing plants about developing and modifying products.

Education Needed:
Nutritional formulators need a bachelor-of-science degree in food science or in a biological science, such as biochemistry or microbiology. A background in operations management and general business is useful, since they often work on product teams involving those divisions.

Your Challenge

Suppose you're a nutritional formulator. Your company plans to increase a product's vitamin content, then claim it is "now more nutritionally complete." You know that a higher dosage isn't better.

1. To what divisions within the company would you take your concern?
2. What would you do if the firm went ahead with its plan?

Food Guide Pyramid

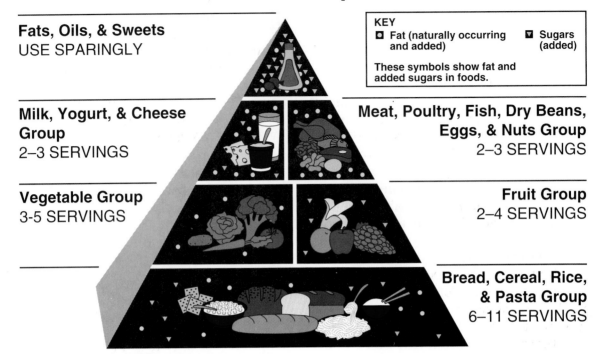

Figure 12-2
The Food Guide Pyramid shows the types of foods you need and in what amount in order to balance your daily food intake. Why was a pyramid chosen to represent these groups?

The RDAs are reconsidered every five years by scientists from the Food and Nutrition Board of the National Academy of Sciences. Occasionally an RDA is revised if research shows that more or less of a particular nutrient is needed for good health.

Health care professionals commonly use RDAs. A dietitian might consult an RDA table to decide whether to suggest iron supplements for a mother in her fourth month of breast feeding. RDAs are currently being used to help develop a new approach to presenting nutrition information, called the Dietary Reference Intakes.

Dietary Reference Intakes

Along with the RDAs, nutrition experts are using various sets of nutrition data to create the **Dietary Reference Intakes (DRIs)**. These *updated recommendations on nutri-*

ent intake are under development through the combined efforts of several agencies in the United States and Canada.

To come up with the new recommendations, researchers are looking at the latest information on how nutrients affect long-term health. They are evaluating data on minimum, maximum, and optimum nutrient amounts that different types of people need.

DRIs for certain vitamins and minerals were established in 1997. Eventually all nutrients will be covered by these updated standards, which are intended to replace the RDAs at some point in time.

Food Guide Pyramid

How do all of the recommendations from experts translate into daily food choices? To answer that question, the Department of Agriculture created the Food Guide Pyramid, shown in **Figure 12-2**. This chart is a visual reminder of what makes up a balanced diet. It outlines what and how much of certain foods a person should eat each day. You may not recall whether sweet potatoes or oranges provide more vitamin C, but you will know how much to eat for good health.

You may have seen the Food Guide Pyramid on cereal boxes, food packages, and bread wrappers. The guide divides foods into the following five groups, according to the major nutrients they supply:

- **Bread, Cereal, Rice, and Pasta.** These foods provide vitamins, minerals, and complex carbohydrates. They are also rich in dietary fiber, which is not a nutrient but is important to digestive health.
- **Vegetables.** Naturally low in fat, vegetables are rich in vitamin C, folate, many minerals, and beta carotene, which the human body converts to vitamin A. They also supply complex carbohydrates and fiber.
- **Fruits.** Lots of beta carotene, potassium, and vitamin C are among fruits' contribution to the diet. Fruits are also low in fat and sodium, while providing a healthy dose of fiber.
- **Milk, Yogurt, and Cheese.** Milk products are high in proteins, vitamins, and minerals, especially calcium. You need to check your choices for fat content, which varies greatly.
- **Meat, Poultry, Fish, Dry Beans, Eggs, and Nuts.** This group supplies proteins, minerals, and vitamins. Like milk products, however, some are high in fat.

Note that groups are placed within the Pyramid according to the number of recommended servings. With this arrangement, the Pyramid reminds you that a healthful diet is based on foods from the Bread, Cereal, Rice, and Pasta Group. Up to 11 servings are recommended. A small section that includes fats, oils, and sweets tops the Pyramid. What do you think that says about the part such foods play in a nutritious eating plan?

Within each group you find many food choices. The most healthful ones are **nutrient dense**. This means they *supply many nutrients that are critical to good health in comparison to their low-to-moderate number of kcalories.* Compare iceberg lettuce and spinach in the Vegetable Group, for example. Two ounces (55 grams) of lettuce supply five kcalories. The same amount of spinach has twice the kcalories (ten) but also provides five times the calcium and iron, eight times the vitamin C, and over 20 times the vitamin A. Clearly, not all vegetables are created equal, nor are all fruits, meats, and cereals.

Nutrient dense foods are typically low in sugars and fats and prepared using methods that keep them that way. A seven-ounce (200-gram) baked potato, at 220 kcalories and a trace of fat, supplies the same amount of protein and calcium as three ounces (85 grams) of potato chips, with their 450 kcalories and 30 grams of fat. In later chapters, you'll learn more about how preparing and processing foods affect specific nutrients.

How can you use the Pyramid to make a real difference in your diet? **Figure 12-3** on page 184 gives some examples of food choices and serving sizes from each group.

Dietary Guidelines for Americans

Choosing nutritious foods doesn't have to be difficult. Two federal government agencies, the

A baked sweet potato has about 115 kcalories, almost no fat, and is a good source of many vitamins and minerals. What additions to the potato could make it higher in fat and kcalories?

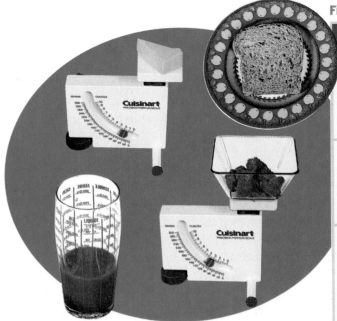

Figure 12-3

Serving Sizes in the Food Guide Pyramid

Breads, Cereal, Rice, and Pasta Group
1 slice bread
1/2 cup (125 mL) cooked cereal, rice, or pasta
1 ounce (28 g) ready-to-eat cereal

Vegetable Group
1/2 cup (125 mL) chopped vegetables, raw or cooked
1 cup (250 mL) raw leafy vegetables
3/4 cup (175 mL) vegetable juice

Fruit Group
1 piece of fruit or melon wedge
1/2 cup (125 mL) canned fruit
1/4 cup (50 mL) dried fruit
3/4 cup (175 mL) fruit juice

Milk, Yogurt, and Cheese Group
1 cup (250 mL) milk or yogurt
1 1/2 ounces (42 g) natural cheese
2 ounces (56 g) processed cheese

Meat, Poultry, Fish, Dry Beans, Eggs, and Nuts Group
2 to 3 ounces (56 to 85 g) cooked lean meat, poultry, or fish
Substitutes for 1 ounce meat: 1/2 cup (125 mL) cooked dry beans; one egg; or 2 tablespoons (30 mL) peanut butter

Fats, Oils, and Sweets Group
Use sparingly

Department of Health and Human Services and the Department of Agriculture, have summarized the best, most current advice on health and nutrition. With input from the U.S. Department of Education, they present this information in the practical, easy-to-understand *Dietary Guidelines for Americans*. The following reflects the most recent update to those guidelines:

Aim for fitness. . .

- **Aim for a healthy weight.** Healthy weights vary among individuals. Amounts outside a recommended range, however, are known to be harmful.

- **Be physically active each day.** Active people are less likely to become overweight. They are less prone to the hazards associated with the extra pounds: greater risk of heart disease, stroke, high blood pressure, diabetes, and certain cancers.

Build a healthy base. . .

- **Let the Pyramid guide your food choices.** Choose variety. No single type of food supplies the energy, protein, vitamins, minerals, fiber, and fluid needed for good health.

- **Choose a variety of grains daily, especially whole grains.** Whole grains supply the complex carbohydrates and fiber you need.

- **Choose a variety of fruits and vegetables daily.** These foods offer the assorted minerals and vitamins needed for a healthy system, while adding little fat and no cholesterol.

- **Keep food safe to eat.** Chapter 25 describes how to avoid the illnesses caused by unsafe foods.

Monitoring Health Claims

An oatmeal canister label notes that "the soluble fiber found in oatmeal, as part of a low-saturated-fat, low-cholesterol diet, may reduce the risk of heart disease." A box of prunes links the food's fiber with possible prevention of some cancers. Bread wrappers make similarly worded claims about folate content and certain birth defects.

These are examples of statements a food may legally make relating nutrients it contains, or does not contain, to health conditions. You may have read claims concerning fat and cancer, fiber and cancer, or sodium and high blood pressure.

Like other parts of food labels, such claims are monitored by the FDA. They must reflect mainstream scientific findings, which so far only link certain nutrients and certain diseases. No direct cause-and-effect relationship has been proved. Stating that "eating oatmeal prevents heart disease" would not be allowed.

The growing number of nutritional supplements has also caught the FDA's attention. Claims on these packages must also be limited. A pill containing an herbal supplement might be described as "a support to the immune system" or "helping maintain eye function."

Applying Your Knowledge

1. Why do you think health claims on food labels have become more common in recent years?
2. How do health claims affect consumers' rights and responsibilities?

By balancing a healthful diet with exercise, an appropriate weight and good health are more easily maintained.

Choose sensibly. . .

- **Choose a diet that is low in saturated fat and cholesterol and moderate in total fat.** Avoiding these substances is shown to reduce the risk of heart disease and certain types of cancer. Maintaining a healthy weight is easier.
- **Choose beverages and foods that limit your intake of sugars.** A high-sugar diet can contribute to tooth decay. A greater

hazard, however, is that it provides very few nutrients for the kcalories, "crowding out" more beneficial foods.

- **Choose and prepare foods with less salt.** This may help reduce your risk of high blood pressure.

Food Labeling

Twenty years ago, the label on a box of raisins said that you were buying "Brand X Raisins," but maybe little else. Even foods that obviously had more than one ingredient may not have told you what those ingredients were. Certain items, including peanut butter and mayonnaise, were covered by "standards of identity."

Manufacturers used one accepted recipe for that product, which was the standard the food met to be identified as peanut butter or mayonnaise. Nutrition information was voluntary, and only two-thirds of all products on the market volunteered any such information.

The 1980s saw a growing consumer rights movement and greater awareness of foods' value to good health. The Nutrition Labeling and Education Act of 1990 was one result. This sweeping piece of legislation requires that almost all foods produced in the United States have labels with specific facts about nutritional content. Now the Nutrition Facts panel is as much a part of the food label as the enticing picture on the front.

Figure 12-4

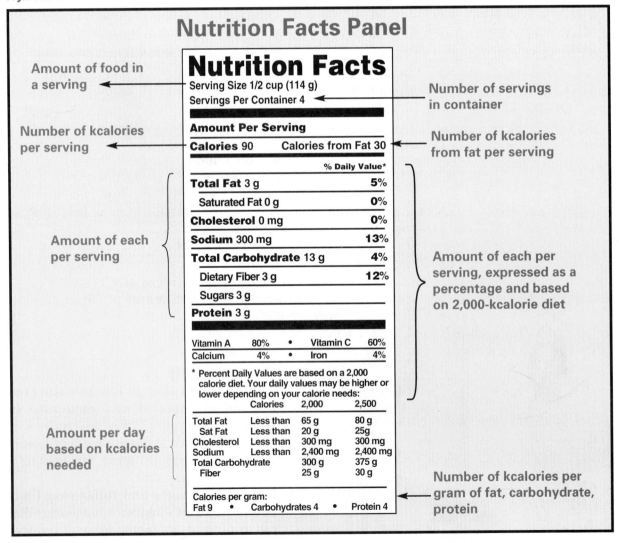

The FDA regulates the meaning of food descriptors. "Light" means that, compared to regular versions, a food has half the fat, or one-third fewer kcalories with no more than half from fat.

The Nutrition Facts Panel

Thanks to the Nutrition Facts panel, almost every food label is a hand-held nutrition guide. As you can see in **Figure 12-4**, the panel has a great deal of information. Amounts (in grams or milligrams) are given for total fat, cholesterol, sodium, total carbohydrate, dietary fiber, sugars, and protein.

You won't find nutrition panels on every food item sold. Legal exceptions include foods meant for immediate consumption, including those bought in restaurants, from sidewalk vendors, or at a supermarket deli. Also exempt are plain coffee, tea, spices, and other products that contain no significant amount of nutrients. Packages measuring less than 12 square inches may carry a nutritional panel; if not, they must give an address or telephone number where you can obtain this information.

Daily Values

With the **Daily Values (DVs)** on the Nutrition Facts panel, you can determine *how the nutrients in a food serving fit with what you can or should have for the day*. DVs are expressed as a percentage. For example, on a box of cereal under "% Daily Values," you might see "2%" beside "fat." That means that if you eat a serving of the cereal, you'll get 2% of the maximum allowable amount of fat for a day (based on a 2000-kcalorie diet). If you see "25%" beside "fat," what does that tell you?

The DV for fat is based on an upper limit. It compares the fat per serving with the maximum amount recommended for good health. The DVs for saturated fat, cholesterol, and sodium are used in the same way. In contrast, the DVs for total carbohydrate, fiber, vitamin A, vitamin C, calcium, and iron are based on a minimum. For each nutrient, they tell how much a serving contributes to the least amount needed for good health.

Cholesterol and sodium DVs are the same for every adult. Those for fat, saturated fat, carbohydrate, and fiber vary according to how many kcalories you need daily.

Food Label Descriptors

Label law reform has also taken the guesswork out of interpreting food descriptions on food labels. The Food and Drug Administration has identified core terms and how they may be used to describe foods.

For example, "free" on a product means it contains none or an insignificant amount of a nutrient. "Low" means a product can be consumed in large amounts without exceeding the Daily Value for the nutrient. A "lean" meat product contains less than 10 g of fat, 4 g of saturated fat, and 95 mg of cholesterol per serving.

Nutrition for Life

Upcoming chapters will detail the properties of nutrients and their effects on the body. What you learn may inspire you to take a closer look at your eating habits. Developing healthful habits is important, not only to how you feel now but also to how healthy you are in the years ahead.

Identifying Basic Nutrients in Foods

Do you know what nutrients are in the bread you toast for breakfast and the banana you eat for lunch? This experiment will help you find out as you test a few common foods for the presence of complex carbohydrates (starches), simple carbohydrates (sugars), fats, and proteins.

Equipment and Materials

5 test tubes	safety goggles	Biuret reagent	bread suspension
test tube rack	brown paper	starch suspension	banana suspension
10-mL graduated cylinder	iodine solution	5% dextrose solution	fat-free milk
400-mL beaker	Benedict's solution	vegetable oil	whole milk
metric ruler	burner or hot plate	egg white	potato suspension

Procedure Part 1

1. Place four clean test tubes in a test tube rack. Add 3 mL of 1% starch solution to test tube 1; 3 mL of dextrose solution to test tube 2; 3 mL of egg white (protein) to test tube 3; and 3 mL of water to test tube 4.
2. To each of these test tubes add one drop of dilute iodine solution. Record any color change in your data table.
3. Wash the test tubes. Then repeat Step 1.
4. To each of the test tubes, add 5 mL of Benedict's solution. Place the test tubes in a 400-mL beaker containing enough water to completely surround the contents of the test tubes. Heat the water to boiling; boil for three minutes. Record any color change you observe in your data table.
5. Again wash and prepare the test tubes as in Step 1.
6. Add 3 mL of water to each of the first three test tubes.
7. Put on safety goggles. Add 5 drops of Biuret reagent to all four test tubes. (Be careful not to get this solution on your hands. If you do, wash it off immediately with plenty of water.) Record any color change you observe in your data table.
8. Fats don't dissolve in water; a different test is needed to check for their presence. Place a drop of the carbohydrate, dextrose, and protein solutions on a piece of brown paper. Place a drop of vegetable oil on the paper. Then hold the paper up to a light. Any sample containing fat will produce a translucent spot on the paper. Describe the appearance of each sample in your data table.

Procedure Part 2

9. **Place 5 clean test tubes in a rack.** Obtain 3-mL samples of the suspensions of bread, banana, and potato flakes prepared by your teacher. Also obtain 3-mL samples of fat-free and whole milk. Place these in the test tubes. Test with iodine as in Step 2 and record results in your data table.
10. Wash and prepare test tubes as in Step 9. Test each sample with Benedict's solution as you did in Step 4. Record results in your data table.
11. Wash and again prepare test tubes as in Step 9. Test each sample with Biuret reagent as in Step 7. Remember to keep your safety goggles on whenever using Biuret reagent. Record results in your data table.
12. Using a small amount of each of the five foods, repeat Step 8 to test for the presence of fat.

1. With what type of food did each reagent react? What color changes indicated this?
2. What can you conclude about the nutrient content of each food tested?
3. Did any foods seem to contain only one type of nutrient? If so, explain.
4. How do whole milk and fat-free milk differ in nutrient content?

SAMPLE DATA TABLE PART 1

Nutrient Tested

	Iodine Solution	Benedict's Solution	Biuret's Solution	Brown Paper
Starch				
Sugar				
Protein				
Water				
Fat				

Reagent Used

SAMPLE DATA TABLE PART 2

Food Tested

	Iodine Solution	Benedict's Solution	Biuret's Solution	Brown Paper
Bread				
Banana				
Potato Flakes				
Whole Milk				
Low-Fat Milk				

Reagent Used

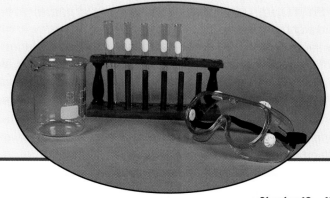

Chapter Summary

- Nutrition is the science of how food is used by the body.
- Nutrients are critical to life. Those that cannot be produced by the body must be consumed in foods.
- Different nutrients perform different functions in the body.
- Dietary Reference Intakes are a range of amounts of nutrients needed for acceptable and optimum health.
- RDAs are recommended amounts of nutrients for specific groups in the population.
- The Dietary Guidelines for Americans offer advice on choosing healthful foods.
- The Food Guide Pyramid illustrates a healthful, daily balance and intake of foods and their nutrients.
- Food package labels are required to give specific, accurate nutritional information.

Using Your Knowledge

1. How did limes play a part in early studies in nutrition?
2. Describe the process that Lavoisier discovered while experimenting with fire in his laboratory.
3. How is respiration related to nutrition?
4. What are the main classes of essential nutrients and why are they called essential?
5. Why is water considered an essential nutrient?
6. Why are organic compounds necessary to nutrition?
7. How are RDAs used, and by whom?
8. Why are the food groups arranged a certain way in the Food Guide Pyramid?
9. Evaluate this dish according to the Food Guide Pyramid: one cup of pasta tossed with a half-cup of fresh, chopped vegetables and sprinkled with one ounce of grated cheese.
10. What problems might arise from choosing foods without paying attention to nutrient density?

Real-World Impact

Symbols of Good Eating

The Food Guide Pyramid is a visual version of a healthful diet promoted in the United States. The Canadian Food Guide to Healthy Eating is a four-banded rainbow. The large, outside band contains foods that should be eaten most. Citizens of the Philippines symbolize a good diet with a star. In Sweden it's a circle. Japan uses the Number 6, and the United Kingdom, a plate. Many food guides are based on bread, cereals, and grains. Canada's, however, is based on fruits and vegetables.

Thinking It Through

1. Why do you think many countries base their food guides on bread, cereals, and grain products?
2. Israel's guide includes water as an important part of the diet. Why?
3. Why do you think the symbols mentioned were chosen to represent eating guidelines?

Using Your Knowledge *(continued)*

11. Which Dietary Guidelines would be especially helpful for someone who wants to manage weight gain?
12. Does the DV for cholesterol mean the same thing as the DV for iron? Explain.

Skill-Building Activities

1. **Math.** Bring in a food label that lists the kcalories supplied per gram of fat, carbohydrate, and protein. Using the rest of the Nutrition Facts panel, compute the number of kcalories each nutrient contributes to one serving of that food.
2. **Decision Making.** Plan a one-day menu you would enjoy, following the recommendations of the Food Guide Pyramid.
3. **Problem Solving.** Use a reliable nutrition resource to find the nutritive and kcaloric value of common foods. With this information, plan a day's menu of nutrient dense foods for each of these individuals: a) someone on a 1500-kcalorie-a-day diet; b) someone who eats no meat or poultry products.
4. **Teamwork.** Working with a group, create your own version of the Food Guide Pyramid. Choose another way to symbolize healthful eating.
5. **Research.** The fortification of folate (folic acid, a water-soluble vitamin) in all cereal grains has been required by law since January, 1998. Locate information and write a report on the importance of folic acid to pregnant females as well as to healthy babies.
6. **Social Studies.** Research the history of the Food Guide Pyramid. Who actually designed the Pyramid? How did they choose the shape, food groups, and servings? What factors besides nutrition influenced their decisions?

Thinking Lab

Unlocking the Door to Nutrition

The Dietary Guidelines are general recommendations for healthful food choices. The Food Guide Pyramid describes a more detailed pattern of eating that helps ensure a nutritionally balanced, adequate diet.

Analysis

The Dietary Guidelines and the Food Guide Pyramid overlap in some aspects. They aren't simply different ways of saying the same thing, however. Instead, each one complements the other, like a lock and key.

Representing and Applying Data

Choose three Dietary Guidelines. Determine how each one relates to each of the groups of the Food Guide Pyramid. How does that guideline affect your choices from that food group? Communicate this information in a chart, table, or other visual form.

7. **Communication.** Express the Dietary Guidelines for Americans in a different form of communication, such as song, verse, visual images, or drama. Aim your message for a specific group, such as children or teens, and choose the tone and medium to best reach your audience.

Understanding Vocabulary

Create fill-in-the-blank questions that use the "Terms to Remember." Keep your quiz for review at test time.

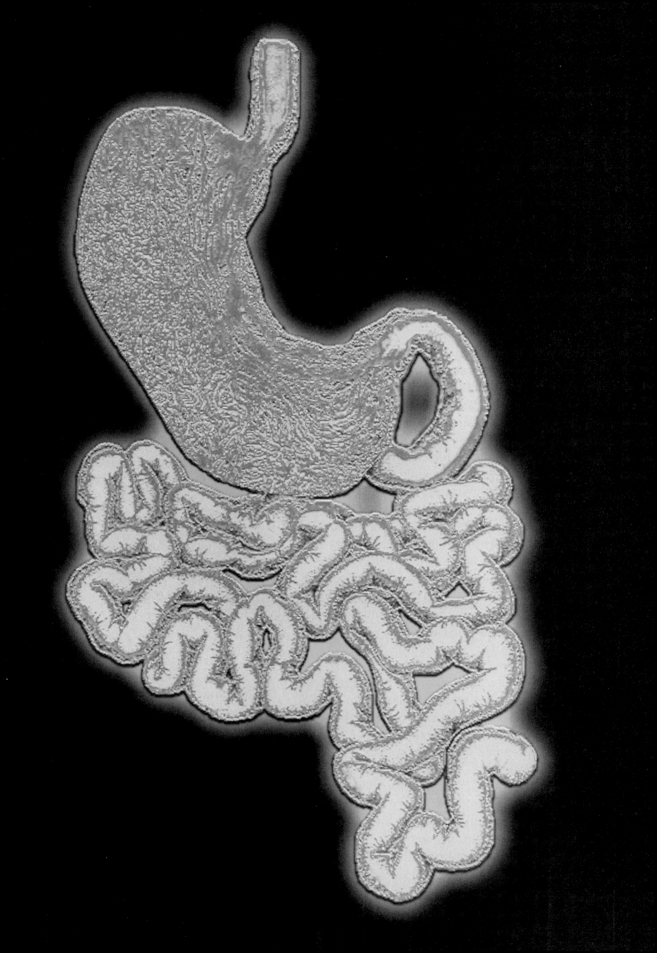

CHAPTER (13) Digestion

Objectives

- Identify in order the parts of the alimentary canal.
- Describe the processes that take place in each part of the digestive tract.
- Explain the function of enzymes in digestion.
- Describe the roles of accessory organs in digestion.
- Explain how nutrients are absorbed.

Terms to remember

alimentary canal
bile
cardiac sphincter
digestion
epiglottis
esophagus
mastication
pancreatic juice
peristalsis
pyloric sphincter
saliva

With every bite of food you take, an amazing process begins. Most of the time, you're not even aware of what's happening.

Body cells are incredibly small. Nutrients, to enter and nourish cells, must be smaller still. Getting nutrients to the cells is the task of the digestive system. This system unlocks nutrients, makes them available to body cells, and even cleans up the leftovers. Start to finish, the system works on a food for 12 to 48 hours, depending on the food's makeup and the final destination for each of its parts.

Digestion of Nutrients

Digestion is *the chemical and mechanical process of breaking down food to release nutrients in a form your body can absorb for use.* Through digestion, nutrients are made available to supply you with energy. **Figure 13-1** on page 194 shows the digestive system in the body.

Digestion occurs in the **alimentary canal**, commonly called the *digestive tract.* The canal is about eight meters of tubing. Coiled and winding, most of it fits within the rib cage. Various sections have different shapes and perform different functions. The tract doesn't work alone, but is helped by accessory organs, which produce substances to aid the digestive process.

Digestion in the Mouth

Digestion begins with the mechanical step of breaking down food into small pieces. **Mastication**, or *chewing*, not only grinds food for easier swallowing but also creates more food surface area for the chemical reactions that are already starting to take place. As you chew, liquids released by the mucous membranes and glands under the tongue soften the food. Your tongue helps push the food along on its way to the next stage of digestion.

Figure 13-1
The digestive system begins in the mouth and proceeds to the stomach and then the small and large intestines. **How is food progress controlled along the way?**

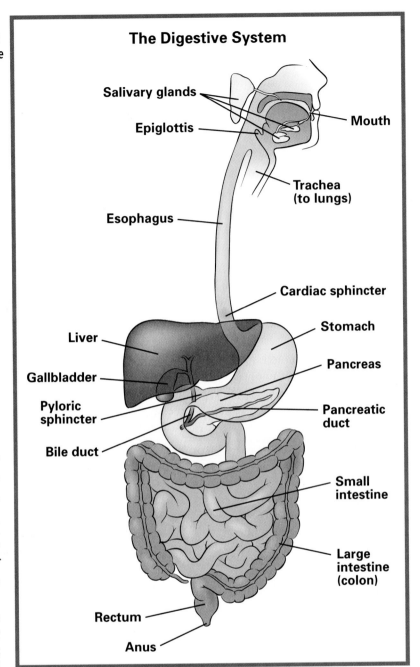

The Digestive System

Salivary glands

Epiglottis

Mouth

Trachea (to lungs)

Esophagus

Cardiac sphincter

Liver

Stomach

Gallbladder

Pancreas

Pyloric sphincter

Pancreatic duct

Bile duct

Small intestine

Large intestine (colon)

Rectum

Anus

At the back of the tongue is a most important structure, the **epiglottis** (e-puh-GLAH-tus). This *thin, elastic flap at the root of the tongue shields the windpipe when you swallow and prevents food and water from entering.* You can interfere with this reflex if you talk or laugh while swallowing and misdirect food "down the wrong pipe."

Action of Saliva

Many chemical processes complement the mechanical actions in the mouth. These processes are aided by enzymes, proteins that control chemical activity in living organisms. (You'll discover more about enzymes in Chapter 19.)

Salivary glands in the mouth start the chemical processes by secreting **saliva**, *a mixture of water, mucus, salts, and the digestive enzyme amylase.* The amount and type of saliva secreted are controlled by your automatic nervous system. Most people produce about one to 1.5 liters of saliva every day.

A vital component of the digestive process, saliva performs the following functions:
- Adds water and salt to foods. This combination helps to dissolve and compress food, making it easier to swallow.
- Lubricates and binds food so it slides easily into the stomach.
- Helps protect and cleanse the teeth and mouth lining.
- Provides a slight alkaline, buffering effect.

Saliva also begins the digestion of carbohydrates, or starch, in the food. The amylase in saliva reacts with the carbohydrates, breaking them down into simpler sugars. Among these sugars is maltose, which tastes something like malted milk. If you slowly chew a soda cracker or a piece of bread, holding it in your mouth as it disintegrates, you begin to taste the subtle sweetness of maltose.

Nutrient Digestion

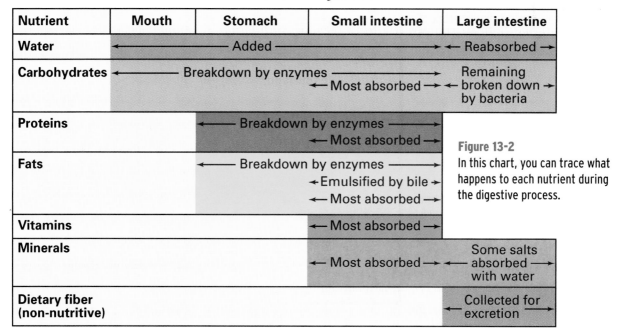

Nutrient	Mouth	Stomach	Small intestine	Large intestine
Water	←———————— Added ————————→			← Reabsorbed →
Carbohydrates	←——— Breakdown by enzymes ———→			Remaining
			← Most absorbed →	← broken down → by bacteria
Proteins		←——— Breakdown by enzymes ———→		
			← Most absorbed →	
Fats		←——— Breakdown by enzymes ———→		
			←Emulsified by bile→	
			← Most absorbed →	
Vitamins			← Most absorbed →	
Minerals			← Most absorbed →	Some salts ← absorbed → with water
Dietary fiber (non-nutritive)				Collected for ← excretion

Figure 13-2
In this chart, you can trace what happens to each nutrient during the digestive process.

In the mouth, chemical action occurs only on carbohydrates. Fats and proteins are dealt with at other points in the digestive process. All three of these are energy-producing nutrients. Vitamins and minerals—the energy releasers—and water don't need to be digested. The body uses those nutrients just as they are. You can see what happens to each of the nutrients in **Figure 13-2**.

Food in the Esophagus

Swallowing food starts the journey down the esophagus. Your **esophagus** (ih-SAH-fuh-gus) is *the tube-like passage connecting the mouth to the stomach*. It is the least complex section of the digestive tract, yet noteworthy in its way. The esophagus is repeatedly exposed to all kinds of abrasive substances—rough, tough, and acid foods. It is protected, however, by a membrane lining that secretes mucus.

Even the acid in fresh tomato salsa and the rough edge of a tortilla chip that wasn't chewed well pass harmlessly.

As food descends your esophagus to your stomach, it's broken down into finer particles through **peristalsis** (PEHR-uh-STAHL-sus), *waves of muscular contractions.* Strong muscles that line the entire digestive tract accomplish peristalsis. These muscles continuously churn and push food along the tract, as shown in **Figure 13-3**.

Figure 13-3
During peristalsis, the muscles pull in two directions. They contract and relax around the diameter of the esophagus. They also contract and relax along the length of the esophagus.

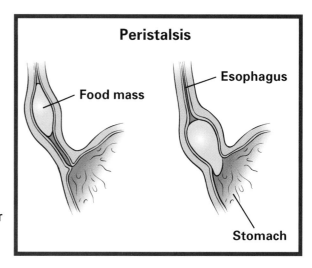

Peristalsis

Food mass

Esophagus

Stomach

Figure 13-4
Food enters the stomach through the cardiac sphincter, where it churns around and turns to chyme. The pyloric sphincter slowly releases the chyme into the small intestine.

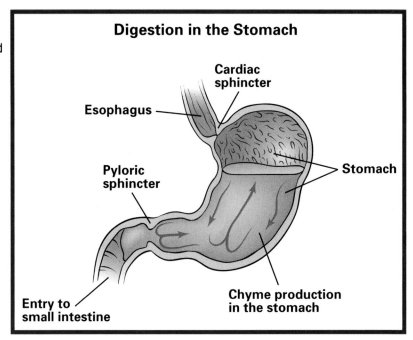

Digestion in the Stomach

- Esophagus
- Cardiac sphincter
- Pyloric sphincter
- Stomach
- Entry to small intestine
- Chyme production in the stomach

The esophagus ends with a ring-like, muscular valve called the **cardiac sphincter** (SFINK-tur). Peristalsis pushes food through this valve into the stomach. The cardiac sphincter then contracts again to keep food from rising back into the esophagus.

Digestion in the Stomach

The stomach, shown in **Figure 13-4**, is basically a pouch, an enlarged section of the digestive tract. In shape it resembles a sweet potato. In size it varies with its human owner. A full stomach can stretch to hold between two and four liters of food, but one liter is the average, more comfortable fit.

The stomach secretes gastric fluid, sometimes called gastric or digestive juices. This water-based fluid contains four main substances that begin the next stage of digestion: the breakdown of proteins and fats.

- **Hydrochloric acid.** Protein digestion requires a high-acid environment. Hydrochloric acid creates such a condition. Due to this acid, gastric fluid has a pH of 2 or below. This low pH not only allows the breakdown of proteins but also prevents harmful bacterial growth in the stomach.
- **Enzymes.** Various enzymes begin the digestion of proteins and fats. Two of these are pepsin and rennin. Acting with the acid environment, these enzymes break down complex proteins into smaller ones. Another enzyme, gastric lipase, starts fat digestion. The amylase, which began digesting starches in the mouth, continues its work in the stomach.
- **Gastrin.** Gastrin is a protein made by the body. It controls acid secretion.

Stress can cause people to have indigestion. Eating slowly without overeating may help. Stress management can too. What techniques are good for relieving stress?

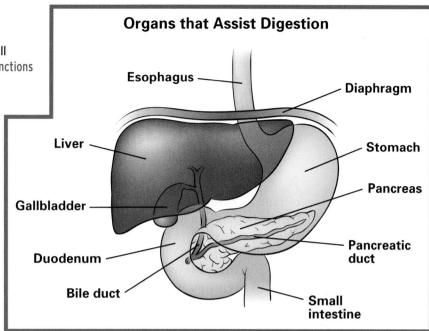

Figure 13-5
The liver, gallbladder, and pancreas all assist the digestive process. What functions do they serve?

Organs that Assist Digestion

Esophagus
Diaphragm
Liver
Stomach
Pancreas
Gallbladder
Pancreatic duct
Duodenum
Bile duct
Small intestine

- **Mucus.** The stomach lining protects itself by releasing a coat of heavy, bicarbonate-rich mucus. This is why your stomach isn't digested by its own acid and enzymes.

The time a food stays in the stomach depends on its nutrients. Some nutrients take longer than others to break down. Carbohydrates may be in and out in one to two hours. Proteins, which are more complex, take three to five hours. Fats are a bigger task for enzymes; they take up to seven hours to break down.

Leaving the stomach, the meal you ate is no longer recognizable. It has been churned into a thick fluid called *chyme* (KIME). Chyme passes through the **pyloric sphincter** (pie-LOHR-ik), *a circular muscle that controls the food's rate of movement from the stomach to the small intestine.* The pyloric sphincter releases food a little at a time. This leisurely pace helps ensure the best possible absorption of nutrients in the small intestine.

Digestion in the Small Intestine

The small intestine is the longest section of the alimentary canal. Straightened out, it would stretch about six meters, roughly the length of your classroom. In diameter, however, it would probably fit through a circle made by your thumb and forefinger. It's a narrow 3.75 cm across. This is why food must be mostly liquefied before leaving the stomach.

Spurred by many complex chemical reactions, the process of nutrient digestion swings into high gear in the small intestine.

Bile and Pancreatic Juice

Digestion in the small intestine gets an assist from three accessory organs, the liver, the gall bladder, and the pancreas (PAN-kree-us). You can see these in **Figure 13-5**. The liver produces **bile**, *a greenish liquid that helps fat mix with the water in the intestine.* By creating this water-fat emulsion, bile helps the body digest and absorb fats.

Bile is stored in the gall bladder until needed. The body continually monitors the amount of fat in the small intestine. It signals the gall bladder to release the needed quantity of emulsifying bile, which enters the small intestine through the bile duct.

Meanwhile, the pancreas secretes **pancreatic juice** (pan-kree-A-tik), *an enzyme-rich fluid that continues to reduce food to smaller molecules.* (There are, in fact, thousands of enzymes at work in the digestive process.) The pancreas also releases bicarbonate in precise amounts to neutralize the strong, acidic fluids carried from the stomach.

FOOD SCIENCE
Careers
Flavorist

Can you make banana pudding without bananas? Ask a . . . Flavorist

Flavorist Catherine McCloud sees her job as creating complicated recipes. "A single flavor may take dozens of substances. Banana flavoring, for instance, has 200 ingredients," she says. "Isolating and blending natural flavors or creating new ones is almost like doing nature one better. If an enzyme in a food contributes to its taste, we have to figure out how to recreate that effect in another food. Some people frown on artificial flavors, but I see the positives. For one thing, they make foods more affordable. And with those butter-flavored sprinkles for vegetables and popcorn, the mouth gets the taste and feel of fat, but the body handles them like protein. That cuts calories."

Education Needed: Flavorists need a bachelor's degree in chemistry or food science. A grounding in organic chemistry and coursework in neuroscience are useful for understanding how the brain decodes sensory responses to food. Flavor chemists who have a Ph.D. can work in food technology research. An advanced degree combined with a

KEY SKILLS
- Attention to detail
- Creative thinking
- Math
- Organization
- Critical thinking
- Keen sense of taste and smell

sales background could lead to the position of vice-president of ingredient sales.

Life on the Job: Flavorists identify the chemical compounds in flavors and then plan and conduct experiments. Likely employers include biotechnology and pharmaceutical firms and government agencies. A flavorist often works with a team.

Your Challenge

Suppose you work for a restaurant chain that wants to locate in a country known for spicy cuisine. As a flavorist, you are asked to develop menu items that will appeal to this new market.

1. How would you learn what tastes and flavors are popular in the country?
2. What factors besides taste would you weigh in reworking a recipe?

Absorption of Nutrients

Breaking food down into its basic components is only half the job of digestion. Through absorption, nutrients are made available to nourish the body cells for their many vital functions.

Absorption in the Small Intestine

Nutrients need no special ducts or tubes to carry them from the digestive tract to the rest of the body. By this time they are mere molecules, passing through the thin, intestinal walls.

The lining of the small intestine is specially fashioned to create the greatest possible surface area for nutrient absorption—an area equal to that of a tennis court, in fact. First, it is pleated into numerous folds. Each of these folds is lined with tiny, finger-like projections called *villi*. You can see them in **Figure 13-6**. The villi, in turn, are covered with more projections called *microvilli*. Each microvillus is specifically designed to aid the absorption of one particular nutrient. Microvilli designated as absorbers for a certain sugar will not absorb a protein, or even another sugar.

As with digestion, foods are absorbed in a set order, the same order in which they are broken down. Simple sugars, which begin to be digested in the mouth, are absorbed earlier, high in the small intestine. Proteins are absorbed further along and lower in the small intestine. Fats are absorbed last.

Once nutrient molecules flow through the intestinal wall, they are transported to the rest of the cells by two systems. The *lymphatic* (lim-FA-tik) system carries off most of the digested fat molecules and the fat-soluable vitamins. Later it "hands off" these nutrients to the *circulatory* system (the bloodstream), which has already picked up proteins, carbohydrates, water-soluable vitamins, and minerals.

Normally, by the time food reaches the end of the small intestine, mostly water, dissolved minerals, and indigestible fiber remain. These substances are dealt with in the last section of the alimentary canal, the large intestine.

Absorption in the Large Intestine

Even with most of the work of digestion complete, the large intestine still has a few important tasks to carry out. The large intestine includes two parts, the colon and the rectum. These segments make up the final 1.5 m of the digestive tract and perform the following three major functions:

- **Bacterial action.** Bacteria that live in the colon complete the breakdown of any carbohydrates that were not digested earlier by enzymes. These bacteria also break down indigestible fiber into simpler compounds. In addition, they synthesize vitamin K and certain B vitamins, creating those nutrients from other, existing chemicals.

Figure 13-6
The lining of the small intestine absorbs nutrients. How is the lining equipped for that purpose?

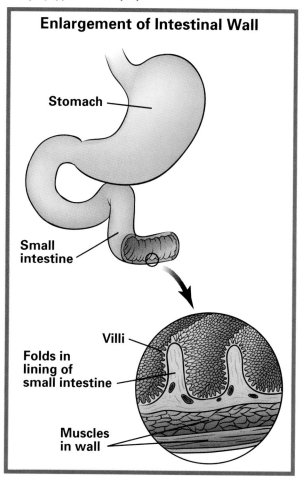

Enlargement of Intestinal Wall

Stomach

Small intestine

Villi

Folds in lining of small intestine

Muscles in wall

Fabulous Fiber

For years, people have been urged to get enough fiber in the diet. Newer research adds weight to the old advice.

People have long known that fiber absorbs water, easing foods through the digestive tract. Some studies indicate that this effect may help prevent colon cancer. By moving food quickly through the body, fiber may limit your exposure to potential cancer-causing substances that some foods contain.

Other research suggests that fiber may lower blood cholesterol levels slightly and help stabilize the production of insulin, a substance that regulates blood sugar levels.

Both the American Dietetic Association and the National Cancer Institute urge a daily intake of about 30 grams of fiber.

With so many varied sources, getting enough fiber is fairly easy. One cup of cooked kidney beans or lentils provides about 10 grams of fiber; a cup of cooked spinach or broccoli adds five more. In general, you can add fiber by eating fruits and vegetables and choosing whole grain products over refined ones. Remember to increase fluids as you increase fiber.

Applying Your Knowledge

1. From the examples of high-fiber foods given here, what nutritional benefits would you expect from choosing them more often?
2. Why do you think fiber consumption is relatively low in the United States?

- **Water recovery.** Water has played a major role in digestion. It has helped break down foods and transport nutrients through the digestive tract and beyond. From the large intestine, much of this water is reabsorbed by the body, along with such mineral salts as sodium and potassium.
- **Collection of waste.** Some parts of foods that cannot be used or digested are stored until elimination from the body.

Once the nutrients have been released into your body, what does your body do with them? How is energy extracted from the breakfast you ate so that you have what your body needs to work efficiently? That story is detailed in the next chapter on metabolism.

By choosing a variety of foods that are high in fiber, you help ensure the health of your digestive system. What are some foods that are high in fiber?

Digestion of Starch

Digestion starts in the mouth. As teeth grind the food, an enzyme in saliva begins the digestion of starch. Other enzymes repeat the process throughout the digestive process. In this experiment, you'll see how starch in food reacts with the salivary enzyme.

Equipment and Materials

5 test tubes	medicine dropper	thermometer
1 g of potato starch	water bath	tincture of iodine

Procedure

1. Label five test tubes A through E.
2. In test tube A, place 1 g of potato starch. Add 15 mL of water. Gently flick the test tube to mix.
3. With a medicine dropper, remove a sample of the water-starch mixture from test tube A. Place 20 drops of mixture in test tube B. Add 2 drops of tincture of iodine. Flick tube gently to mix.
4. Record the color of the mixture in each test tube.
5. Heat a water bath until the thermometer registers 37ºC.
6. While the water bath is heating, add approximately 1 cm of saliva to test tube A. Flick the tube to mix. Then place in the water bath.
7. When test tube A has been in the water bath for 5 minutes, remove a sample of the starch-saliva mixture with a medicine dropper. Place 10 drops of the mixture in test tube C. Add 1 drop of tincture of iodine. Gently flick the tube to mix. Then record the color of the mixture.
8. When test tube A has been in the water bath for 10 minutes, remove another sample of the starch-saliva with a medicine dropper. Place 10 drops in test tube D. Add 1 drop of tincture of iodine and flick tube to mix. Record the color of the mixture.
9. When test tube A has been in the water bath for 15 minutes, remove a final sample of the starch-saliva mixture with a medicine dropper. Place 10 drops in test tube E. Add 1 drop of tincture of iodine and flick the test tube to mix. Record the color of the mixture.

Analyzing Results

1. What color is the mixture when starch is present?
2. Was starch present in all the samples tested? If not, when did it disappear?
3. How do you think heat affects enzyme action? How is heat a factor in digestion?

SAMPLE DATA TABLE

Mixture	Color
Test tube A: starch-water	
Test tube B: water-starch-iodine	
Test tube C: starch-saliva (5 mins.)	
Test tube D: starch-saliva (10 mins.)	
Test tube E: starch-saliva (15 mins.)	

Chapter Summary

- Digestion is the process by which nutrients are made available for the body to use.
- Food is broken down by both mechanical and chemical means, especially by the work of enzymes.
- Digestion begins in the mouth.
- Peristalsis moves food through the esophagus to the stomach.
- The major breakdown of food into nutrients begins in the stomach.
- Most food is thoroughly digested and absorbed in the small intestine.
- The large intestine completes the digestive process.

Using Your Knowledge

1. Describe the mechanical process that begins digestion.
2. How does saliva aid in digestion?
3. Are all nutrients digested? Explain.
4. Why is the esophagus relatively sturdy?
5. What special function do the muscles of the esophagus perform?
6. A friend is alarmed to learn that the stomach contains hydrochloric acid. How do you respond?
7. Name two enzymes and explain their role in digestion.
8. How long does a food remain in the digestive tract? What factors determine the length of its stay?
9. In what chemical reaction is bile involved? Why is this reaction important to digestion?
10. How are nutrients absorbed and transported from the small intestine?
11. Should you be concerned about bacterial growth in the colon? Why or why not?

Real-World Impact

Taking Medication

Medications that are taken by mouth can affect the digestive system. These include prescriptions and over-the-counter products. Some medications can cause harmful effects when taken together. To prevent problems, never take medicine that was prescribed for someone else. When you need medication for illness, read labels and warnings and carefully follow directions for use. Any questions or concerns you have should be referred to a doctor.

Thinking It Through
1. Why do some people use prescriptions written for others?
2. Overuse of aspirin has caused ulcers in some people. What other examples of problems caused by medications can you cite?

Skill-Building Activities

1. **Communication.** Prepare a short presentation (about five minutes) to teach third-graders about the digestive process. Include diagrams or other visual aids and comparisons to help make concepts clearer to your audience.
2. **Social Studies.** Using the Internet, explore the impact of lifestyle factors on digestive system diseases. How do eating habits, tobacco use, and alcohol use relate to various ulcers and cancers? If possible, compare illness rates among cultures with different lifestyles. What conclusions do these findings suggest?
3. **Research.** Find information on screening methods that provide early detection of colon cancer. Write a report on the screening methods used.
4. **Demonstration.** Plan a short demonstration for the class that visually depicts measurements mentioned in the chapter. Include the length of the digestive tract, average daily saliva production, and stomach capacity.
5. **Management.** Suppose you're an assistant Little League coach. You have games scheduled for 11 a.m. and 2 p.m. Based on your knowledge of nutrition and digestion, what suggestions would you give to the team for eating that day?
6. **Critical Thinking.** Suppose a friend complains about continual stomach discomforts. You know that the friend has been having problems at home. What would you suggest?

Understanding Vocabulary

Create a word puzzle that uses most or all of the "Terms to Remember." Exchange puzzles with a classmate for solving.

Thinking Lab

Do Antacids Spell Relief?

To digest proteins, an environment must be acidic. The body provides such an environment in the stomach. The stomach lining secretes gastric fluids to maintain a stomach pH around 2.

Analysis

Antacids were originally marketed as an antidote to the acid or burning sensation of heartburn or indigestion. The active ingredient in these products is calcium carbonate, which neutralizes acids, thus offering some relief. Some people routinely reach for the antacid bottle whenever they feel an indigestion attack coming on.

Recently, antacids have been promoted for their calcium content as well. Package labels strictly limit how much of the product should be taken over a period of time. However, some consumers may take excessive doses to make sure they meet their calcium allowance.

Thinking Critically

1. How might overreliance or prolonged use of antacids make them less effective?
2. For what other reasons might manufacturers warn against overusing antacids?

Objectives

- Explain the purpose of metabolism and the conditions needed for it to occur.
- Explain the role of energy in metabolism.
- Compare anabolism and catabolism.
- Explain the process that stores and transfers energy in the body.
- Explain how cells maintain chemical balance.
- Relate the influence of various factors to metabolic rate.
- Relate basal metabolism and voluntary activity to energy needs.
- Evaluate weight-loss diets and exercise habits in relation to metabolism and health.

Terms to remember

adenosine triphosphate
anabolism
basal metabolic rate
basal metabolism
catabolism
cytoplasm
glycogen
homeostasis
lactic acid
membranes
metabolism
metabolic rate
osmosis
semipermeable
voluntary activities

Have you ever heard someone say, "I just look at food and gain weight," yet someone else says, "I eat all day and never gain an ounce"? These comments may be exaggerations, but they do bring up some interesting questions about how the human body uses nutrients. Why do people gain and lose weight at different rates? Why do you and a friend need different amounts of food to stay healthy?

The answer to these questions lies in the process of metabolism. Through **metabolism**, *living cells use nutrients in many chemical reactions that provide energy for vital processes and activities.* Metabolism is the sum of all chemical reactions occurring within a cell or organism. Understanding this interrelated series of events can give you a good foundation for making wise food choices.

The Main Function of Metabolism

Through all its functions, the body strives for a state of **homeostasis** (hoe-mee-oh-STAY-sis), *a healthy and relatively constant internal environment.* To maintain homeostasis, the body regulates its systems to avoid dangerous lacks or excesses. For example, you breathe to take in needed oxygen and expel unwanted carbon dioxide. If you hold your breath, your brain eventually senses a lack of oxygen. It signals you to breathe deeply and your heart to beat quickly to speed oxygen to the cells.

Normal metabolism helps make homeostasis possible. It allows the body to maintain or regain an "even keel," a safe range of functioning.

The Metabolic Process

Metabolism might be likened to the chemical reactions that keep a car running. Both processes rely on the following four conditions:

- **Chemical balance.** Gasoline must have a specific ratio of hydrogen to carbon to run a car. For metabolism to occur, the body needs the right balance of nutrients and other substances dissolved in the blood and cells.

Certain chemical reactions enable a car to run. How is that similar to your body?

eat animals that eat plants, they gain the nutrients that provide energy for metabolism.

Of the six groups of essential nutrients, the ones that produce energy are carbohydrates, fats, and protein. As the body uses these nutrients to create energy, metabolism takes place.

Carbohydrates are the body's preferred source of energy. They are quickly broken down into simple sugars, which are available for immediate use. Some are reserved as **glycogen** (GLY-kuh-jun), *the form of carbohydrates stored in the muscles.* Remaining carbohydrates are converted to fat. They, along with fats in the diet, are stored in fat tissue until needed for energy.

Proteins are mainly used to build the body's own protein. Excess amounts are also converted to fat. As you'll see later, under extreme conditions protein itself may be broken down for energy.

Catabolism and Anabolism

Metabolism is actually two separate and opposite processes, as shown in **Figure 14-1**. One is **catabolism** (kuh-TA-buh-li-zum), *breaking down complex molecules into simpler ones during chemical reactions.* Through catabolism, the nutrients made available by digestion are broken down into even simpler materials that can enter cells. Catabolic (ka-tuh-BAH-lik) reactions, which include digestion, release energy.

Some of the energy released by catabolism is used during anabolism. **Anabolism** (uh-NA-buh-li-zum) is *the combining of molecules during chemical processes in order to build the materials of living tissue.* Nutrient molecules broken down by catabolism are reconstructed into body cells, from the brain to the toenails. The protein in peanut butter becomes protein in your muscle though anabolic (a-nuh-BAH-lik) reactions.

In the cell, this molecular tearing down and building up produces cytoplasm. **Cytoplasm** (SY-tuh-pla-zum) is *a colloidal substance consisting of organic and inorganic substances, including proteins and water found in a living cell. Cytoplasm is the main component of both animal and plant cells.* Catabolism breaks down food to make cytoplasm, which

- **Oxygen.** Like the "burning" of gasoline that runs a car engine, many metabolic processes require oxygen to take place. This is why you breathe: to supply cells with oxygen.
- **Temperature.** A car won't start if its internal temperature is too low. It stops running if it overheats. An organism's body temperature must be within a certain range too. A fever of 50°C, for example, would destroy the enzymes that participate in human metabolism (and the rest of your systems).
- **Removal of waste products.** Cars have an exhaust system to get rid of water vapor and carbon monoxide, the gases produced by the combustion of gasoline. The waste products of metabolism are water and carbon dioxide. These are carried in the blood to the lungs, where you exhale them.

Energy for Metabolism

During metabolism, energy is both used and produced. In all cells, life processes constantly move and rearrange atoms, ions, and molecules. This takes energy. Energy originates from the sun. Plants, but not humans, are able to trap the sun's energy for use. When humans eat plants or

When you eat carbohydrates, your body can produce energy. What other nutrients can produce energy?

the body uses for maintenance during anabolic processes.

The ATP Cycle

As nutrients break down in cells, they produce more energy than can be immediately used. Metabolic reactions take only a small amount of energy. Therefore, some of the energy needs to be stored and transferred among cells. *Certain molecules serve as energy warehouses.* They are **adenosine triphosphate** molecules (uh-DEH-nuh-seen try-FOS-fayt), or ATP.

ATP molecules combine the compound adenosine with three phosphate groups. The three phosphate groups form a chain. Thus, ATP can also be written as A-P-P-P.

Phosphate groups contain one atom of phosphorus and four atoms of oxygen and are written $-PO_4$. Energy is carried in the bonds between phosphate groups. These high-energy bonds are primarily between the second and the third groups. When a cell needs energy, the bond between these two phosphate groups is broken and the third group transfers to another molecule. With only two phosphate groups remaining attached, ATP becomes a̲denosine di̲phosphate, or ADP (A-P-P). This process is shown in **Figure 14-2** on page 208.

Because ATP is constantly needed, it must be remade—from ADP. As nutrients break down, ADP uses the energy to link with another phosphate group, re-forming into ATP.

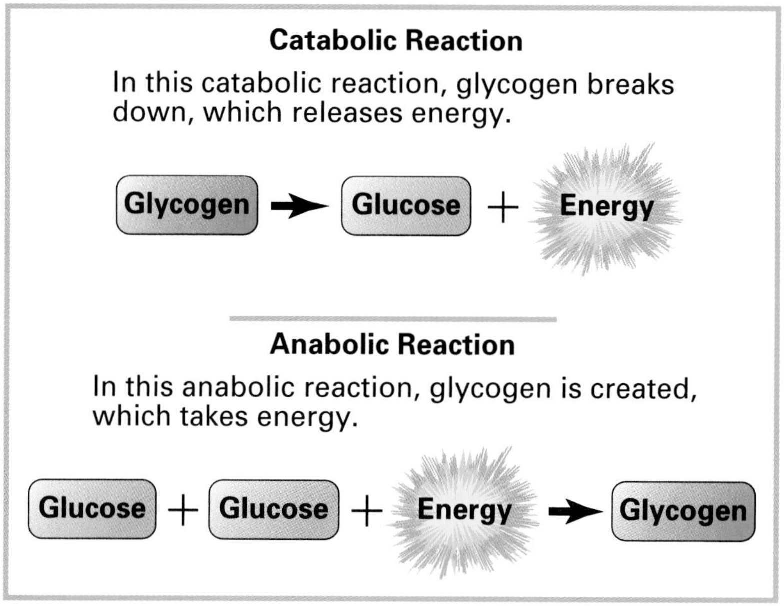

Chemical Balance during Metabolism

Normal metabolism can't take place without the right concentration of chemicals inside and outside the cell. Certain cell characteristics make this balance possible.

The cells in the body are mostly cytoplasm walled in by

Figure 14-1

Catabolic Reaction

In this catabolic reaction, glycogen breaks down, which releases energy.

Glycogen ➡ Glucose + Energy

Anabolic Reaction

In this anabolic reaction, glycogen is created, which takes energy.

Glucose + Glucose + Energy ➡ Glycogen

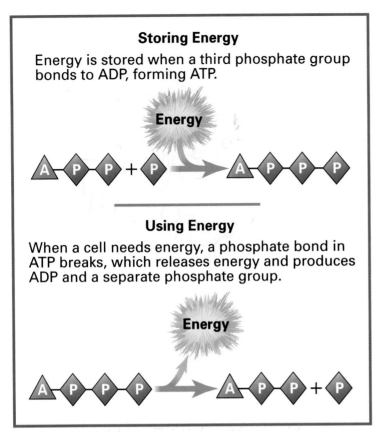

Storing Energy

Energy is stored when a third phosphate group bonds to ADP, forming ATP.

Energy

Using Energy

When a cell needs energy, a phosphate bond in ATP breaks, which releases energy and produces ADP and a separate phosphate group.

Energy

Figure 14-2

thin layers of tissue called **membranes**. These membranes are **semipermeable** (se-mee-PUR-mee-uh-bul). *They allow varying amounts of specific substances to pass through them.* The result is a constant interchange of molecules across the membranes of every cell in the body.

The "open door policy" of semipermeable membranes can lead to chemical imbalances between cells, however. Some kind of control mechanism is needed—and one exists. It's called **osmosis** (ahz-MOH-sus), *the movement of fluid through a semipermeable cell membrane to create an equal concentration of solute on both sides of the membrane.* See **Figure 14-3**.

By controlling the movement of fluid—in this case, water—the body balances the concentration of the substances dissolved in it. This is done with the help of ions of four minerals. Sodium and chloride ions are the principal

minerals in blood. Potassium and phosphate ions are the most abundant minerals in cell cytoplasm. If the number of sodium ions in the blood increases to a level that's too high, the body responds by drawing water from the cells into the blood. This lowers the sodium ion concentration. At the same time, it raises the potassium concentration inside the cells to an appropriate level.

Influences on Metabolism

Metabolism is a combined effort of your body's systems. Together, the body's systems regulate metabolic processes and **metabolic rate** (met-uh-BAHL-ik), or *how fast the chemical processes of metabolism take place.* Metabolic rate is similar among similar creatures in similar conditions. Variations are many, however, due to the influence of several factors.

Body Temperature

Changes in body temperature affect an organism's metabolic rate. You and other warm-blooded creatures have mechanisms to keep body temperature fairly stable in most surroundings. In a cold room, for instance, you shiver to increase body heat. Such responses keep your metabolic rate relatively constant.

In cold-blooded creatures, such as lizards and turtles, body temperature is more dependent on environmental temperature. A lizard's metabolic rate rises as it lies in the sun, where the heat speeds up its metabolic processes, such as heartbeat. Thus, a healthy lizard's metabolic rate varies more in a given day than a healthy human's rate.

Environmental extremes can have a limited effect on metabolism. For example, since metabolic processes release energy, and thus heat, the rate may slow in very hot weather to help cool the body.

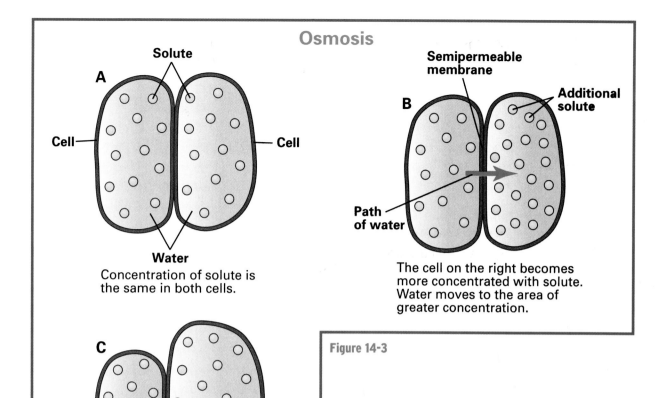

Osmosis

A

Cell — Cell

Solute

Water

Concentration of solute is the same in both cells.

B

Semipermeable membrane

Additional solute

Path of water

The cell on the right becomes more concentrated with solute. Water moves to the area of greater concentration.

C

With less water on the left, the cells change size, and the concentration of solute equalizes.

Figure 14-3

A polar bear can be active in many different climates. Why is that possible?

Body Size

If you tried to take a rabbit's pulse, you might not be able to count fast enough. Small animals have more surface area compared to body mass than larger animals, so they lose body heat more quickly. Staying warm is harder for the rabbit than for you. Thus, its metabolic rate is much higher.

This principle holds for all living things. In one minute, a mouse breathes about 150 times; an elephant, six times; and a human, 16 times. Your own metabolic rate varies from the average depending on your inherited body type.

Geriatric Food Designer

What are the nutrition concerns of older adults?
Ask a ... Geriatric Food Designer

To geriatric food designer Gretchen Wells, an aging population presents a challenge. "In twenty years, six-and-a-half-million Americans will be age eighty-five and older," she notes. "People want to stay healthy and enjoy life, not just live longer. To design foods for older adults, you need to know how aging affects the body. Metabolism slows. The senses of taste and smell weaken and so does the immune system. Since older people are more prone to illness, they need good nutrition.

"We put a lot of work into nutritional supplements, coming up with different formulas for healthy seniors and for those who are ill or taking treatments. We're working on a peach pudding, fortified with vitamin B_{12} and calcium. The taste is too strong for me, but my ninety-year-old grandmother says it's the best thing she's tasted in years."

Education Needed: Geriatric food designers typically have a degree in food science, emphasizing nutrition. Courses in neuroscience are valuable for understanding how aging affects food perception. Some medical knowledge is also an asset.

KEY SKILLS
- Laboratory skills
- Creative thinking
- Relationship skills
- Teamwork
- Communication skills
- Keen sense of taste and smell

Life on the Job: Altering foods to make them more pleasing to older adults takes much laboratory work. Geriatric food designers often consult with flavorists and nutritionists. They may also work with sensory evaluation panels to test new products.

Your Challenge

Suppose you're reviewing a new processing method to make foods more digestible for elderly people. The process is expensive and may be too costly for many older adults.

1. What are pros and cons of recommending the new process?
2. If you recommend this process, how could you keep the foods affordable?

One person may be naturally tall and thin; another is short and stocky. How would you expect their metabolic rates to compare? **Figure 14-4** will help you answer this question.

Physical Development and Age

Scientists believe that cells function rapidly in young individuals in order to build the new material needed for development. Thus, metabolic rate is high. As a fully grown adult, however, a person's metabolism drops to a lower, "maintenance" level. The next time someone tells you, "I can't believe how you've grown since I last saw you," you can give much of the credit to metabolism.

Body Composition

Lean tissue (muscle) takes more kcalories to maintain than other types. Males, who tend to have a greater proportion of lean tissue, generally have a higher metabolic rate than females, who have a higher percentage of body fat. As people age and lose lean tissue, metabolism often slows.

Energy Supply

Like a careful consumer, your body can slow its spending when it has less to spend. In other words, when the energy supply is scarce, the body automatically slows down metabolism. With less food available, the body stretches the food supply by burning it up at a slower rate. Such a survival mechanism may have saved your early ancestors in times of famine. This ability is found in other organisms also.

Metabolism and Weight Management

Discussions of energy needs may have you wondering whether metabolism figures in weight control. It does—significantly. One foundation of good health is maintaining a desirable weight. Knowing how metabolism affects your kcalorie needs can help you reach that goal.

Basal Metabolism

About two-thirds of the energy your body produces is spent on basal metabolism. **Basal metabolism** is *energy used by a body at rest to maintain automatic, life-supporting processes.* Regulating heartbeat, breathing, and body temperature are all part of basal metabolism. Also included are the metabolic reactions going on in the cells.

Basal metabolism is expressed as **basal metabolic rate**, or BMR. BMR is *a measure of heat given off per time unit*—usually, as kcalories per hour. Basal metabolism varies among individuals, affected by the factors influencing overall metabolism described earlier. In the Nutrition Link on page 212, you learn how to use BMR to estimate your daily kcalorie needs.

Voluntary Activities

Basal metabolism accounts for only part of your energy needs. The remaining one-third of your kcalories are spent doing **voluntary activities**, *conscious and deliberate actions.*

The number of kcalories a voluntary activity uses depends on the physical work, the time spent on it, and your level of fitness.

A *sedentary activity* such as reading burns 80–100 kcalories per hour. Slow walking is a

Figure 14-4
Here are two structures made from the same number of bricks. How many brick surfaces are exposed on each structure? Even though they weigh the same, the tall, thin structure has a greater surface area. If these represented a tall, thin person and a short, stocky one, what would you conclude about their metabolism?

Nutrition*Link*

Metabolic Rate Math

Calculating your basal metabolic rate (BMR) can help you gauge your nutrition and energy needs. Measuring BMR precisely takes special equipment and procedures, but the shortcut below gives a useful estimate.

Find your weight in kilograms. Divide your weight in pounds by 2.2. Since Brianna weighs 130 pounds, she does the following calculation:

$$\frac{130 \text{ lb.}}{2.2} = 59 \text{ kg}$$

Find your basal metabolic rate. Calculate the kcalories you use per hour. If you are female, multiply your weight in kilograms by 0.9. If you are male, multiply by 1. (These numbers are called the BMR factor.) Brianna multiplies 59 by 0.9 to learn her BMR.

$$59 \times 0.9 = 53$$

Determine kilocalories used per day. Multiply your BMR by 24. Brianna multiplies 53 by 24 to find her daily basal kcalorie needs.

$$53 \times 24 = 1272$$

Thus, Brianna uses about 1272 kcalories per day just breathing, digesting food, growing new cells, and other basic processes.

Applying Your Knowledge

1. Use the above method to find your minimum kcalorie needs.
2. How can you use your BMR with the RDA for nutrients to maintain good health?

Basal metabolic rates aren't the same for everyone. Of these individuals, who is likely to have the lowest BMR? Who might have the highest? Which of the two parents would likely have the higher BMR? Why?

light activity that burns 100–170 kcalories per hour. A *moderate activity* like a brisk walk might burn 170–250 kcalories in the same time. Waxing the car is a more *vigorous activity* that could burn 250–350 kcalories. Swimming or bicycling, on the other hand, is a *strenuous activity* that can burn 350 kcalories or more per hour.

Like BMR, the type of activity that predominates in your life affects your kcalorie needs. To estimate how much energy you need for voluntary activities, figure your daily kcalorie needs for basal metabolism. Then use the following information to complete the calculation:

- If your activities are mostly sedentary, multiply by 20 percent.
- For light activities, use 30 percent.
- If moderately active, multiply by 40 percent.
- Multiply by 50 percent for vigorous activities.
- For strenuous activities, multiply by at least 50 percent, depending on your activities.

To find the total kcalories you need each day, add your basal kcalories to those from your voluntary activities. Consuming this number of kcalories daily should help you maintain a steady weight.

Metabolic rate goes up when muscles are toned. This means you burn more kcalories when you're in good condition.

Weight-Loss Diets

Would it surprise you that very low-kcalorie, "crash" diets are not the best way to lose weight? The body doesn't take kindly to being denied the energy it needs for survival. It defends itself by slowing metabolism, thus saving energy. The fewer kcalories you consume, the fewer the body tries to get by on.

Crash diets are not just ineffective. Like any type of malnutrition, they take a heavy toll on health. After exhausting its blood sugar and glycogen supplies for energy, the body breaks down its fat and protein deposits. Neither tissue can provide the glucose needed, and both have other, more important functions in the body. Protein catabolism reduces muscle mass, which is eliminated as solid waste and urine. People on starvation diets are literally "wasting away."

The catabolism of fats leaves ammonia and ketones as waste products. Ammonia is a base that is poisonous in cells. It, too, is excreted in urine, again stressing the kidneys. Ketones are compounds consisting of one oxygen atom and three carbon atoms. As they accumulate, ketones can lower blood pH, thus upsetting blood chemistry and the vital cell functions that depend upon it. You may recall this condition as acidosis.

Deprivation diets can cause a host of other harmful effects, from hair loss, to a weakened immune system, to sudden death. In the extreme, such dieting is more dangerous than carrying excess weight, and certainly too high a price to pay for trying to fit a societal ideal of physical beauty.

Current research shows that the most healthful eating pattern consists of five or six small meals evenly spaced throughout the day. This is better for both metabolic rate and energy levels than even the traditional three larger meals. Positive weight management may mean limiting, but never eliminating, food intake.

Watching television is a sedentary activity. How many hours a week do you think you spend in sedentary activity?

Kcalories in Food

In this chapter, you've seen how food converts into energy during metabolism. Sometimes it's difficult to imagine that a food can actually produce energy. In this experiment, you'll see how burning releases the heat energy in a nut.

Equipment and Materials

electronic balance	long needle	coffee can or large juice can with top and bottom removed
laboratory thermometer	soup can	stirring rod
large cork		

Procedure

1. Create a nut assembly: stick the eye of the needle in the narrow end of the cork, then on the point of the needle mount the shelled nut assigned by your teacher.

2. Determine the mass of the nut assembly. Record it in your data table.

3. Remove both ends of a large can, and punch holes in the sides near the bottom. This will serve as a chimney to minimize heat loss during the experiment.

4. Remove one end of a small aluminum can. Punch two holes, opposite each other, in the sides of the can near the open end.

5. Pour exactly 100 mL of tap water into the small can. Record the temperature of the water in your data table.

6. Insert a stirring rod through the holes in the sides of the small can. Use the rod to balance the small can within the large can.

7. Place the nut on a nonflammable surface, and ignite it with a match. Immediately place the large can around the nut assembly so the small can of water is above the nut.

8. Allow the nut to burn for 2 minutes or until it goes out.

9. Stir the water with the thermometer. In your data table, record the water's highest temperature.

10. Mass the nut assembly, and record it in your data table.

11. Write your results on the chalkboard. Copy the results for the other kinds of nuts in your data table.

A small but nutritious mini-meal could be just what you need to fill the gap between lunch and the evening meal.

Physical Exercise

Severe dieting slows metabolism. Exercise, whether hiking in the mountains or mowing a lawn, increases it. Muscle cells first use the energy supplied by catabolizing simple sugars in the blood, then dip into their glycogen stores.

Also, as muscles are toned, they become more "active," requiring more kcalories to maintain. Thus, a muscular body has a higher basal metabolism than a less developed one.

Analyzing Results

1. Using the following equation, calculate the calories of heat from the burning nut. The 100 mL of water has a mass of 100 g.

$$calories = \frac{grams}{of\ water} \times \frac{degrees}{of\ temperature} \times cal/g^oC$$
$$change$$

2. Divide the figure from question 1 by the change in mass of the nut. This gives the calories released per gram of nut burned. Record this on the chalkboard in kcalories.

3. Which kind of nut released the most heat per gram? The least? Do your results agree with the standard calorie tables provided by your teacher? If not, how do you explain any differences between your calculated values for calories per gram and the values listed in the calorie table?

SAMPLE DATA TABLE

Kind of Nut	Mass			Temperature		
	Original	Final	Change	Original	Final	Change

Have you ever exercised or worked until your muscles "burned" or gave out? That sensation is caused by a buildup of lactic acid in the muscles. **Lactic acid** is *a waste product formed when carbohydrates are not completely metabolized.* The problem occurs when the cells don't get enough oxygen to completely break down the carbohydrates. Metabolism, remember, requires oxygen.

Rest, which allows the body to replenish its oxygen supply, is one antidote to lactic acid buildup. The oxygen combines with lactic acid to complete metabolism, and blood carries the acid to the liver. Regular exercise also helps prevent lactic acid buildup. Well-conditioned muscles build a supply of the enzymes needed to quickly process carbohydrates and fat, effectively drawing on both nutrients for energy.

Chapter Summary

- Metabolism is how the body releases energy to maintain and use all of its systems.
- Metabolism occurs only under certain conditions.
- Metabolism involves two processes: catabolism and anabolism.
- Energy is carried to cells by ATP molecules.
- Osmosis helps maintain the balance of chemicals in the blood cells needed for metabolism.
- Metabolic rate is influenced by a variety of factors.
- Basal metabolism is the amount of energy needed by a body at rest to maintain the automatic activities that support life.
- Basal metabolism and voluntary activities together use the energy released by metabolism.
- Voluntary activities use varying amounts of kcalories.
- Severely restricted diets can damage health by disrupting normal metabolic processes.
- Physical exercise affects metabolism.

Using Your Knowledge

1. How is metabolism related to homeostasis?
2. Describe an environment in which metabolism can occur.
3. How do humans get energy?
4. How are carbohydrates metabolized?
5. How are anabolism and catabolism related?
6. How is energy transferred from nutrients to the body's cells?
7. How does a body maintain its supply of ATP?
8. Why is osmosis necessary for metabolism?
9. Why is your metabolic rate different from an alligator's?
10. How does your metabolic rate compare to small animals?

Real-World Impact

Set-Point Theory

In some people, no amount of diet and exercise causes weight loss. Some scientists believe that such people may be fighting their "set point," the body's inborn metabolic rate. According to set-point theory, your metabolic rate is genetically determined. Trying to substantially change the rate at which you burn kcalories is useless and possibly dangerous. Instead, you should practice healthful eating and exercise habits and be comfortable with the body weight and shape that result.

Thinking It Through

1. What are some positive and negative consequences of accepting set-point theory?
2. A friend is excited about a weight-loss plan. You lose weight by eating certain foods that contain fewer kcalories than the body uses to metabolize them. How do you respond?

Using Your Knowledge *(continued)*

11. Do you use energy while sleeping? Explain.
12. How will your body respond if you start skipping meals?
13. How can you prevent lactic acid buildup? Why do these techniques work?

Skill-Building Activities

1. **Math.** Running a fever can raise BMR as much as 4 percent for every °C that body temperature increases. Imagine that a flu bug raises your normal 37°C temperature to 38.5°C. If your usual BMR is 62, what is it while you're sick with the flu?

2. **Management.** Estimate your total daily kcalorie needs. Suppose you want to increase that sum by 20 percent. Using the information on pages 211-212 as a guide, suggest three activities that would meet this goal. How long would you have to do these activities? How would you fit them into your schedule? What factors would you weigh when choosing activities, besides the time and kcalories they require?

3. **Critical Thinking.** Jack and Jill each weigh 60 kg. Jill is 1.8 m tall and Jack, 1.5 m tall. According to the formula for calculating BMR found in the chapter, these people have the same BMR. Do you agree with this calculation? Why or why not?

4. **Consumer Skills.** Visit a local or Internet bookseller. Obtain enough information on some popular diet books to summarize each weight-loss plan. Based on what you've read in this chapter, evaluate the safety and effectiveness of each plan. Share your ratings with classmates.

Thinking Lab

Feed a Fever, Starve a Cold?

An increase in body temperature raises basal metabolism. A higher metabolic rate uses more kcalories.

Analysis

When you're sick with a viral or bacterial infection, you may have a fever and lose your appetite.

Thinking Critically

1. What problems might the symptoms described above eventually lead to?
2. How can the problems you've identified be lessened?

5. **Problem Solving.** Record the foods you eat and the activities you participate in for two days. Calculate how much more you would need to eat to gain weight at your current activity level or how much less you would need to eat to lose weight at the same activity level.

Understanding Vocabulary

With a partner, choose one of the "Terms to Remember" that is suitable for the following: plan a lesson in which you use the chalkboard to explain the term. Present your lesson to the class.

Objectives

- Explain the chemical reaction by which plants produce carbohydrates.
- Describe the molecular structure of simple and complex carbohydrates.
- Describe properties of sugars.
- Summarize how glucose is made available to the body.
- Contrast healthy blood glucose regulation to the complications of diabetes.
- Discuss caramelization.
- Compare the structures of amylose and amylopectin and how these structures affect cooking properties.
- Define the terms gelatinization, paste, retrogradation, and syneresis as used in starch cookery.

Terms to remember

amylopectin
amylose
caramelization
carbohydrates
gelatinization
glucose
hormone
hydrolysis
hydroxyl group
inversion
photosynthesis
polymer
retrogradation
saccharide
supersaturated
syneresis
viscosity

Suppose you're late for a part-time job. With a quick burst of energy, you sprint down the street to get to where you work. What fueled your body for that sprint? The energy you used came primarily from the carbohydrates in food. Your body changed carbohydrates into glucose, a form that you can use for immediate energy. It also stored some of the glucose in your muscles and liver so you have energy available when you need it.

What Are Carbohydrates?

A **carbohydrate** is *an organic compound that is the body's main source of energy.* If you break down the word "carbohydrate," you'll find part of the name of two of its main elements: carbon and hydrogen. A carbohydrate molecule also includes oxygen.

Carbohydrates are found mainly in foods from plant sources, such as fruits, vegetables, grain products, dry beans, and peas. People typically think of carbohydrates as sugars and starches. More technically, they are simple and complex carbohydrates. Simple carbohydrates consist of one or two sugars in very small molecules. Complex carbohydrates, which are starches and fiber, are very large molecules made of many simple carbohydrate units. Together, these two types of carbohydrates are the most abundant organic molecules on earth.

Plants use chlorophyll to make carbohydrates. The plant then becomes your source of carbohydrates.

219

How Carbohydrates Form

Carbohydrates are produced by green plants. *Using the sun's energy, plants convert carbon dioxide and water into glucose and oxygen. The carbon dioxide comes from the air, and water is taken up by a plant's roots.* Known as **photosynthesis** (foe-toe-SIN-thuh-sus), this process is shown in **Figure 15-1**. The green pigment that must be present in plants for photosynthesis to occur is called *chlorophyll* (KLOR-uh-fill).

As you see in the equation, the carbohydrate produced by photosynthesis is glucose (GLOO-kohs). **Glucose** is *the basic sugar molecule from which all other carbohydrates are built.*

A plant can convert glucose molecules into other sugars, starches, and fiber. It usually starts by changing glucose into other sugars. These become starches as plants mature. Green peas from a young plant taste sweeter

Figure 15-1

Through photosynthesis, plants make carbohydrates. How does each part of the equation relate to the illustration?

Photosynthesis

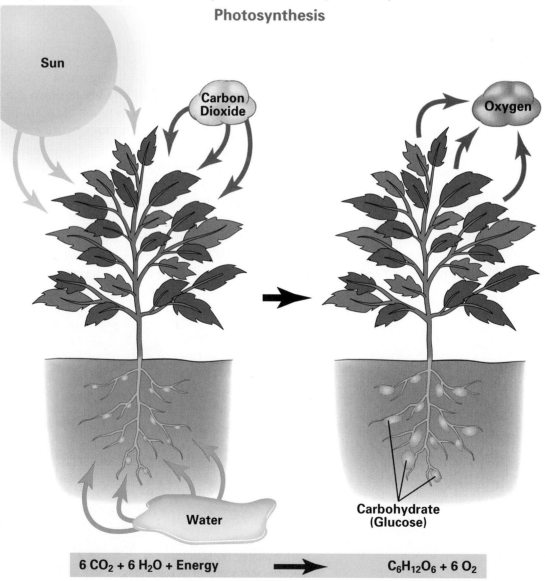

Sun

Carbon Dioxide

Oxygen

Water

Carbohydrate (Glucose)

$$6\ CO_2 + 6\ H_2O + Energy \longrightarrow C_6H_{12}O_6 + 6\ O_2$$

Both the orange and the orange drink contain carbohydrates. **Which one is the more nutritious food choice? Why?**

Sugars can also be extracted from plants and used to sweeten such foods as candy, soft drinks, and baked products. Overeating these products can lead to health problems.

Simple carbohydrates exist in pure form as crystalline solids. The basic sugar molecule is a six-sided ring made of 6 carbon atoms, 12 hydrogen atoms, and 6 oxygen atoms. The hydrogen and oxygen atoms are arranged in hydroxyl groups. A **hydroxyl group** is *a combination of hydrogen and oxygen, containing one atom of each element.* It is written –OH. The dash represents the chemical bond that attaches the group to the rest of the molecule.

Sugars are also called saccharides. A **saccharide** (SAK-uh-ride) is simply *a sugar or a substance made from sugar.* Sugars have the following two basic molecular structures, as shown in **Figure 15-2**:

• Monosaccharides (mah-nuh-SA-kuh-ryds), or single sugars, include glucose, the "building block" of all other sugars and thus of all carbohydrates. The monosaccharide fructose (FRUHK-tohs) can be found in fruits and tree sap. Galactose (guh-LACK-tohs) is not found free in nature, but is always bonded to something else. Small amounts occur in such milk products as yogurt and aged cheese.

• Disaccharides (dy-SA-kuh-ryds) are sugars made of two monosaccharides bonded together. These include the two most common sugars found in food. Sucrose (SOO-krohs) is made of glucose and fructose. Lactose (LAK-tohs), which is found in milk, is com-

than peas from an older one because they are higher in simple carbohydrates, or sugar. Peas from older plants contain a higher level of starch. The fiber formed from glucose strengthens and supports the plant's cell walls.

Simple Carbohydrates: Sugars

Simple carbohydrates, or sugars, are a natural part of many foods. They have a sweet taste, easily noticed, for example, in the apple or orange you eat for a snack. Foods with naturally occurring sugars often contain other nutrients too.

Figure 15-2
Glucose and fructose are both monosaccharides. When combined, they produce a disaccharide called sucrose. Why are these simple carbohydrates? Which one is the basis for forming all carbohydrates?

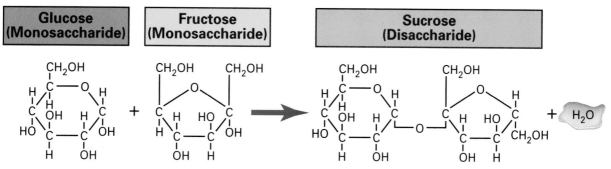

posed of glucose and galactose. A less common disaccharide, maltose (MAWL-tohs), occurs in cereals and sprouting grains. Maltose molecules consist of two glucose units.

Properties of Sugar

While all simple sugars have the same chemical formula, they differ in how the atoms are positioned in the molecular ring. This difference in arrangement causes a difference in sweetness, solubility, and other qualities.

Sweetness

It may sound strange, but some sugars are sweeter than others. In taste tests of pure sugars, fructose is generally found sweetest. Sucrose, glucose, galactose, maltose, and lactose follow in decreasing order of sweetness.

Honey is an example of a very sweet sugar. When bees make honey, they collect nectar from flowers that contain fructose, glucose, and sucrose. An enzyme in their body converts most of the sucrose into fructose and glucose. This makes honey a highly concentrated solution of two of the sweetest sugars.

Sweetness depends in part on how a certain sugar's molecules "fit" with the taste bud sites that register the sweet taste. Variations in taste bud patterns affect how a person rates sweetness.

Such factors as concentration, consistency, temperature, and pH level can also affect how sweetness is judged. For example, fructose seems sweeter when used in cold foods and in slightly acidic drinks. Adding other ingredients affects a sugar's molecular structure, and so its taste. Food technologists consider all of these factors when choosing sweeteners for recipes.

Caramelization

Some mysteries remain in food science. **Caramelization** (kahr-muh-ly-ZAY-shun), *the browning reaction that can occur with any kind of sugar*, is one of them. This complex chemical process is not fully understood.

Caramelization requires either low or high pH. As sugar is heated, some of the water leaves its molecules in the form of hydrogen and oxygen. At high temperatures, the molecule "remnants" join to form larger molecules. The new molecules have a higher concentration of carbon, creating the distinctive caramel color.

Sugars differ in the degree to which they caramelize and the temperature at which they react. Sucrose, galactose, and glucose all caramelize at 170°C. Fructose, on the other hand, caramelizes at 110°C.

Solubility

Sugars are highly soluble in water because they contain many hydroxyl groups, which form hydrogen bonds with water molecules. Interestingly, the solubility of sugars in water mirrors their order of sweetness: fructose is most soluble, followed by sucrose, glucose, galactose, maltose, and lactose in descending

Most flowers produce nectar, a liquid made of proteins and sugars. The nectar collects in the cuplike base of the petals. Why is the honey produced by bees very sweet?

The ability of sugars to carmelize can be applied to many cooking processes, including candy making. What gives peanut brittle its brownish color?

to boil. Water evaporates, increasing the concentration of the sugar in the solvent. When sugar reaches a certain concentration, crystallization occurs. Sugar crystals separate from the solution, each sugar in a unique organized pattern of molecules.

Crystallization from a supersaturated solution begins when particles enter the solution. Generally these particles are very tiny sugar crystals that form on the sides of the container and fall into the solution. However, a drifting speck of lint or dust can have the same effect. The sugar in the solution crystallizes around these particles. These newly formed crystals can trigger additional crystallization. Under the right circumstances, this process continues until a solid forms.

The size of the crystals depends on how many particles are present and how quickly the crystals grow around them. The more particles you have, the smaller the crystals are. Rapidly growing crystals tend to be smaller.

Different sugars form crystals of different sizes. Sucrose crystals are among the largest. Glucose and fructose crystals are relatively small.

solubility. Water temperature affects solubility of sugar. At 25°C, the solubility of sucrose is 211 g per 100 g of water. Water that's too chilly for a shower can dissolve twice its weight in sucrose.

Suppose, however, that you want to dissolve 275 g of sucrose in 100 g of water. Sugar's solubility increases dramatically with water temperature. Thus, you can dissolve more sugar if you heat the solution. Eventually, you may get all the sugar in solution. If you then cool the solution very carefully, with absolutely no agitation, you'll have a **supersaturated** (SOO-pur-SA-chuh-ray-tud) solution. The solution *contains more dissolved solute than it would normally hold at that temperature.*

Crystallization

What happens if you add more sugar to a solvent than it can dissolve even at high temperatures? As you raise the heat, the solution starts

By controlling the cooking process, a boiled sugar solution can produce both crystalline and noncrystalline candies. Crystalline candies are usually soft, smooth, and creamy, with very small crystals. Noncrystalline candies are typically chewy or hard.

Complex Carbohydrates: Starch and Fiber

When people talk about starch and fiber in the diet, they're speaking of complex carbohydrates. Both are found in dry beans, peas, and lentils; vegetables, such as potatoes and corn; and such grain products as rice, pasta, and breads. These foods also contain other nutrients.

Structurally, simple and complex carbohydrates are different. You saw that simple carbohydrates have one or two monosaccharides per molecule. In complex carbohydrates, glucose forms compounds called polysaccharides (pah-lee-SA-kuh-ryds). A polysaccharide is an example of a **polymer** (PAH-luh-mur), which is *a large molecule formed when small molecules of the same kind chain together*. Polysaccharides are made of chemically linked monosaccharides—from ten to several thousand.

The cell walls of a plant have a firm structure that protects the cell and gives it shape. The fibrous nature of cellulose is shown in this magnified photo.

Structure of Starches

Starches are literally plant food. They are to plants what glycogen is to people: reserve energy. Starches are stored in granular form, largely in the seeds and roots.

Although they are chains of sugar molecules, starches taste rather bland. They are made of a form of glucose called alpha-D-glucose. A single molecule of starch can include anywhere from 400 to several hundred thousand alpha-D-glucose units.

Starch molecules can have two structures, as shown in **Figure 15-3**. **Amylose** (A-muh-lohs) *molecules are linear; they're long and narrow like a line*. Starch molecules known as **amylopectin** (A-muh-loh-PEK-tun) *have multiple branches, like the veins in a leaf*. Most starches found in foods contain both molecular structures, but in differing proportions. Amylopectin is typically more abundant in starches than amylose. The cereal starches of corn, wheat, and rice have more amylopectin (about 80 percent) than the root starches of tapioca and potato (about 75 percent).

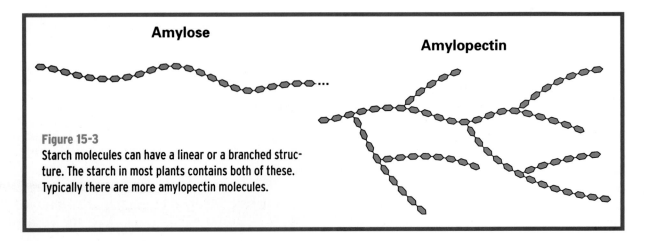

Amylose

Amylopectin

Figure 15-3
Starch molecules can have a linear or a branched structure. The starch in most plants contains both of these. Typically there are more amylopectin molecules.

Carbohydrates in a Healthful Diet

Carbohydrates that occur naturally in foods deserve starring roles in your diet. They bring with them the plant's energy and nutrient store. Healthful carbohydrate choices are as varied as they are nutritious.

- Whole grains, whether rice, oats, buckwheat, or barley, are rich in vitamin E and several B vitamins. They supply such minerals as iron, zinc, potassium, and phosphorus. *Whole* is the key word; processing methods that strip away the grain's bran also remove many nutrients.
- Fruits and vegetables are almost your sole source of vitamin C. Vitamins A, E, and a range of B vitamins and minerals are plentiful. Some nutrients are found just under the skins; eat those when possible. Vegetable leaves, roots, tubers (underground stems such as potatoes), and flowers (including broccoli), are most nutrient dense.
- Legumes, including dry peas, beans, and lentils, are rich in protein, B vitamins, iron, calcium, and potassium.

In contrast, refined or added sugars offer only kcalories that must be used now or stored, often as fat. Such energy contributes little to your overall, long-term health.

Applying Your Knowledge

1. Some people think of carbohydrates, such as pasta and potatoes, as fattening foods. Why do you think this is so?
2. In what ways might the mild taste of some carbohydrate-rich foods be a health advantage and disadvantage?

Structure of Fiber

Like starches, fiber is related to table sugar but is worlds apart in taste. Fiber is what gives plants their structure. The main plant fiber found in food is cellulose (SELL-yuh-lohs), a polymer made of a form of glucose called beta-D-glucose. These molecules are chained in straight, parallel lines to strengthen the plant's cell walls and woody parts.

Other basic structures of edible fiber include hemicellulose (heh-mih-SELL-yuh-lohs), pectins, and algal polysaccharides. These also contribute to a plant's structural framework. Bran, the most concentrated fiber, forms the outer layers of grains.

Carbohydrates in the Body

Carbohydrates are used and stored by the body in various ways to help supply it with a steady stream of energy. That's why your body and mind function around the clock, although you may eat only a few times a day. Experts recom-

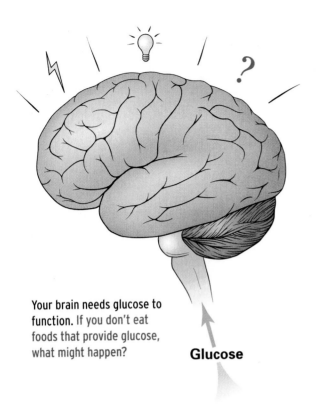

Your brain needs glucose to function. If you don't eat foods that provide glucose, what might happen?

Glucose

Glucose is the main carbohydrate found in blood, where it is called blood sugar. It affects the health and functioning of all the body's cells but is most critical to the brain and nervous system. In fact, it takes about twice as much glucose to run your brain as to power the rest of your body.

Hydrolysis of Sugars

In the small intestine, glucose is "freed" for energy use by **hydrolysis** (hy-DRAH-luh-sus), which is the *splitting of a compound into smaller parts by the addition of water*. This reaction breaks down carbohydrates until they yield the sugars from which they are formed. The word equation below shows how the process splits sucrose into glucose and fructose.

Sucrose + Water → Glucose + Fructose

Hydrolysis of sugars results from the action of enzymes, certain proteins that control chemical activities. The enzyme sucrase (SOO-krase), also called invertase (in-VUR-tase), hydrolyzes sucrose. The enzyme lactase (LAK-tase) breaks down lactose into two monosaccharides, glucose and galactose. Galactose and fructose are chemically changed to glucose in the liver. **Figure 15-4** shows how maltose hydrolyzes into two molecules of glucose.

mend that 55–65 percent of your daily calories come from carbohydrates, mostly complex. Your body needs the nutrients in complex carbohydrates, and dietary fiber helps in digestion.

Energy Production

Carbohydrates are efficient fuel for the body. Each gram of digested carbohydrate—sugars and starches—produces four kcalories of energy.

Blood Glucose Levels

As one of its digestive duties, the pancreas monitors the flow of glucose to the cells. During digestion, the pancreas secretes the

Figure 15-4
The hydrolysis of maltose yields two molecules of glucose. How does this equation show hydrolysis?

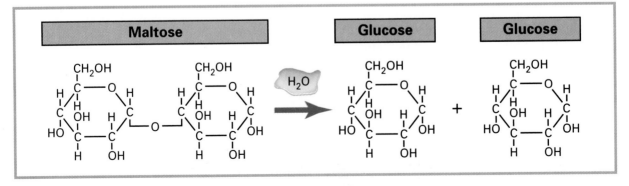

The pancreas produces insulin, shown here microscopically. How is this hormone related to the disease diabetes?

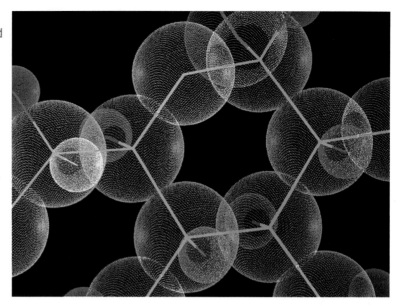

hormone insulin. A **hormone** is *a chemical messenger that affects a specific organ or tissue and brings forth a specific response.* In this case, insulin signals the body's cells to accept the surge of nutrients that have entered the blood, including glucose.

Insulin also helps keep glucose in the blood at a normal level. If blood glucose is too high, insulin triggers the liver and muscle cells to remove glucose from the blood and store it as glycogen. When blood glucose drops below a certain level, glycogen is retrieved from the liver, changed back into glucose, and released in the blood.

Diabetes

Diabetes is a condition in which the body cannot regulate blood glucose levels. It has two forms. In Type I diabetes, which usually occurs in children and young adults, the pancreas secretes little or no insulin. Type II diabetes, which usually occurs in people over age forty, is the most common form of the disease and accounts for over 90 percent of all cases. In this form of the disease, the pancreas does produce insulin, but either not enough or the insulin can't be used effectively. Sometimes a cell defect prevents the body from using the insulin.

Both types of diabetes cause an abnormally high blood glucose level, a condition called hyperglycemia (hy-pur-gly-SEE-mee-uh). The kidneys filter some glucose from the blood and excrete it with urine. This glucose overload can seriously strain and damage the heart and kidneys. Meanwhile, little glucose reaches the cells.

Untreated, diabetes can lead to other serious complications. In Type I, where no insulin is produced, cells cannot use the energy in food. They respond as they do to a starvation diet: they break down proteins and fats and throw off the body's chemistry. A diabetic coma may result. In Type II, cells get enough glucose

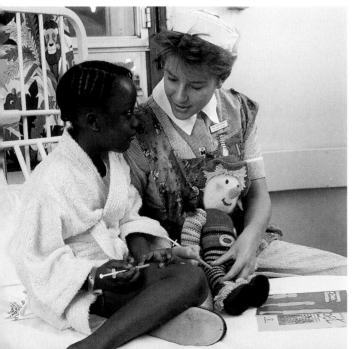

A person who has diabetes may need insulin shots. What type of diabetes do some children get?

to avoid tapping into fats and proteins. A person may become obese, however, trying to feed the glucose-hungry cells. Added fat makes cells even less sensitive to insulin, leading to a dangerous cycle of overeating and overworking internal organs.

Effective treatment for Type I diabetes at this time includes daily insulin injections plus exercise and a planned diet. Through regular exercise, people with Type I diabetes may be able to reduce their daily dose of insulin.

Type II diabetes can often be controlled by diet and exercise alone, without any medication being needed. Exercise may make cells more responsive to the insulin the body produces.

Over half of all diabetics are able to control the disease through diet alone. Diabetic diets are individualized but similar to other healthful eating plans. They include limiting fat, sodium, and protein, with increased intake of complex carbohydrates, especially high-starch foods, such as dry peas and beans. Adding the indigestible bulk of fiber steadies the digestion and absorption of glucose. In addition, eating regularly to avoid excessive hunger steadies the flow of glucose to the bloodstream.

Hypoglycemia

A condition opposite of diabetes is hypoglycemia (hy-po-gly-SEE-mee-uh), an abnormally low level of blood glucose. Without enough glucose pumping to the brain, the person may feel dizzy, weak, and nauseous. Only a physician can determine whether someone has hypoglycemia, which is often corrected by a prescribed diet and eating schedule.

Digestion

The body cannot digest fiber. The parallel chains of molecules in cellulose are not easily broken into monosaccharides by human digestive enzymes. Fiber does promote digestive health, however. High-fiber foods absorb water as they pass through the intestines, helping ensure an easy transit of foods. Fiber's indigestible bulk also exercises and tones intestinal muscles.

The American Dietetic Association recommends 20–35 g of dietary fiber per day for adults. An adolescent can estimate fiber need by adding 5 to his or her age. Thus, a thirteen-year-old needs 18 g of fiber daily. High-fiber foods include whole wheat and other grain products, cereals such as oatmeal and shredded wheat, fresh fruits with skin, dried fruits, raw vegetables, legumes, and nuts.

Carbohydrates in Food Production

Your body uses carbohydrates in one way; the food industry uses them in others. Carbohydrates add taste and texture to many food products.

Simple carbohydrates, not surprisingly, are valued for sweetness. Glucose is the most common monosaccharide in foods, as it is in nature. This compound occurs in varying amounts in all fruits and vegetables. In numerous other foods, sugars are added in different combinations and proportions for the right degree of sweetness. A survey of food labels will likely reveal the simple carbohydrates sucrose, fructose, corn syrup, and honey among the ingredients.

Fruits and vegetables contain fiber. Many people eat the skin of a baked potato because it tastes good and has fiber. With low-fat toppings, a baked potato is a nutritious addition to a meal.

Careers

Peace Corps Volunteer

What is the key to fighting poverty?
Ask a . . . Peace Corps Volunteer

Teaching people to use and care for their resources, including themselves" is how Dominique Chism sums up her work as a Peace Corps volunteer in the West African nation of Senegal. "In agriculture you literally start from the ground up. Most people here raise wheat, but grain farming alone can't lift people out of poverty. While I show farmers a steel blade for plowing and ways to improve soil naturally, other volunteers show how to build solar cookers. Families can sell the wheat plus some of the bread they bake in the cooker. That's two sources of income. It's a small step toward self-reliance, but one that works."

Education Needed: The Peace Corps normally requires a bachelor's degree. However, experience isn't overlooked, especially in agriculture and forestry. In fact, older adults make up an increasing number of volunteers. Knowledge of a foreign language is extremely helpful.

Life on the Job: From Africa to the South Pacific, agriculture and forestry volunteers

KEY SKILLS
- Organization
- Teamwork
- Patience
- Leadership
- Communication
- Problem solving

work with small farmers to increase food production and income, while promoting environmental conservation. They teach organic farming and soil improvement techniques, such as composting. Most of the over 150,000 people who have served in the Peace Corps say the work was rewarding, yet challenging, and one of the most satisfying experiences of their lives.

Your Challenge

Suppose you're an ag specialist in the Peace Corps. Soil tests show nutrient loss caused by planting the same crops every year. The farmers resist your suggestion of crop rotation.

1. How would you persuade the farmers to try a new crop?
2. What small businesses might suit a community in a developing nation?

The average American consumes about 64 pounds of sugar per year. For many teens, the figure is much higher. How does your sugar consumption compare to this statistic?

Complex carbohydrates, too, are used whole and refined. An ear of corn and a baked potato are bundles of carbohydrate energy. The starch from these and similar carbohydrate-rich foods may be used to texturize other products. The Nutrition Facts panel on food labels tells the total grams of carbohydrates, dietary fiber, and sugar per serving.

Sweeteners

When you think of sugars in the food industry, the first thing that comes to mind may be ordinary sugar. Sucrose extracted from sugar cane or sugar beets is the source of many popular sweeteners in the supermarket. Granulated sugar, or table sugar, is almost pure crystalline sucrose. Brown sugar is granulated sugar flavored with molasses. Molasses is a thick, dark syrup produced during the refining of sugar cane. Confectioners' sugar is powdered granulated sugar with an added anti-caking agent such as cornstarch. What type of food would you say cornstarch is?

Sucrose is a main disaccharide in two other natural sweeteners: maple syrup, the highly condensed sap of the sugar maple tree, and honey.

Simple carbohydrates are also added by manufacturers to many foods, from candy bars to ketchup. Some are added for sweetness. Corn syrup contains glucose derived from the enzymatic breakdown of cornstarch. It's often found in ready-to-eat cereals. High-fructose corn syrup is the most commonly used sweetener in food manufacturing.

Other saccharides, including invert sugars, are used as preservatives. These sugars are a mixture of fructose and glucose that results *when sucrose is hydrolyzed by acid and heat, rather than water*. This process is called **inversion**.

Candy Making

Many popular candies, including fondant, fudge, penuche (puh-NOO-chee), and divinity, are called crystalline because they're made by crystals formed from supersaturated sucrose solutions. Smaller crystals produce a smoother texture and a superior product. Sucrose, the sugar used for candy making, tends to produce larger crystals. In the highest quality candy, a small part of the sucrose is hydrolyzed to small-crystal glucose and fructose. Skilled candy makers know how to influence the process to control crystal size.

Simple carbohydrates are in many foods—even ketchup. By reading labels, you can tell what's inside. Look for words that say "sugar," such as sucrose and fructose.

Controlling Crystallization

Crystal formation starts when the solution, also called the syrup, achieves the proper concentration of sugar. Concentration is measured by boiling point. The higher the concentration, the higher the boiling point is. That's why candy recipes call for a syrup to be boiled until it reaches a certain temperature: to show that it has reached the desired concentration. That's also why a candy thermometer should be calibrated on the day it's used.

Substances called *interfering agents* may be added to a sugar syrup to control crystal growth. For example, cream of tartar and vinegar contain acids that can cause sucrose to hydrolize into glucose and fructose. Corn syrup, another interfering agent, is high in glucose. Both fructose and glucose form small crystals. Also, the fats and proteins in butter and egg white can prevent crystallization completely if enough of either is added. They literally get in the way of crystal formation.

Agitation, that is, stirring or beating the solution, also limits crystal growth. The temperature at which agitation occurs is a controlling

In order to get a quality product, manufacturers must control the processes used to make candy very carefully. What effect does crystallization have on candy making?

factor too. Stirring should start when the solution cools to about 104°C and continue until crystallization stops.

Ripening

In order to produce a soft, smooth confection, such crystal candies as fondant are ripened—stored in an airtight container for 12–24 hours. During this time, the smallest crystals in the candy dissolve, allowing large crystals to move more freely. Meanwhile the candy absorbs substances that further interfere with crystal growth. The result is a smooth, pliant candy that's easy to knead into desired shapes.

Noncrystalline Candies

Some candies depend on avoiding sugar crystallization altogether. Taffy, for instance, is made by adding enough interfering agents to invert some of the sucrose. Toffee results from heating and caramelizing sugars before crystals can form. Caramel making uses both techniques.

Starches

Like the human body—your personal food processor—commercial food processors look to cereal grains and legumes for their starch supply. The starch is separated from the oils and proteins in these foods by grinding and screening.

Starches are very important in food preparation as thickening agents. In their role as food thickeners and also stabilizers, starches go by names ranging from arrowroot to xanthan gum. They are matched to a certain use based on their properties. If no existing starch satisfies a manufacturer's needs, food scientists may create a new one.

Gelatinization

When a starch is heated in water, it thickens the mixture in a process that is really quite remarkable. Unlike sugars, starches are not water-soluble. Their molecules are too big to form true solutions. If starches are heated in water, however, energy from the water molecules can loosen the bonds between the starch molecules. *An irreversible thickening process* called **gelatinization** (juh-LA-tun-uh-ZAY-shun) begins. *Hydrogen bonds form between starch and water molecules, causing starch granules to absorb water and swell.*

As a hot starch mixture thickens, it becomes what is called a paste. You've seen this process when watching someone make gravy. The temperature at which a paste forms varies with the type of starch. For example, wheat starch thickens at a slightly lower temperature than cornstarch.

Hot starch mixtures typically flow. However, as gelatinization thickens a mixture, viscosity (vis-KAH-suh-tee) increases. **Viscosity** is the *resistance to flow.* The greater the concentration of starch in a mixture, the more viscous the resulting paste becomes.

Thickening Properties

A starch's thickening ability is affected by its structure. Its use in foods depends on the ratio of amylose molecules to amylopectin molecules.

A basic fondant is made with water, cream of tartar, lots of sugar, and perhaps flavoring. Cooks ripen and knead the dough. Then they roll it into shapes and sometimes cover it with chocolate. Fondant can also be melted and used to cover other candy centers.

To make a lemon sauce, cornstarch is first mixed with a small amount of the liquid before gradually adding the rest of the liquid. Once the sauce thickens, adding more dry cornstarch to the mixture could cause lumps. How could you prevent lumps from forming?

Amylopectin mixes less easily with water than amylose, apparently due to its branched form. This quality gives the desired thickness to such foods as ketchup and gravies, which should stay a bit runny.

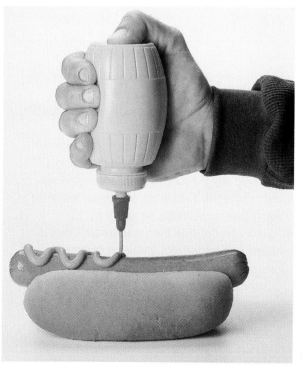

The first squeeze from a mustard container is sometimes watery. What does this demonstrate?

Forming a Gel

When some starch pastes are cooled without stirring, they change to gels. At this point the mixture loses the flow properties it formerly had. A fruit pie that holds its shape when you cut a piece is an example. Amylose starch molecules are most effective at turning from a paste to a gel.

After a period of time, gels have a tendency toward **retrogradation** (reh-tro-gray-DAY-shun). *The amylose molecules shift and orient themselves in crystalline regions, which forms a somewhat gritty texture in the mixture.* A pudding that has been thickened with starch can develop this condition if held in the refrigerator for a few days.

During retrogradation, a mixture may "weep." That is, *water leaks from the gel as it ages.* This is called **syneresis** (suh-NEHR-uh-sus). As amylose molecules pull together more tightly, the gel network shrinks, and water is pushed out of the gel. You might see liquid forming on a pudding that has been kept in the refrigerator for a while.

Modifying Starches

Have you seen the ingredient "modified food starch" on a food label? Such starches are "made to order" to maximize desired qualities and minimize unwanted ones. Food scientists might rework the molecular structure or genetic makeup of an existing starch, or develop a new one. The result is a starch that is more stable after heating or freezing, or in acidic foods.

Chemical tinkering can "fine tune" starches for exact effects. For example, some new strains of corn, rice, and barley contain mainly amylopectin. These "waxy starches" give

EXPERIMENT 15

Thickening Agents

SAFETY FIRST
Review these safety guidelines before you begin this experiment.

Viscosity is a desirable physical property in many foods. To obtain the desired viscosity in a given recipe, different starches can be used as ingredients. Each of these behaves differently. In this experiment, you'll compare how thickening agents used in food can affect the final product.

Equipment and Materials

electronic balance	saucepan	2 heat-diffusing rings	clear pie plate
2, 400-mL glass beakers	100-mL graduated cylinder	(for electric ranges)	muffin tins
stirring rod	line-spread test sheet	plastic ring	paper muffin tin liners

Procedure

1. Obtain a starch from your teacher. Record the type of starch in your data table.
2. Mass 16 g of the starch.
3. Place 60 mL of cold water in a 400-mL beaker; stir in the starch sample. Add 220 mL more water; stir again.
4. Heat the beaker over moderate heat. Stir slowly but constantly until the mixture boils. Wear safety goggles throughout the heating process.
5. Place a glass plate over the line-spread test sheet provided by your teacher. Place a plastic ring in the center of the plate; fill the ring with some of the starch mixture. Lift the ring and allow the mixture to flow for 2 minutes. Count the lines covered at each of 4 points around the circle. Average your readings. Record the result in your data table and on the chalkboard.
6. Cool the rest of the starch mixture to room temperature by placing the beaker in a pan of cold water.
7. Repeat the line-spread test described in Step 5, using the cooled mixture.
8. Label two paper muffin cups with your name, and the name and amount of starch used in the sample. Fill each cup with the starch mixture and place in separate muffin tins. Refrigerate one cup and freeze the other until the next class period.
9. On the next day, thaw the frozen sample. Check the thawed and refrigerated sample for retrogradation and syneresis. Record your observations in the data table and on the chalkboard. In the data table, also record the results from the other starches written on the chalkboard.

ready-made gravies a good consistency but also have a stringiness that you wouldn't want in a fruit pie. By rearranging molecules, scientists have created a modified starch that has the same thickening ability plus a smooth texture.

Another technique has created starches that swell in cold liquids. A starch is cooked, and the granules then flaked into fragments. You can whip up an instant pudding for dessert without waiting for it to cool to eating temperature.

Analyzing Results

1. Compare the findings. Then explain which starches, if any, you would choose to:
   a) Make a molded chicken salad.
   b) Thicken a fruit pie that would be refrigerated before serving.
   c) Make a gravy.
   d) Thicken a cream pie that would be frozen and thawed before serving.
2. What may have caused the different results produced by different starches?

SAMPLE DATA TABLE

Name of Starch	Line-Spread Average		Appearance of Refrigerated Sample	Appearance of Frozen Sample
	Hot	Cold		

Chapter Summary

- Carbohydrates are formed in green plants through the process of photosynthesis.
- Carbohydrates include simple sugars, starches, and fibers.
- Glucose is the basic sugar unit around which all carbohydrates are formed.
- Different types and arrangements of molecules affect the properties of carbohydrates.
- The body regulates levels of blood glucose, which is critical to cell functions. Otherwise, serious health complications can develop.
- Simple carbohydrates play a major role in commercial food production.
- Candy is made by controlling crystallization in sugars.
- Food manufacturers use starches' thickening properties to stabilize food products.

Using Your Knowledge

1. How is the term "carbohydrate" related to the nutrient's composition?
2. How are carbohydrates involved in transferring solar energy to people?
3. Compare monosaccharides and disaccharides, using examples to illustrate.
4. In developing a new recipe, why might you use fructose instead of sucrose?
5. How would a chemist describe a complex carbohydrate?
6. What would happen if carbohydrates in your body were not hydrolyzed?
7. Describe the process that ensures your body a steady flow of glucose.
8. Compare the complications of Type I and Type II diabetes.
9. What is nutritionally significant about fiber?
10. Suppose you forgot to calibrate your thermometer before making candy. It registers 10°C less than the actual temperature of the syrup. What may result?

Real-World Impact

Rice in Japan

Rice is so central to Japanese history, culture, and diet that it's almost sacred. Indeed, Japanese folklore honors Indari as the rice god. Even samurai warriors of ancient times were paid with the grain. Today, rice is part of almost every meal, including breakfast, and is even made into beverages. Sushi, those picture-perfect rolls of rice, seaweed, fish, and vegetables, are a traditional Japanese favorite.

Thinking It Through
1. Why do you think samurai were paid in rice?
2. The traditional Japanese diet is known as a very healthful one. How do you think rice contributes to this reputation?

Using Your Knowledge *(continued)*

11. How do amylose and amylopectin affect thickening?
12. Why do food scientists modify starches for food production?

Skill-Building Activities

1. **Management.** Using a reliable chart of nutritive values, identify five high-carbohydrate foods that you would recommend for a weight-loss diet. Give reasons for your selections.
2. **Consumer Skills.** Bring in labels from two different types of food, such as cereal and canned tuna. List the sweetening agents and starches (including gums) included among the ingredients. Combine your findings with classmates' in a master chart. What conclusions can you draw about the kinds, amounts, and proportions of sweeteners and starches used in various types of commercially available foods?
3. **Communication.** Design a poster showing the progression of a glucose molecule into at least five disaccharides, starches, and fibers. Include examples of foods containing these molecules.
4. **Critical Thinking.** Review the Food Guide Pyramid on page 182. Explain how the information in this chapter relates to the guidelines suggested by the Pyramid.
5. **Communication.** Plan a demonstration using an everyday meal preparation or serving situation to illustrate a chemical process described in this chapter.

Thinking Lab

Diabetic Decisions

Simple or complex, carbohydrates are unavoidable in the typical diet, and certainly so in a healthful one. If they aren't present naturally, they are added as sweeteners or texturizers.

Analysis

People with diabetes, who cannot rely on normal insulin production, must monitor their diet to control blood glucose. Besides limiting obviously sugary foods, they need to watch their entire food intake for foods containing starches or "hidden" sugars that the body breaks down into glucose.

Representing and Applying Data

1. List ten foods you commonly eat for breakfast, lunch, dinner, and as a snack.
2. Evaluate the carbohydrate value of each food. Using the information in this chapter, try to decide whether it would significantly affect your blood glucose level. Identify those foods that might have a positive or negative impact.
3. Communicate your conclusions in a table. Include a column for any comments you wish to make on a food. Should it be included or avoided in a diabetic eating plan? How could you include a beneficial food more often? What food would you substitute for one that could be a problem?

Understanding Vocabulary

Choose one term from the "Terms to Remember." Draw a visual that depicts the meaning of the term. Put your drawing on the bulletin board with others created by classmates.

CHAPTER 16 Lipids

Objectives

- Explain the three categories of lipids.
- Describe how fatty acids form triglycerides.
- Compare the structures of saturated and unsaturated fat.
- Describe the properties of triglycerides.
- Relate the composition of lipids to their functions in foods and in the body.
- Explain the relationship between cholesterol and heart disease.
- Develop an eating plan that keeps dietary lipids within healthful levels.

Terms to remember

adipose tissue
atherosclerosis
carboxyl group
cholesterol
double bond
fatty acids
hydrogenation
lipids
lipoproteins
plaque
rancid
saturated fat
single bond
smoke point
solidification point
triglycerides
unsaturated fat

"Fat-free" and "low-fat" have become common selling points on food labels in recent years. For health reasons, people want to manage their fat intake. To help them, food manufacturers continue to design foods that have little or no fat. Improvements have increased the popularity of these products. The fat-free brownie that once tasted like cardboard is now an enjoyable treat.

Although fat carries the role of villain, the bad reputation isn't all deserved. Fat is a nutrient that your body needs. You just don't need too much.

What Are Lipids?

You're probably already familiar with the term "fats," but science offers a view that you might not have explored before. Fats are actually part of a larger grouping called lipids. **Lipids** *are a family of chemical compounds that are a main component in every living cell.* They include the following three categories:

- **Triglycerides (try-GLI-suh-ryds). Triglycerides** make up *the largest class of lipids. They include nearly all of the fats and oils people typically eat.* Of the lipids that are in foods and stored in the body, most are fats and oils—about 95 percent, in fact. This chapter focuses primarily on these particular lipids because of their link to foods. Any references to fats are actually talking about this lipid category.

- **Phospholipids (fahs-foh-LIH-puds).** These versatile lipids have a structure that enables them to dissolve in both fat and water. For this reason, the food industry uses them as emulsifiers that mix fats with water in such products as mayonnaise. They are also found naturally in some foods, including eggs and peanuts.

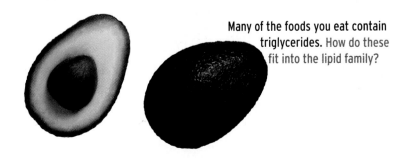

Many of the foods you eat contain triglycerides. How do these fit into the lipid family?

- **Sterols (STIHR-awls).** Sterols include important compounds. Bile acids and certain hormones, for example, perform vital functions. Both plant and animal foods contain sterols, but only animal foods contain the most well-known sterol—cholesterol (kuh-LESS-tuh-rawl).

Triglycerides

The main function of triglycerides, or fat, is to fuel the body and keep it warm. Triglycerides help keep body temperature relatively constant. In the body, fat is a poor conductor of heat. When triglycerides are stored in the body, they collect in **adipose tissue** (ADD-uh-pohs), *pockets of fat-storing cells.* See **Figure 16-1**. The thin layer of fat under your skin, similar to the fat around a cut of beef, is natural insulation.

Fats serve other functions as well. The adipose tissue under each kidney forms a cushion to help protect it from impact. Natural body oils help keep skin supple and hair glossy and healthy. Fat also carries certain vitamins in the body.

Figure 16-1
As it stores fat, an adipose cell grows larger. Triglycerides gather at the outer edge of the cell and eventually merge with the globule in the center.

An Adipose Cell

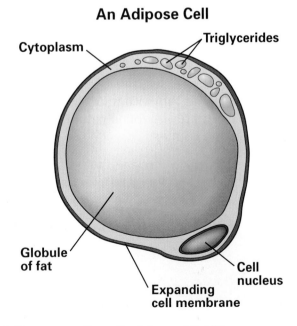

Structure of Triglycerides

Triglycerides are composed of carbon, hydrogen, and oxygen. They form when an alcohol reacts with organic acids. Here's the chemical explanation of how that happens.

Fatty Acids

The organic acids in triglycerides are called **fatty acids**. These are *organic compounds that have a carbon chain with attached hydrogen atoms and a carboxyl group* (kar-BAHK-sul) *at one end.* The carboxyl group is what makes fatty acids organic.

A **carboxyl group** consists of *carbon bonded to oxygen by a double covalent bond, and to a hydroxyl group with a single bond.* (Single and double bonds are described in the pages ahead.) In chemical formulas, carboxyl groups are written as –COOH. **Figure 16-2** shows the carboxyl structure. The "R" in the formula is chemical shorthand. Think of it as meaning "the rest" of the compound.

Not all fatty acids are the same. The length of the carbon chain makes the difference. Most fatty acids found in foods have an 18-carbon chain. The simplest type, however, is acetic acid. It has a two-carbon chain, as shown in **Figure 16-3**.

To form triglycerides, three fatty acids react with the alcohol glycerol (GLI-suh-rawl), as shown in **Figure 16-4**. Glycerol contains three hydroxyl groups (–OH). A hydrogen atom from

Figure 16-2
Fatty acids contain a carboxyl group. How can you tell that a fatty acid is organic?

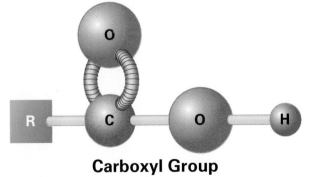

Carboxyl Group

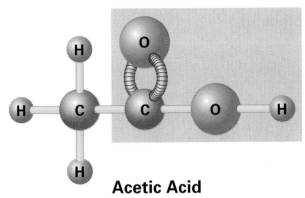

Acetic Acid

Figure 16-3
Acetic acid is a fatty acid with a two-carbon chain. Note the carboxyl group on the right.

each fatty acid molecule attaches in place of the hydrogen in each hydroxyl group. The name triglyceride is a combination of "tri," for three fatty acids, and "glyceride," a glycerol compound. Most triglycerides contain a mix of fatty acids.

Essential Fatty Acids

The human body can make all but two fatty acids. These are linoleic (lin-oh-LEE-ik) and linolenic (lin-oh-LEN-ik). Both are needed for normal growth and development. These essential fatty acids must be obtained from such foods as vegetables, grains, nuts, seeds, and soybeans.

Saturated and Unsaturated Fat

You've probably heard about the health effects of saturated versus unsaturated fats. These terms describe how carbon atoms are bonded in fatty acids. The bonds can be single or double.

A **single bond** is *a covalent bond in which each atom donates only one electron to form the bond.* Thus, two atoms share one pair of electrons between them. A **double bond** is *a covalent bond in which each atom donates two electrons to form the bond.* Thus, two atoms share two pairs of electrons between them. Look at **Figure 16-5** while you read how these bonds define saturated and unsaturated fat.

- **Saturated fat.** In **saturated fat**, *most of the fatty acids are saturated too. In other words, the fatty acids contain all the hydrogen atoms their molecular structure can hold.* As long as no hydrogen atoms are missing, each bond between a carbon atom and its four partnering atoms will be single. Thus, there will be four single bonds.

- **Unsaturated fat.** In **unsaturated fat**, *most of the fatty acids are also unsaturated. That is, they're missing hydrogen atoms.* Where hydrogen atoms are missing due to a chemical change, single bonds can't form. To make up the difference for a missing single bond, a double bond forms. Therefore, two carbon atoms that are each missing a hydrogen atom will bond to each other with a double bond. Forming this dou-

Figure 16-4
When three fatty acids combine with glycerol, they form a triglyceride. Where are the hydroxyl groups in the glycerol structure? Where are the carboxyl groups in the fatty acids? Where does the resulting water on the right side of the equation come from?

Forming a Triglyceride

$$
\begin{aligned}
&H-\overset{\overset{\displaystyle H}{|}}{C}-OH + HO-\overset{\overset{\displaystyle O}{\|}}{C}-R_1 && H-\overset{\overset{\displaystyle H}{|}}{C}-O-\overset{\overset{\displaystyle O}{\|}}{C}-R_1 + H_2O \\
&H-\overset{|}{C}-OH + HO-\overset{\overset{\displaystyle O}{\|}}{C}-R_2 && H-\overset{|}{C}-O-\overset{\overset{\displaystyle O}{\|}}{C}-R_2 + H_2O \\
&H-\overset{|}{\underset{\underset{\displaystyle H}{|}}{C}}-OH + HO-\overset{\overset{\displaystyle O}{\|}}{C}-R_3 && H-\overset{|}{\underset{\underset{\displaystyle H}{|}}{C}}-O-\overset{\overset{\displaystyle O}{\|}}{C}-R_3 + H_2O
\end{aligned}
$$

Glycerol   **Three Fatty Acids**   **Triglyceride**

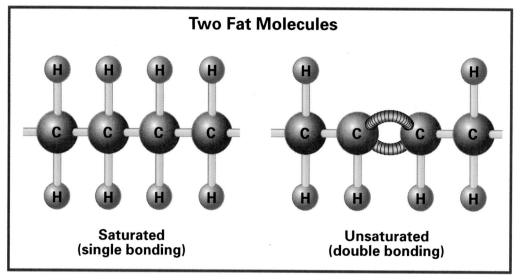

Two Fat Molecules

**Saturated
(single bonding)**

**Unsaturated
(double bonding)**

Figure 16-5

A saturated fat molecule contains all single bonds. Why does the double bond form between carbon atoms in an unsaturated fat?

ble bond fulfills an interesting requirement of nature: all carbon atoms must have four bonds.

Unsaturated fats are classified by the number of hydrogen atoms that "drop out" and produce double bonds. A *monounsaturated* fat lacks two hydrogens, which creates one double bond between carbons. Oleic acid (oh-LEE-ik) is an example of such a fat. A *polyunsaturated* fat has two or more double bonds between carbons and, thus, lacks four or more hydrogen atoms. Linoleic acid has two double bonds, for example, while linolenic acid has three double bonds. How many hydrogens are missing from each of these polyunsaturated fats?

Properties of Triglycerides

The number and arrangement of atoms influence how triglycerides behave. Understanding this relationship between structure and properties helps you see what triglycerides do for you, and what you can do with them.

Energy Value

The structure of triglycerides makes them bundles of energy. Recall that energy is released by oxidation, the reaction of a substance with oxygen. Triglyceride molecules are high in carbon and hydrogen. They offer many sites for oxidation reactions. In contrast, carbohydrate molecules are oxygen-rich. They can't be further oxidized. Triglycerides supply over twice the energy of glucose—nine kcalories per gram compared to four.

Solubility

Unlike sugars, triglycerides are only slightly soluble in water. Because triglyceride molecules are nonpolar, they resist the hydrogen bonding that dissolves sugars in water. That's why you have to shake a bottle of vinegar-and-oil dressing before pouring it over your salad. Triglycerides are quite susceptible to organic, nonpolar solvents, such as chloroform and acetone.

Phase Differences

As pairs, the words "animal fat" and "vegetable oil" seem to go together. In referring to a fat's phase, these expressions are more or less accurate. Fats, which usually come from animals, are saturated. Their single bonds allow a full set of hydrogens, tightly packed like the teeth in a

Some fats are solid and some are liquid. Which are more likely to be saturated? Why?

fat, remember, is an assortment of fatty acids, with both long and short carbon chains. Thus, each fat melts gradually, over a range of about 5 to 10°C, as more of its long-chain acids melt. In a pat of butter on your plate, for example, some fats are in liquid phase and some are solid crystals. The butter softens as it warms to room temperature and melts if spread on a hot potato.

Saturation also affects melting. As with liquidity, fats with a greater proportion of unsaturated fatty acids melt at lower temperatures.

Solidification Point

If you refrigerate leftover buttered carrots, you may find them coated with firm butter flecks the next day. The butter chilled to its **solidification point**, *the temperature at which a melted fat regains its original firmness.* Unlike pure crystalline substances, which melt and freeze at the same temperature, a fat's solidification point is lower than its melting range. Fats with more similar fatty acids solidify with a grainier texture.

zipper. They are denser than oils and in a solid phase at room temperature.

Most plant oils are unsaturated. The molecule "kinks" wherever a double bond replaces two hydrogens. This slight bend makes oils less compact, so they are liquid at room temperature. Plant fats tend to have shorter carbon chains, which also promotes liquidity.

Not all plant oils follow this model. Two "tropical" oils, coconut and palm kernel oil, are over 80 percent saturated.

Melting Range

Molecular structure has a related effect on how fats change phase. Most substances have a melting point and a freezing point, the temperature at which they change from a solid to a liquid and vice versa.

Fats are a different case. Fatty acids with more carbon have higher melting points. Each

Saturation affects melting. Which do you think would melt faster on a hot ear of corn, a pat of margarine or butter? Why?

Triglycerides in Food

Most foods contain some fat. Whipping cream and olive oil contain almost nothing else. A handful of mixed nuts is 50 to 70 percent fat, depending on whether you scoop up more walnuts, peanuts, or pecans. At the low end of the scale, most grains are between three and five percent fat.

Triglyceride Subgroups

As **Figure 16-6** shows, triglycerides are often divided into seven subgroups that include the various types of fats and oils found in food. As you read the chart, look for connections to the previous discussion on triglyceride structure and properties. Likewise, some of the points will be worth reviewing as you read on about the role of triglycerides in foods and health.

Functions of Fats in Food

Why do people drizzle butter over baked potatoes or beat shortening with sugar in a cake batter? Whether it's flavor, texture, or mouth-feel, fats add qualities that can make foods appealing.

Tenderizing

One of the most important uses of fats in foods is tenderizing baked goods. Fats coat the flour particles in baked products, creating a flaky, delicate, lighter texture. Cakes would be crumbly without the moistness provided by fats. Pizza dough and piecrusts would be tough, chewy blocks of baked starch.

Aeration

Fats serve another function in some baked goods. Through aeration, they add air or gas to batters and doughs. The fat forms a bubble around air molecules, which are incorporated in the batter. This fine dispersion of air bubbles decreases viscosity, making the batter flow more easily. Baked goods rise, in part, due to the air trapped by fats.

Emulsions

Lipids play two roles in creating emulsions. They might be part of an emulsion or they might influence one to form. Oils, for example,

Figure 16-6

Triglyceride Subgroups in Foods

Triglyceride Subgroups	Description	Examples
Milk fat	Contains some short-chain fatty acids	Cow's milk
Fats from the coconut palm	Low melting range; usually low in unsaturated fatty acids	Coconut palm fat
Fats containing linolenic acid	Highly unsaturated	Soybean oil, wheat germ oil
Vegetable butters	Fats from seeds of tropical trees; highly saturated; very narrow melting range	Cocoa butter
Fats containing oleic or linoleic acids	Unsaturated; probably most common in American diet	Corn, peanut, sunflower, olive, and sesame oils
Marine (fish) oils	Contain long chains of polyunsaturated fatty acids; some very high in vitamins A and D	Cod liver oil
Animal fats	Usually high in saturated fat; contain oleic acid; high melting range	Fat on edge of meat and marbling throughout

are frequently one of the liquid phases in an emulsion, such as mayonnaise. Triglycerides can be broken down into *monoglycerides*. These have one fatty acid attached and are stable enough to emulsify other fats with liquids. As mentioned earlier, phospholipids are another lipid, also used as emulsifiers.

Flavor

As the aroma of frying doughnuts tells you, fats carry flavor. They dissolve aromatic molecules in foods and distribute their essence throughout the recipe. Some fats add their own characteristic taste. Olive oil is valued to mildly flavor salads, while bacon fat is unmistakably salty and smoky. Others, such as corn oil, are preferred because they lack distinctive flavor.

Oxidation

The oxidation of fats occurs as the surface of foods react with oxygen. When fatty acids oxidize, they lose electrons by combining with oxygen. This complex, somewhat mysterious, chemical chain reaction is a main reason that high-fat foods spoil.

Many people cook with olive oil today. It's highly unsaturated and doesn't easily oxidize. Olive oil labeled as "virgin" was obtained from the first pressing of the olives without further processing. By spraying the oil on the pan, you can use a smaller amount.

Imagine what a piece of pie would taste like without fat of some type in the crust. Some people use oil in their crust recipe, and some use lard. Which contains more saturated fat?

Oxidation involves the loss of a hydrogen atom from a single-bonded carbon next to a double bond. Unsaturated fats, with more double bonds, are more prone to oxidization.

Heat speeds the rate of oxidation. Cooling the environment, even to freezing, slows but doesn't stop the process. Water inhibits oxidation. Water molecules interfere with contact between air and fatty acids. The oxygen in water is not free to react with fats. Just as water is used to put out fires, which are the most rapid form of oxidation, water prevents slow oxidation as well by preventing contact between substances and elemental oxygen.

What does nutrition research mean to real life? Ask a . . . Dietetic Technician

As a dietetic technician for a community health program, Margarita Tillman works with low-income mothers enrolled in the government nutrition program called Women, Infants, and Children (WIC). "A dietitian teaches the class, and I work with the moms one-on-one, helping them develop menus based on what we talk about in class. Our main goal is to help them feed their families nutritiously on a tight budget, but we accomplish something else too. We teach problem solving and decision making. As our clients build skills, it's a big boost to their confidence. They gain a sense of power."

Education Needed: Preparing for a career in dietetics can start in high school, with classes in biology, chemistry, and nutrition. In college, candidates complete a two-year program, earning an associate's degree. Courses in psychology and communication can help dietetic technicians understand how people learn and how to best present unfamiliar topics.

KEY SKILLS
- Communication
- Food preparation
- Organization
- Ability to work with people
- Problem solving
- Applying concepts

Life on the Job: Dietetic technicians teach principles of food and nutrition and provide dietary counseling under the direction of a licensed dietitian. They help guide people in selecting and preparing foods and planning menus based on nutritional needs. Dietetic technicians often work in hospitals; however, opportunities exist in other locations.

Your Challenge

Suppose you're a dietetic technician for a school district. Tired of seeing cafeteria lunches end up in the garbage, administrators may let a fast-food chain operate in the schools. The school board wants public input.

1. What are your concerns about the school board's plan?
2. What actions should the board take?

Rancid is the *term that describes the unpleasant flavors that develop as fats oxidize.* Because foods contain different fats, however, oxidation can be a different chemical change in each product. Thus, rancid walnuts taste different from rancid potato chips and certainly from rancid meat. Rancidity produces a distinctive flavor, one you know when you taste it.

Controlling Oxidation

Food spoilage is a loss to both producers and consumers, so food scientists continually research ways to minimize oxidation during processing. Potato chips are packaged in pure nitrogen. Bacon is vacuum-sealed. Oils may be bottled in dark glass. Dried foods, which lack protective water molecules, may be treated with substances that prevent oxidation. These substances are called antioxidants.

Commercial Uses of Fats

Some fats don't occur naturally in foods but are extracted and processed from natural sources. Some are tailored for specific uses. Commercially important fats fall into two main groups: animal fats and plant oils.

Among animal fats, butter, lard, and beef fat are most important. Of these, butter is most familiar to consumers. Like milk, butter is a natural emulsion, 80 percent fat and 18 percent water. The remaining two percent is the protein that binds the other two ingredients.

Lard, or rendered hog fat, may not sound very appetizing. It's popular in the baking industry, however, for use in pastries, cakes, and frostings. Modifying lard produces different effects. In baked goods, for instance, the fatty acids are shifted among triglycerides. This "jumbling" creates smaller crystals for a more uniform, creamy texture.

Beef fat is usually combined with vegetable fats for use in foods.

Plant lipids are derived from certain oil-rich seeds. In liquid form, they're often called vegetable oils. Corn, olive, canola, and oil blends are popular choices on grocery shelves. These and other oils are also extensively used in processed foods.

Hydrogenated Oils

Another fat that you may have used is solid vegetable shortening. This is a plant oil that has undergone **hydrogenation** (hy-DRAH-juh-NAY-shun), *a chemical process in which hydrogen is added to unsaturated fat molecules, breaking some double bonds and replacing them with single bonds.* Partial hydrogenation changes liquid oil to spreadable, semisolid fat. What effect do you think complete hydrogenation would have?

Deep-fried foods are totally immersed in fat, producing a browned, crispy coating on the food. If the fat is not hot enough, how might that affect the amount of fat in the food?

Besides imparting texture, hydrogenation adds stability by eliminating some of the double bonds in fatty acids. Hydrogenated oil resists rancidity better than liquid oil. It doesn't develop a stale flavor and odor as quickly. Hydrogenation helps make margarine a less delicate substitute for butter, for example.

Cooking with Fats

Cooking with fats often means frying. Fats form so many bonds between carbon chains that it takes temperatures of at least 260°C to separate the molecules. Thus, the boiling point of fats is more than twice that of water, making them excellent heat conductors.

Heating fats to frying temperatures produces a range of chemical changes that affect the quality of the cooked food.

Effects on Fats

Frying may add appeal to foods, but it's tough on fats. Repeated exposure to intense heat causes a decomposition similar to oxidation. Also, foods release some of their own fat, water, and other substances into the frying oil. These factors contribute to the deterioration of fat, a process called cracking. Cracking can discolor oil and produce off flavors and odors, qualities that are passed to fried foods.

Another effect of cracking is to lower the **smoke point**, *the temperature at which a fat produces smoke.* Reaching a smoke point is part of a cycle of fat breakdown: smoking makes the fat less stable, and the less stable the fat, the lower its smoke point. Vegetable oils generally have a higher smoke point than animal fats, making them more useful for frying.

Frying Safety

Safe frying starts with using the right equipment. A properly designed deep-fat fryer is well insulated, with high sides and a reliable thermostat. A frying pan used on the range top should be of heavy-duty metal, with sides high enough to contain spills.

Food should be thoroughly dry before it's placed in hot fat. Recall that fat and water don't mix. In "fleeing" the water, fat will spatter. Fat droplets that hit you can cause serious burns. Fat that comes in contact with a heating element can start a fire. Again, dousing with water will only spread the flames. Instead, smother a fire in the pan with a frying pan lid, another pan, or baking soda.

Frying demands constant monitoring. Use a deep-fat thermometer to keep the temperature in the correct range. Keep the fryer filter clean and change the oil often. These steps help ensure the food's quality as well as your safety.

Triglycerides in the Diet

It sounds like common sense: the fats you eat become fats in your body. Basically this is true. Fats in food are broken down into fatty acids and glycerol, then absorbed by villi in the small intestine. Some of these fatty acids are stored in the liver, either converted to glycogen or left as fat. Remaining fats end up as adipose tissue throughout the body.

Did you ever eat fried chicken that was fried in overused fat? Why was the flavor off?

Making meal selections can be a challenge. Some fat in food is easily recognizable, but some isn't. Where might "invisible" fat be in one of these meals? Which meal is the best choice nutritionally?

While some dietary fat is essential, most Americans eat too much. A typical American diet gets 45 percent of its calories from fat—an all-time high, and well over the 30 percent maximum recommended by nutrition experts. This trend alarms health professionals, for excess dietary fat, especially saturated, is one of several factors suspected in heart disease.

Low-Fat Options

According to recent surveys, as many as 56 percent of American adults are trying to limit their fat intake. They are helped by a greater variety of low-fat foods in the supermarket: fat-free ice cream made with skim milk; lean cuts of meat; and baked, rather than fried, snack foods. To keep weight under control, however, they must also watch calorie consumption.

Nutrition Link

Limiting Fats and Cholesterol

While heredity is a leading factor in heart disease, evidence linking heart disease to saturated fat and cholesterol is mounting. You can't choose your genes, but you can choose an overall low-fat diet that stresses monounsaturated oils. To help ensure your heart's health for many years, use the following suggestions:

- Choose legumes, lean meats, poultry, and fish as protein sources.
- Trim excess fat from meats.
- Eat liver only occasionally. Limit eggs to four a week.
- Use fat-free milk and cheeses.
- Use butter, cream, shortening, and solid (hydrogenated) margarine sparingly.
- When possible, choose liquid oils instead, especially olive oils.
- Bake, boil, and broil foods rather than fry them.
- Read labels to become familiar with the fat content of foods. See **Figure 16-7** on page 250.
- Learn how to substitute other food ingredients for fats in recipes.
- Make careful choices when eating out. Ask how foods are prepared. Request that high-fat sauces and dressings be left out or served on the side. Learn terms for high-fat preparation methods.

Applying Your Knowledge

1. If someone said that heart disease is only a problem for overweight people, how would you respond?
2. Do you think foods labeled low or reduced fat are helpful in limiting dietary fat? Explain.

Approximate Fat Content in Selected Foods

Foods	Quantity	Total Fat (g)	Saturated Fat (g)
Bacon	3 pieces	9	3.3
Bagels, plain	1	1	.2
Beef, roast (lean and fat)	4 ounces	35	15.8
Beef steak, t-bone (lean and fat)	4 ounces	9	4
Bread, wheat	1 slice	1	.2
Butter	1 tablespoon	11	7.1
Cereal, corn flakes	1 cup	<1	.1
Cheese (American)	1 ounce	9	5.5
Cheese (Cheddar)	1 ounce	9	5.9
Cheeseburger	1	20	9.2
Cheesecake	1 piece	18	9.2
Chicken breast (roasted; skinless)	1	6	1.8
Chicken breast (batter-fried)	1	37	9.9
Chocolate, plain milk	1 ounce	9	5.2
Cookies, chocolate chip	4	15	4.5
Corn chips	1 cup	9	1.2
Croissant	1	12	6.7
Cupcakes	1	7	2
Eggs (whole)	1	5	1.5
Granola bar, hard	1	5	.6
Ice cream, vanilla	1 cup	14	9
Mayonnaise (regular)	1 tablespoon	11	1.7
Milk, nonfat	1 cup	<1	.4
Milk, whole	1 cup	8	5.1
Pasta, spaghetti	1 cup	1	.1
Peanuts (oil roasted)	1 cup	71	9.8
Pie, cherry	1 piece	29	7.2
Pizza, combination	1 piece	5	1.5
Popcorn (air popped)	1 cup	<1	Trace
Pork chops (lean and fat)	1	12	4.5
Potatoes, mashed	1/2 cup	1	.4
Pretzels, thin sticks	1 ounce	1	.2
Pudding, vanilla	1 cup	5	.8
Salad dressing, ranch	1 tablespoon	8	1.2
Trail mix	1 cup	47	9.3
Tuna, in water	1 cup	1	.4

< = Less than

Low-fat and fat-free foods are often high in calories.

Fat Replacers

Fat replacers, substances that mimic fats in foods but not in the body, are adding low-fat choices. With fats' complex nature and many attributes, food technologists have yet to develop one product to meet every need. Instead different replacers fill different functions.

Fat replacers are loosely grouped as either fat, protein, or carbohydrate based. Fat-based replacers are manufactured from very short and long carbon-chain acids, which supply fewer kcalories. Protein-based replacers use milk or protein particles to stabilize and texturize dairy products and some baked goods, sauces, and soups. Digested as protein, they also provide fewer kcalories per gram. Carbohydrate fat replacers include cellulose, gums, and modified food starch. They imitate fat's mouthfeel at half the kcalories.

A newer fat replacer consists of fatty acids attached to a sucrose molecule. This "cross-breed," called sucrose polyester, is the first artificial fat that is suitable for frying. Since it cannot be digested, it is calorie-free.

Cholesterol

Another lipid found in humans and animals is cholesterol (kuh-LESS-tuh-rawl). **Cholesterol** is not a triglyceride but *a sterol, a fatty alcohol made from glucose or saturated fatty acids.* The cholesterol molecule is complex; its formula is $C_{27}H_{45}OH$. Cholesterol is vital in producing vitamin D and some hormones; it strengthens cell membranes.

The liver makes all the cholesterol the body needs. Dietary cholesterol, found only in animal products, is useless and possibly harmful. Cholesterol in the blood is thought to contribute to the manufacture of **plaque** (PLAK), *a mound of lipid material mixed with calcium and smooth muscle cells.* Plaque can lodge in artery walls in and around the heart, reducing blood flow. This may lead to **atherosclerosis** (a-thuh-ro-skluh-RO-sus), or "hardening of the arteries," *a buildup of plaque along the inner walls of the arteries.* Atherosclerosis raises the risk of heart attack and stroke.

Research shows that fat, especially saturated, is related to blood cholesterol levels. To avoid saturated fats, many people now substitute margarine for butter in their diets and avoid foods made with lard. The long-term impact of such dietary changes is still under study. Limiting full-fat dairy foods and high-cholesterol meat products probably has a greater impact on lowering blood cholesterol.

LDL and HDL

In the bloodstream, cholesterol and other lipids travel to where they're needed. For the journey, they are packaged as **lipoproteins** (lip-oh-PRO-teens), *large, complex molecules of lipids and protein that carry lipids in the blood.* See **Figure 16-8**. The more lipids in a

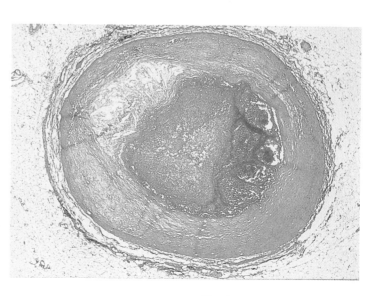

Artery walls can become clogged with cholesterol. A blocked artery won't allow blood to pass through. You can manage cholesterol better by limiting servings of red meat to 3 ounces and having three or four servings or less per week.

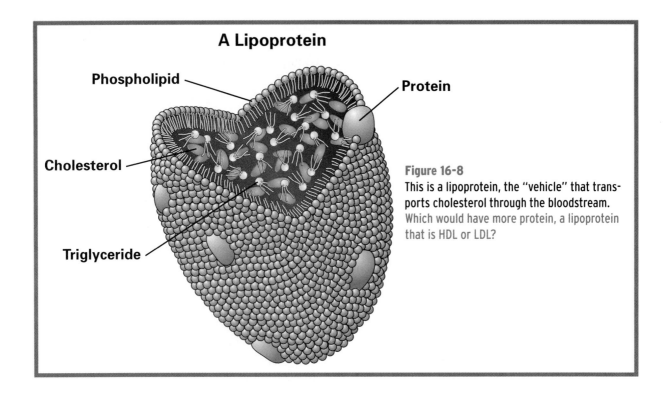

A Lipoprotein

Phospholipid

Protein

Cholesterol

Triglyceride

Figure 16-8
This is a lipoprotein, the "vehicle" that transports cholesterol through the bloodstream.
Which would have more protein, a lipoprotein that is HDL or LDL?

lipoprotein, the less dense it is. Lipoproteins that transport cholesterol from the liver to other tissues are, thus, known as low-density lipoproteins, or LDL. They carry about 75 percent of the cholesterol in the blood.

On the return trip to the liver, lipoproteins are higher in protein than lipids, making them more dense. These high-density lipoproteins (HDL) are thought to return cholesterol to the liver for breakdown and disposal (**Figure 16-9**).

Tests that measure lipoprotein levels can determine whether cholesterol is a problem. Since LDL has a greater percentage of cholesterol, an LDL level that is too high warns of increased risk of heart disease. An HDL level that is too low also indicates risk.

Omega-3 Fatty Acids

Why do natives of the Arctic region, whose diet is high in fat and cholesterol, have a low rate of heart disease? It may be they're protected in part by a steady diet of fish and the marine oils they contain. Scientists believe that omega-3 fatty acids, found in some fish, promote heart health in two ways. First, they make it more dif-

ficult for plaque to form or clump. They also make plaque less sticky and less likely to collect in the arteries.

Sardines, salmon, tuna, herring, and other ocean fish are highest in omega-3 fatty acids. Cod stores fatty acids in its liver, so doses of cod-liver oil are more beneficial than the fish itself.

Figure 16-9
Cholesterol travels in lipoproteins from the liver, to body cells, and back. Why are HDLs often called "good cholesterol"?

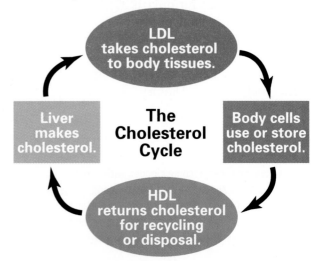

The Cholesterol Cycle

LDL takes cholesterol to body tissues.

Body cells use or store cholesterol.

HDL returns cholesterol for recycling or disposal.

Liver makes cholesterol.

Effect of Light on Flavor

SAFETY FIRST

Review these safety guidelines before you begin this experiment.

To choose an appropriate packaging method for a food requires knowing the properties of the food, the microorganisms (if any) that affect the food, and how the food changes as its quality deteriorates. As you know, over time, fat oxidizes to become rancid, causing undesirable flavors and odors. These flavors sometimes develop in high-fat foods, such as peanut butter, potato chips, and crackers. How these foods are packaged can affect whether they become rancid, since both light and oxygen accelerate oxidation and rancidity. In this experiment, you will study the effect of light on the flavor of potato chips stored in different environments for a specified period of time.

Equipment and Materials

2, 400-mL beakers	plastic wrap	masking tape
aluminum foil	30 g of potato chips	marking pen

Procedure

1. Wrap the outside of one beaker completely in aluminum foil. Leave the other unwrapped.
2. Mass the potato chips. Place 15 g in each beaker.
3. Close the foil-wrapped beaker with additional foil to seal it completely.
4. Tightly seal the unwrapped beaker with plastic wrap.
5. Label the beakers with your name and class period. Place them in a location specified by your teacher.
6. Sample the potato chips every other day for nine days. Rate their flavor, using the following scale:
   - 1—Strongly dislike flavor
   - 2—Slightly dislike flavor
   - 3—Neither like nor dislike flavor
   - 4—Like flavor
   - 5—Greatly enjoy flavor
7. Chart your ratings over the course of the experiment on a graph. Let the y-axis (the side) represent the flavor score; use the x-axis (the bottom) to represent the days.

Analyzing Results

1. As shown on your graph, when did the flavor of each sample begin to deteriorate? When did the difference become pronounced? What caused this change in flavor?
2. Did either sample stay completely fresh? Why do you think this is so?
3. Based on this experiment, how would you store foods with a high risk of rancidity?

SAMPLE DATA TABLE

Day	Rating of Chips Exposed to Light	Rating of Chips Protected from Light
1		
3		
5		
7		

Chapter Summary

- The body needs fat, but only in limited amounts.
- Lipids are a family of chemical compounds that are a main component in every living cell.
- Triglycerides make up the largest category of lipids and are commonly referred to as fats.
- Triglycerides perform a number of necessary functions in the body.
- Fatty acids are the organic acids that make up triglycerides.
- Saturated fats have single bonds between carbon atoms and contain the maximum number of hydrogen atoms. In unsaturated fats, double bonds between two or more carbon atoms occur when hydrogen atoms are missing.
- The structure of triglycerides influences their properties and, thus, their functions in foods and cooking.
- Reducing intakes of total fat, saturated fat, and cholesterol can help prevent heart disease.

Using Your Knowledge

1. Describe the three categories of lipids.
2. What functions do triglycerides serve in the body?
3. Describe the structure of triglycerides.
4. Why does the body need to obtain linoleic and linolenic acids from plant sources?
5. How are unsaturated fats classified? Give examples.
6. How do fats compare to carbohydrates as an energy source?
7. How does carbon bonding affect the phase of a fat?
8. Why do fats have a melting range rather than a melting point?
9. From a chemical standpoint, why might stale potato chips become rancid?

Real-World Impact

A Controversial Fat Substitute

In 1996, the sucrose polyester Olestra became available in the United States. This synthetic fat withstands the high heats of frying without adding kcalories to such foods as corn chips and potato chips. Olestra is under fire from nutritionists and consumer advocates, however, because it also picks up fat-soluble vitamins as it passes through the system, preventing their absorption. In some tests, Olestra caused diarrhea and other digestive problems in some people. Labels of foods fried in Olestra must warn of these effects, but its use remains controversial.

Thinking It Through

1. Do you agree with the Food and Drug Administration's approval of Olestra? Why or why not?
2. What precautions should be taken by someone who eats products with Olestra?

Using Your Knowledge (continued)

10. How does hydrogenation affect oils?
11. If you fried chicken in a restaurant, what signs might tell you the frying oil needed changing?
12. Compare low- and high-density lipoproteins.
13. How do marine oils differ from animal fats?

Skill-Building Activities

1. **Math.** Nutrition experts recommend that you get no more than 30 percent of your kcalories from fat. If each gram of fat provides nine kcalories, how many fat grams could you eat without exceeding this limit on a daily diet of 1600 kcalories? 1900 kcalories? 2200 kcalories?

2. **Nutrition.** Bring in three different food labels. Read the Nutrition Facts panel for information related to fat content. Compare your findings with those of classmates. Are you surprised at the amount of fat in certain foods? How might you account for high levels of fat in foods that don't seem particularly fatty?

3. **Teamwork.** In a group of three or four, create a meal plan for two days, aimed at balanced overall fat intake. For reference, use a cookbook to find recipes and obtain a chart that lists the fat content of foods.

4. **Communication.** On the Internet, find and print one article that deals with fats in foods or in health. Summarize your article to share and discuss with the class.

5. **Critical Thinking.** Devon cut down on fat by choosing fat-free snacks, which he really likes and eats often. Now he is gaining weight. What might the problem be? What should he do?

Thinking Lab

Low-Fat Fats

Nutritionists advise limiting saturated fats and cholesterol to reduce your risk of heart disease. For some people, this means the reluctant decision to give up some favorite high-fat foods.

Analysis

To enjoy some of the flavor and texture of fat while reducing their health risk, many people have switched from butter to margarine, and especially "diet" spreads.

Representing and Applying Data

Compare labels from a package of butter, regular margarine, and a reduced-fat and reduced-calorie spread. Note the amounts of fat, saturated fat, cholesterol, and kcalories. Compare prices also. Present this data in chart form.

1. Based on your findings, how effective is changing from butter to another fat in cutting saturated fats and cholesterol?
2. Do price differences or restrictions in use make any option more or less worthwhile?
3. Some newer research says that eating small amounts of butter is better than using margarine. Locate up-to-date information on this topic and report to your class.

Understanding Vocabulary

Take one term from "Terms to Remember" and plan a short lesson on that term. Present your lesson to the class and take notes on the lessons that class members provide on other terms.

CHAPTER ⑰ Protein

If you think of proteins primarily in terms of building muscles, you have only part of the story of this interesting nutrient. The human body contains an estimated 10,000 to 50,000 different proteins. Only about a thousand of these have been studied. Proteins have specialized functions in the body, which affect your health. As you will see, their role in foods is also interesting, both chemically and in real life.

What Is Protein?

Protein molecules are very large and complex. They are made of hydrogen, carbon, oxygen, nitrogen, and sometimes other elements. Because of their large size, proteins are often called **macromolecules** (mak-ro-MAHL-uh-kyools). A macromolecule is *a large molecule containing many atoms*. Protein molecules can make up to 50 percent of the dry mass of a body cell.

Nitrogen is a crucial player in proteins. This element sets protein apart from the nutrients you've studied so far; nitrogen occurs in neither carbohydrates nor lipids. Nitrogen is also the site of the chemical bond that gives proteins their variety and, thus, their versatility.

Structure of Protein

Protein is made of chains of substances called amino acids (uh-MEE-no). **Amino acids** are *a type of organic acid*. You will recall that organic acids are molecules that contain a carboxyl group (–COOH). In addition to the carboxyl group, amino acids also contain an amine group (uh-MEEN). An **amine group** is composed of *two atoms of hydrogen and one atom of nitrogen and is written* $-NH_2$.

As you can see in **Figure 17-1** on page 258, both the carboxyl group and the amine group are attached to a central carbon in the amino acid. A third bond to this carbon is to a single hydrogen.

With four bonds needed around the carbon atom, one is left. This is the variable side group, or side chain, that makes one amino acid different from another.

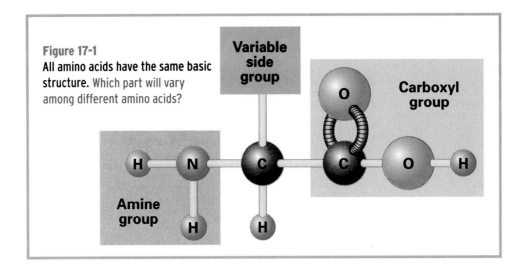

Figure 17-1
All amino acids have the same basic structure. Which part will vary among different amino acids?

In **Figure 17-2**, you see the simplest amino acid, glycine. It contains all the components just named. The side group in glycine, however, is only one atom of hydrogen. In other amino acids, the side group has different structures.

Peptide Bonds

How do amino acids link to form a protein molecule? This attachment occurs through **peptide bonds**, *bonds between the nitrogen of one amino acid and the carbon of a second amino acid.* An example is in **Figure 17-3**.

Suppose that in glycine, the amine group reacts with the carboxyl group of a second amino acid. One hydrogen in the amine group joins with the –OH of the carboxyl group to form a

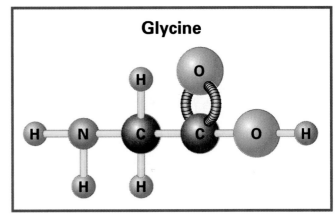

Figure 17-2
Glycine is an amino acid. How does its structure compare to the basic structure of an amino acid?

Figure 17-3
Peptide bonds hold amino acids together in protein. How does a peptide bond form?

Creating a Peptide Bond

familiar compound: H_2O, or water. The water molecule is released, leaving a bond between the amine group's nitrogen and the carboxyl group's carbon. This is a peptide bond.

Through peptide bonds, amino acids chain together. Eventually they may create a **polypeptide**, *a single protein molecule containing ten or more amino acids linked in peptide chains.*

Some protein chains contain only a few amino acids. Most molecules, however, contain 100 to 500 amino acids, and some have thousands.

Hydrogen Bonds and Structure

Peptide bonds hold chains of amino acids together. Rather than stretching out like railroad tracks, however, the chains coil, fold, and tangle—more like a wadded-up piece of string. How do these varied structures form?

Hydrogen bonds between parts of the peptide chain help fold the protein into shapes and hold it together. The shape that results depends on the particular amino acids in the chain and their order. Some chains form spirals called helixes, shown in **Figure 17-4**. The helixes may also entwine to form rope-like structures.

Polarity also influences a protein's structure. Some amino acids are attracted to, or repelled by, other amino acids and water. Amino acids with side groups that are attracted to water push to the outside of the structure to get closer to water. Those that are repelled by water avoid it by moving toward the inside.

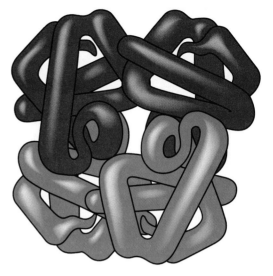

Hemoglobin

Figure 17-5

When amino acid chains fold, they form globular structures. Hemoglobin is a globular structure.

This rearrangement, shown in **Figure 17-5**, can cause a structure to fold over on itself, becoming more compact and three-dimensional.

Structure and Function

A protein's shape helps determine its function. Protein molecules that form rope-like fibers are called *fibrous* protein. This structure strengthens them to serve as connective tissue in the body. Collagen (KAHL-uh-jun) and elastin (ih-LAS-tin) are such fibrous proteins.

The more compact protein molecules have a structure that can be compared to a ball of steel wool. The rounded shape of these *globular* (GLAHB-yuh-lur) proteins makes them convenient carriers. The protein hemoglobin (HE-muh-glo-bun), for example, transports oxygen in the blood.

Other functions of proteins are detailed later in the chapter.

Denaturation of Protein

The size and folded shape of protein molecules help prevent them from dissolving and forming true solutions with water. They are prone to a

Figure 17-4

When amino acids chain together in a protein, some form rope-like spirals. Where would such protein be found in the body?

Collagen

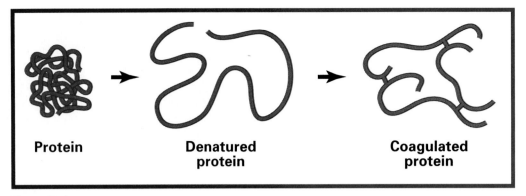

Figure 17-6
When protein denatures, the molecules unfold. New bonds produce the coagulated state as protein molecules bind together and produce a gel or a solid mass.

different type of chemical transformation, however, called **denaturation** (dee-nay-chuh-RAY-shun). *This process changes the shape of a protein molecule without breaking its peptide bonds.* Denaturation breaks the hydrogen bonds that create the twists and turns of a protein molecule. The result is a looser, less compact structure. As the molecule literally unfolds, some of the protein's original properties are diminished or lost.

This kind of change is unique to protein. Also, because each protein is unique, the denaturation process varies from one molecule to the next.

Denaturation is the first step in the process of **coagulation** (ko-ag-yuh-LAY-shun), as shown in **Figure 17-6**. Coagulation *changes a liquid into a soft, semisolid clot or solid mass.* It occurs when polypeptides unfold during denaturation, then collide and clump together to form a solid.

An example of these processes is found in scrambled eggs. Beating the egg denatures its protein. The protein coagulates as the egg cooks.

Denaturation by Heat

Heat is the most common agent in denaturing protein. The temperature is significant. The rate and the degree of denaturation increase 600 times for every 10°C rise in temperature. The structure of the protein affects the process also. Most proteins denature at temperatures between 47° and 67°C, as in eggs and milk. Denaturation in beef, however, takes a much higher temperature.

Other Means of Denaturation

Denaturation is caused not only by heat. Protein molecules may unfold in reaction to the following:

- Freezing, pressure, sound waves, and the addition of certain compounds.

A good example of coagulation is scrambling an egg. What happens to protein during this process?

- Mechanical treatment, as in beating eggs and kneading bread.
- Very high or very low pH. Adding lemon juice, which lowers pH, can sour milk. Milk proteins denature, coagulate, and separate from the liquid.
- Certain metal ions. Sodium and potassium ions are most commonly involved in this process, but copper and iron can have the same effect.

Before coagulation occurs, denaturation is sometimes reversible. That's why whipped egg whites that are left to stand will subside again into a viscous pool. Coagulation is not reversible.

Fish, milk, eggs, nuts, and legumes are all protein sources. What other foods provide protein?

Protein in Foods

Protein in the diet can come from many varied sources. Eggs, meat, fish, and poultry score highly as protein sources. Legumes provide some proteins as well. A closer look at some of these foods shows how diverse protein can be.

Eggs

Eggs are complex biological systems. They contain almost every vitamin and mineral you need. The yolk is rich in iron, phosphorus, vitamin A, and several B vitamins. All that eggs lack is vitamin C, which chickens make in their own bodies, and calcium. Calcium is present only in the shells, and there are tastier ways to get calcium than by eating egg shells. For all their nutritional offerings, however, eggs are foremost considered a protein food.

Proteins in Eggs

Inside and out, protein is the foundation of an egg. You can see the parts of an egg in **Figure 17-7** on page 262. The shell contains protein molecules interwoven with calcium carbonate crystals. Pores in the shell and the inner membrane lining allow a developing chick to breathe.

Cracking open an egg, you first encounter the **albumen** (al-BYOO-muhn), or *egg white.* The substance that makes up 54 percent of the inner contents of an egg is named ovalbumin, one of an egg white's six main proteins.

The thickest part of the albumen is the **chalaza** (kuh-LAY-zuh), *a twisted, ropelike structure that keeps the egg yolk centered.* Due to its thickness, the chalaza is the last part of the egg to coagulate and may remain slightly watery.

The egg yolk is very different from the white in composition. Main components in the yolk include the globular protein livetin and both high- and low-density lipoproteins. Thus, the yolk contains all of the fat, along with most of the other nutrients, found in eggs.

Effects of Storage

If you want eggs to retain their quality, the tray on the refrigerator door is not the best place to keep them. It exposes eggs to light and temperature change every time you open the door.

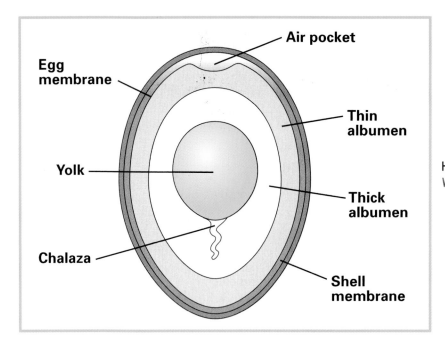

Egg membrane

Air pocket

Thin albumen

Yolk

Thick albumen

Chalaza

Shell membrane

Here you see the parts of an egg. What is the albumen typically called?

Such storage hastens chemical changes that cause physical changes that ultimately spell egg deterioration.

Eggs lose quality as compounds in the egg white break down, forming water. The white grows less viscous. Some of this water enters the yolk, making it thinner also. At the same time, carbon dioxide and more water escape through the porous shell. This can increase the pH, from a normal 7.6 to over 9. One result of this change is that proteins begin to break down. What's more, odors and flavors from other foods can enter the egg through its shell, giving it an off taste.

To delay these changes, a thin film of natural oil in the shell protects eggs. Shells may also be lightly coated with mineral oil immediately after eggs are laid. Storing eggs closed in their carton helps preserve their quality after purchase. The best temperature for long-term storage is -1°C. This temperature is below the freezing point of water but well above the freezing point of the liquids found in eggs.

To protect egg quality, store them in the refrigerator and in the carton they came in. When you crack an egg, how can you tell whether it's fresh?

Meat

In this text, meat refers to the edible portion of mammals, including both muscle and fatty tissue. Muscle tissue, the lean part of meat, is about 15–20 percent protein. Other components include water, fat, and minerals. Muscle is composed of fibrous proteins called actin (AK-tin) and myosin (MY-uh-sin). These proteins form bundles of fibers, which are held together by connective tissue made of collagen and elastin, two proteins with long, strong molecules.

The Steak and Eggs Diet

Eat all the meat and eggs you want—and lose weight. That's the promise of high-protein diets, which are based on getting 30 to 40 percent of kcalories from protein, while strictly limiting carbohydrates.

Many people do lose weight on such diets. Protein-rich foods, such as lean meat, are not always high in kcalories. They are filling, so dieters may consume fewer kcalories voluntarily. Does this disprove the accepted advice that you get 15 percent of kcalories from protein and 60 percent from carbohydrates?

From your studies, you know that carbohydrates are your best source of glucose, the fuel especially needed by the brain. Denied carbohydrates, the body relies on inadequate stores of glucose in body tissue. Ketones, a waste product of protein breakdown, accumulate in the blood, lowering the pH of body fluids. Nutrients can't work as intended in this unnaturally acid environment. Weight loss includes useful fat, protein, and water. Excreting this added volume stresses the kidneys. A buildup of uric acid may lead to kidney stones and gout, a painful joint inflammation. Meanwhile, you're likely eating unhealthful amounts of saturated fats and cholesterol.

Obviously, you can't stay healthy on such a diet. It's also hard to maintain. Once you resume your usual eating habits, the body rebuilds itself and the weight returns. Basing your diet on grains, fruits, and vegetables is a more sound (and economical) way to maintain a healthful weight and a healthy body.

Applying Your Knowledge

1. A friend who has lost 10 kilograms after a month on a high-protein diet claims to have more energy than before. How do you explain this fact?
2. What are some disadvantages of diets that require unusual foods or eating patterns?

Meat cookery is very complex due to variations in fat and in types of muscle fibers and connective tissue. In a beef stew, for example, moist heat softens some of the collagen, tenderizing the meat. Elastin is heat-resistant, however. It must be broken down mechanically, as by grinding or pounding.

Fish

Recipes for fish often say "flake with a fork" to test for doneness. This demonstrates a basic difference in fish proteins compared to those in meat. In fish, shorter, segmented muscle fibers are layered between thin sheets of connective tissue. There is far less connective tissue, and it's a type that liquefies easily. This protein structure also explains why fish cooks much more quickly than meat.

Nuts and Legumes

Among foods from plants, nuts and legumes—dry beans, peas, and lentils—are considered good sources of protein. Like eggs, these foods carry a rich store of nutrients. While nuts and legumes are cholesterol-free, their fat content ranges from almost zero to very high. Nuts are generally too high in fats to be a main source of protein in a healthful diet.

Fish flakes easily when cooked. Why doesn't a steak do the same thing?

Protein in Cooking

Like carbohydrates and lipids, protein's structure and traits give it many different roles in food preparation. Proteins lend their "special effects" to foods, ranging from a simple buttermilk biscuit to a towering soufflé.

Emulsifiers

The polarity of amino acids makes some protein foods good emulsifiers. Certain proteins contain amino acids that are polar. One end of these amino acids is attracted to water, where it forms hydrogen bonds, while the other end avoids water and bonds with oil. In this way, protein can stabilize an oil-and-water mixture. This is how egg yolk, the most common emulsifier used in home cooking, keeps oil and lemon juice blended in mayonnaise.

Foams

Technically, the fluffy meringue on a cream pie is a **foam**, *air bubbles incorporated and trapped in a protein film by whipping.* Foams add volume and lightness to a recipe, and proteins provide the chemical ability to make foams.

Soybeans, on the other hand, are a nutrition standout. They're composed of 40 percent protein, and their protein is equal to that in foods from animals. Soybeans are also high in fiber and polyunsaturated fatty acids. Recent research shows that soy foods in the diet may reduce the risk of coronary heart disease. Other possible health benefits are also being researched.

Soybeans are used to create many food items, including meat substitutes. Improved tastes and textures make soy foods easy to incorporate in the diet today.

Many healthful recipes today call for tofu, the product shown here. Just as cheese is made from cow's milk, tofu is made from soy milk. Soybeans produce soy milk, which is separated into curds and whey to form tofu.

Meringues are mainly sugar and egg white. More sugar means a stiffer meringue. Have you ever noticed syrupy beads on a meringue? A high baking temperature may have caused proteins to coagulate too fast, forcing out water that didn't have time to evaporate.

Foam formation begins when a protein-containing liquid, such as egg white or cream, is whipped. Whipping introduces air and denatures the protein molecules. The protein molecules coagulate to form a fine film around the pockets of air. As the protein molecules in this film unfold during denaturation, they form new bonds. Polarity determines the way bonds form with either the liquid phase on one side or the gas phase on the other. The new bonds help give the foam stability and structure.

Promoting Factors

Some fluids make better foams than others do. Milk has too little protein for its water content, and milk proteins resist denaturation. Thus, bubbles in milk don't last long. Egg white is better suited on both counts. Large protein molecules make the white more viscous. Some of its proteins are easily denatured by mechanical means. Ovalbumin, one of the proteins in egg white, coagulates nicely when heated, helping to set macaroons and other puffy baked goods.

Egg whites foam best at a pH of 4.6–4.8. Since pH rises as eggs age, fresh, quality eggs produce better foams. Cream of tartar, which lowers pH, is often added to stabilize the protein and produce a higher quality foam.

Inhibiting Factors

Fats are trouble in protein foams. Fat molecules, especially those in egg yolk, keep protein molecules from bonding with each other and with water. Even a trace of fat can drastically reduce a foam's volume and stability, or prevent foaming entirely. That's why a cook is careful to keep the yolk out when separating whites to whip. Fats tend to cling to plastics, their chemical relatives, so recipes often specify using glass bowls for whipping egg whites.

Sugar partially interferes with coagulation, delaying foam formation. However, sugar molecules form hydrogen bonds with the water in the fluid, improving foam stability. Adding sugar gradually, after the whites have started to foam, tends to get the best results.

Whipping can itself cause problems. A foam gains volume from air but is most stable just before reaching maximum volume. Past that stage, the proteins unfold to the point that they lose elasticity and water. Such overbeaten foams are called "dry." You can see that they are less moist and more fragile.

Gelatin

The gelatin used to set desserts and thicken meat sauces is an animal protein. It's made by using heat and water to hydrolyze the collagen on the inner layer of hides and bones. Gelatin is highly purified, making it safe to eat. Due to very strict conditions for manufacturing, it contains no harmful bacteria.

Suppose you were separating eggs to get whites for a meringue. A few spots of yolk land in the whites. **What would you do? Why?**

Is a cure for cancer growing in the forest?
Ask a . . . Senior Protein Biologist

Rachel Bergstrom, senior protein biologist for a private research foundation, says, "Using plants to heal is as old as recorded history, but now we understand more about how they work. We need to look seriously at the many plants in the world so that we don't miss a possible cure for AIDS or cancer. I worry about what life-saving treatments may be destroyed along with the rain forests and wetlands.

"I also worry about seriously ill people who are frustrated when conventional medicine can't cure them. They often grasp for so-called cures that are unproven. It takes years to find effective therapies. I hope researchers like me can continue to expose the myths as well as find what works."

Education Needed: Protein biologists leading research efforts generally need a Ph.D. in biochemistry, chemistry, or biochemical engineering. However, many positions exist for researchers with bachelor's degrees to validate testing, carry out basic experiments, and assist senior researchers with keeping records and preparing reports.

KEY SKILLS
- **Problem solving**
- **Data analysis**
- **Patience**
- **Communication**
- **Record keeping**
- **Curiosity**

Life on the Job: Producing curative proteins from plants takes long hours in the lab. Protein biologists separate, analyze, and purify these nutrients, then test their effectiveness against diseases. Highly complex techniques are employed. When therapeutic proteins are found, senior protein biologists plan production.

Your Challenge

Suppose as your team of protein biologists makes slow but steady progress, the media mistakenly reports you're nearing discovery of a treatment for AIDS. As senior researcher, you must respond.

1. How would you explain your advance without raising false hopes?
2. What would you say to those who criticize the slow pace of your work?

As a connective-tissue protein, gelatin has long molecules, many of which are polar. This structure allows gelatin to dissolve in water and to create a more viscous solution. As it cools, the protein forms a network that locks up the liquid in a firm, semisolid mass. Gelatin is what causes refrigerated meat drippings to congeal.

Gelatin can bind 100 times its weight in water. Unflavored gelatin may be used to stabilize liquids in food preparation. The sweetened gelatin powders you find in stores are mostly sugar, coloring, and flavoring.

Gluten

Gluten (GLOO-tun) is *an elastic substance formed by mixing water with the proteins found in wheat.* Gluten includes both fibrous and globular proteins. Extremely complex reactions between the water and strong, insoluble proteins make gluten both stretchy and springy. As a component of flour, gluten gives baked goods their structure and shape.

Gluten is developed as dough is kneaded, mechanically denaturing the protein molecules. The molecules align themselves lengthwise, while linking crosswise. The result is a smooth, elastic dough with enough "give" to stretch as the air and other gases trapped within it start to expand. At the same time, it is firm enough to form walls around the gases, creating an elastic mesh that keeps them from bursting through the surface. The heat of baking coagulates the gluten, setting the structure it has created.

Protein in the Body

How do you build body protein from food protein? During digestion, proteins are denatured by hydrochloric acid in the stomach, making the peptide bonds easier for enzymes to break.

They are hydrolyzed into smaller chains in the small intestine, then absorbed through the villi. Most proteins enter the bloodstream as individual amino acids.

In the blood, amino acids travel to cells. There, special genetic material assembles them into the specific proteins needed for the many functions they perform. At various points in life, some needs outweigh others—building bone in your teens, for instance. Like an auto maker, cells respond by making more of those proteins that are in highest demand.

The body creates all of its proteins from 20 different amino acids. It combines these compounds in thousands of ways to make many different proteins. For comparison, think of all the different words you can spell from the same letters of the alphabet. At the same time, if you lack even one letter, think of all the words you can't spell. The body relies just as heavily on those 20 amino acids.

Gluten gives dough its stretchy, springy qualities. What makes protein denature when a dough is prepared?

Your body can't grow and function without protein. How does protein make a difference?

Functions of Body Proteins

Proteins have earned their title as "nature's body builders." Their complexity makes them useful throughout the body. A few major functions are described below.

- Structural protein is needed by every cell in the body. For example, collagen helps build your bones, the ligaments that bind them, the tendons that connect them to muscles, and the muscles themselves.
- New growth requires a continuous supply of protein to replace and repair cells. Red blood cells must be replaced every few months. Those lining your intestinal tract last only three days. Your hair and nails couldn't grow without new cells.
- The enzymes and hormones that are responsible for most body processes are largely proteins. A single cell may contain over 1000 different enzymes, some of which break down other proteins. Various hormones aid growth, balance body fluids, and regulate metabolism.
- Some proteins pick up, deliver, and store nutrients in cells. Lipoproteins transport lipids in the blood, for example.
- Proteins called antibodies help you ward off disease. **Antibodies** are *very large proteins that weaken or destroy foreign substances in the body.* True to their specialized nature, each antibody is developed as a custom-made response to a specific substance.
- Proteins are crucial for stabilizing pH levels. Bodily functions continually produce acids

and bases that must be in balance. The blood, for example, can't carry oxygen unless its acid-base balance is correct. Proteins attract and release hydrogen ions in order to control the acidity of a solution.

- Proteins can supply energy if needed, but this is an expensive, and even dangerous, way to fuel your body. It results when a body starved of carbohydrates breaks down its own protein in search of glucose.

Essential Amino Acids

Of the 20 amino acids the body needs to build protein, it manufactures 11. The other nine are considered **essential amino acids**. *The body needs them but cannot make them itself.* Essential amino acids, which are listed in **Figure 17-8**, must be included in the diet. Otherwise the body dismantles its own protein to find them.

Figure 17-8

Essential Amino Acids

- **Histidine (HISS-tuh-deen)***
- **Isoleucine (eye-suh-LOO-seen)**
- **Leucine (LOO-seen)**
- **Lysine (LYE-seen)**
- **Methionine (muh-THIGH-uh-neen)**
- **Phenylalanine (fen-ul-AL-uh-neen)**
- **Threonin (THREE-uh-neen)**
- **Tryptophan (TRIP-tuh-fan)**
- **Valine (VAY-leen)**

*Because histidine is essential only for children, some scientists don't consider it an essential amino acid.

Combining grains with seeds, nuts, or legumes offers variety and other nutrients along with essential amino acids. Legumes and grains are an especially efficient pair, since each food contains the amino acids that the other lacks. Including both of these in the diet, often called complementing proteins, is how people in many cultures typically get complete protein.

Protein Allowances

Protein is not only a workhorse in the body, but also an efficient one. It accomplishes all of its vital work in relatively small amounts.

The RDA for a healthy adult is 0.8 g of high-quality protein per kilogram of ideal body mass (0.9 g for males under age eighteen). A teen female whose ideal body mass is 55 kg needs about 44 g of protein a day (55 kg × 0.8 g/kg = 44 g).

A protein that contains all the essential amino acids is called a **complete protein**. Logically, then, an **incomplete protein** is *lacking one or more essential amino acids.* Animal protein is generally complete. Among plants, only soybeans provide complete protein.

Another concern about dietary protein is its quality. A **high-quality protein** *contains all the essential amino acids in proportion to the body's need for them.* Egg protein is the traditional standard for quality by which all other proteins are measured.

Vegetarian Diets

Vegetarians who eat no foods from animals can still get all the essential amino acids. Many foods from plant sources contain some of these amino acids. Eating a variety of these foods over the day is likely to supply all that you need.

Beans and rice make a delicious meal whether you're vegetarian or not. How does this combination help fulfill the need for protein?

The RDA for children ages eleven to fourteen is slightly higher, at 1.0 g of protein per kilogram of body mass. A five-year-old child needs almost twice as much per kilogram as an adult—1.5 g of protein. Women who are pregnant or nursing require an extra 20 to 30 g of protein daily to keep themselves and their babies healthy.

Most people in the United States eat more than enough protein. Depending on your food choices, you might meet your protein requirement more easily than you realize. For example, 100 g of light-fleshed fish and light chicken meat each supply 25-30 g of complete protein. Many people eat larger portions than those, in addition to other protein foods throughout the day.

Learning the protein content of a variety of foods makes nutritional and economic sense. **Figure 17-9** shows you the protein content in a few selected foods. How do you think your protein intake compares to the recommended daily amount?

Figure 17-9

Approximate Protein Content in Selected Foods

Foods	Quantity	Protein (g)	Foods	Quantity	Protein (g)
Avocados	1	4-5	Frankfurter	1	5-7
Beans, black	1/2 cup	8	Macaroni	1 cup	7
Beef, ground	4 ounces	32	Milk	1 cup	8-9
Bread	1 slice	2-3	Oatmeal cereal, instant	1/2 cup	4
Cake	1 piece	2-5	Peanuts	1 ounce	7
Cheese, cheddar	1 ounce	7	Peanut butter	2 tablespoons	8
Cheese, cottage	1 cup	28	Pear	1	1
Cheese, ricotta	1 cup	28	Pork chop	1	23
Cheese, swiss	1 ounce	8	Potato, baked with skin	1	5
Chicken, breast, roasted	1	53	Pudding, chocolate	1 cup	9
Eggs	1	6	Raisins	1 cup	5
Egg substitute, liquid	1/2 cup	16	Rice, brown	1 cup	5
English muffin, whole wheat	1	6	Tofu	1/2 cup	8-9
Fish, cod	4 ounces	26	Tomato, raw, whole	1	1
Fish, tuna	1 cup	40	Yogurt	1 cup	13

The Effect of Acid on Protein

SAFETY FIRST

Review these safety guidelines before you begin this experiment.

In this experiment you'll compare the behavior of egg white in pure water and in an acid vinegar, as temperature increases. You will be able to draw some conclusions about the effect of acids on protein in egg white during cooking.

Equipment and Materials

egg white	2, 100-mL beakers	2 large saucepans
white vinegar	metal spoon	

Procedure

1. Obtain an egg white sample from your teacher.
2. Pour white vinegar into a 100-mL beaker to a depth of about 3 cm. Pour room-temperature water into a second 100-mL beaker to exactly the same depth as the vinegar.
3. Scoop up 5 mL of egg white. Avoid the chalaza, if possible. Gently add the white to the beaker of water, with a minimum of dripping and stirring. Try to keep the egg white intact. Add 5 mL of egg white to the beaker of vinegar in the same manner.
4. Pour tap water into a large saucepan to a depth of about 3 cm. Heat until the water boils. Wear safety goggles during heating.
5. Pour cold tap water into a second large saucepan to a depth of about 2 cm. Place both 100-mL beakers in this saucepan. Stir the cold water as you slowly add the boiling water. Do not pour any water in the beakers. Stop adding boiling water when the water in the pan is about twice as deep as the liquid in the beakers.

6. Observe the egg samples after they have warmed for 1 minute and again after 3 minutes. Record your observations in your data table.
7. After 5 minutes, remove both beakers from the water bath. Hold them up to examine the samples from the side. Record your observations in your data table.

Analyzing Results

1. What differences did you observe between the egg white cooked in water and the egg white cooked in vinegar? What caused these differences?
2. What substances might you substitute for vinegar to get a similar effect?
3. How might you apply your findings when preparing food?

SAMPLE DATA TABLE

Time	Egg White in Water	Egg White in Vinegar
1 minute		
3 minutes		
5 minutes		

Chapter Summary

- Protein molecules are very complex.
- Amino acids, the building blocks of protein, consist of a carboxyl group and an amine group.
- Proteins are made up of chains of amino acids joined by peptide bonds.
- Denaturation changes a protein's shape and properties.
- Protein is found in a variety of foods, affecting their storage and preparation.
- Proteins serve many purposes in food preparation.
- Proteins are indispensable to the body's functioning.
- Complete proteins are needed in the diet to provide the nine essential amino acids.

Using Your Knowledge

1. Why are amine groups important to proteins?
2. Compare peptide bonds and hydrogen bonds in protein formation.
3. Describe the two basic protein shapes. What is the main function of each?
4. What occurs during denaturation?
5. Can proteins coagulate without denaturation? Why or why not?
6. How does the egg yolk compare in composition to the egg white?
7. A friend refrigerates eggs still in their carton immediately after returning from the supermarket. Explain whether this is a good idea.
8. Compare the protein structure of meat to that in fish.
9. Suppose your egg whites aren't making a very impressive foam. What might be some reasons?
10. Why is kneading the dough an important step in making bread?

Real-World Impact

When Protein Is Missing

Protein-energy malnutrition (PEM), the most common type of malnutrition in the world, graphically illustrates the body's need for protein. PEM affects mostly children, contributing to tens of thousands of deaths daily. It slows physical and mental development, causes muscle wasting and diarrhea, weakens the immune system, and prevents the body from using other nutrients. It is most common among people facing natural disasters, war, and large-scale displacement.

Thinking It Through

1. Why do you think most PEM victims are children?
2. Why do you think PEM is more common than other types of malnutrition?

Using Your Knowledge *(continued)*

11. Why might injury and illness be more serious for someone with a long-term protein deficiency?
12. Should you be concerned if your diet lacks just one essential amino acid? Explain.
13. Can a diet of fruits, vegetables, and grains supply all the essential amino acids? Explain.

Skill-Building Activities

1. **Math.** How much protein would be needed by a sixteen-year-old male whose ideal body mass is 67 kg?
2. **Social Studies.** Research well-known national and regional dishes that are based on complementing proteins, such as tabouli, a Lebanese salad made with chick peas and wheat kernels, and Louisiana red beans and rice. What plant proteins are used? How is the dish typical of a culture? Does it have any special cultural significance?
3. **Management.** Plan a one-day vegetarian menu. Compare your eating plan with those of classmates. Does each provide complete protein? Are they otherwise nutritious? Varied? What are some advantages and disadvantages of getting protein only from plant sources?
4. **Critical Thinking.** Why do you think that cream, which is mostly fat, whips into a better foam if chilled, while egg whites produce a better foam at room temperature? Review Chapter 16 for help if needed.
5. **Science.** Further research proteins, using a biology or chemistry text. How does the new information you find relate to the facts in this chapter? Choose one aspect of proteins discussed in this chapter and prepare a more detailed explanation for the class based on your research.

Thinking Lab

Gluten in Bread Making

Gluten is the elastic, stretchy protein in wheat that gives bread its final shape and structure. As the cell walls of the gluten stretch during baking, heat coagulates the gluten and forms holes in the dough. But when does that structure become strong enough to keep a bread from falling?

Analysis

If you've ever been around when an oven door slammed while bread was baking, you probably know what can happen. Often the slamming door causes the bread to fall flat. The slamming door causes an impact that puts pressure on the fragile gluten structure.

Practicing Scientific Processes

Will a slamming door have an effect early in baking? Is there a point at which the bread can withstand the force of a slamming door? Develop an experiment to test when in the baking process gluten is most vulnerable to collapse.

Understanding Vocabulary

Create a crossword puzzle that uses as many of the "Terms to Remember" as possible. Exchange puzzles with a classmate for completion.

CHAPTER (18) Vitamins and Minerals

Objectives

- Explain in general how vitamins and minerals function in the body.
- Describe the basic structure of vitamin molecules.
- Explain specific contributions of different vitamins and minerals.
- Evaluate foods as sources of various vitamins and minerals.
- Relate vitamin and mineral deficiencies to the diseases that result.
- Explain some interrelationships among vitamins and minerals.

Terms to remember

beriberi
beta carotene
deficiency disease
fat-soluble vitamins
major minerals
megadoses
minerals
osteomalacia
osteoporosis
pellagra
phytochemicals
precursor
rickets
trace minerals
vitamins
water-soluble vitamins

Scurvy, beriberi, rickets—you may never encounter these odd-sounding illnesses. Each is a **deficiency disease**, *a disease caused by a lack of a specific nutrient.* At one time, they were quite common.

Scientists now know that such conditions result from a serious lack of a certain vitamin or mineral. Severe deficiencies are fortunately rare in developed countries. In the many areas afflicted by malnutrition, however, deficiency diseases are still quite common.

How can vitamins and minerals, which you need in such small doses, make such a huge difference in health? This chapter introduces you to the valuable contributions made by these two classes of nutrients.

The History of Vitamins

Vitamins, *complex organic substances vital to life,* are a relatively new discovery. The body uses them in such tiny amounts—sometimes only a millionth of a gram a day compared to hundreds of grams of energy nutrients. Until the early 1900s, scientists didn't have the techniques to identify them or the knowledge of chemistry and biology to understand their role.

Scientists knew of vitamins' importance to life long before then, however. You'll recall how experiments done in the mid-1700s showed that an ingredient in citrus fruits, now known as vitamin C, protected British sailors from scurvy.

In the 1880s, the Japanese navy's surgeon general found that adding meat and vegetables to the sailors' diet seemed to prevent **beriberi** (ber-ee-BER-ee), *a disease of the nervous system that causes partial paralysis, weakness, mental confusion, and death.* Soon after, a Dutch army doctor showed that eating rice with the hulls also prevented the disease. Because the substance in the hulls was first thought to be an amine, and was critical to life, it was called a vitamine. (*Vita* is Latin for "life.") Today, it's called vitamin B_1, or thiamine.

What other sources of vitamin C are there besides citrus fruits?

With the discovery of vitamin A in 1915, vitamins were recognized as a class of nutrients. Since then, a dozen more vitamins necessary to humans have been identified, ending with vitamin B_{12} in 1948. Scientists believe that still more await discovery.

The Function of Vitamins

Vitamins don't provide energy or build the body physically. Instead, they perform specific functions in the body by working with compounds called enzymes. You read about digestive enzymes in Chapter 13, and you'll study them in detail in Chapter 19. A quick explanation here will help.

Basically, enzymes speed up reactions that would otherwise proceed too slowly to sustain cell life. They work in metabolism and cellular reproduction, for instance. Enzymes include two parts, a protein molecule and a coenzyme. An enzyme needs a specific coenzyme to function fully.

As you can see in **Figure 18-1**, a vitamin may form part of a coenzyme. If you don't get enough of a certain vitamin, the enzyme that needs that vitamin as its coenzyme gradually slows its work. Cells continue to function, but less efficiently. The cells—and the organs and tissues they make up—begin to die.

This process takes time, which is fortunate. Otherwise, a short-term lack of a vitamin could be serious. On the other hand, when the signs of the deficiency do appear, weeks or even months later, the damage to your body may be hard to repair.

The Structure of Vitamins

Unlike the other nutrients you have studied, vitamin molecules are single, self-contained units. They don't link to form chains as glucose units form polysaccharides, for instance.

Each vitamin has its own structure, as shown by the structure of vitamin C in **Figure 18-2**. Most vitamins have similar basic components that include the following:

- A six-sided ring that includes carbon or carbon and nitrogen atoms. These are joined by a variety of single or double bonds. Depending on the specific vitamin, a molecule may have several of these rings.

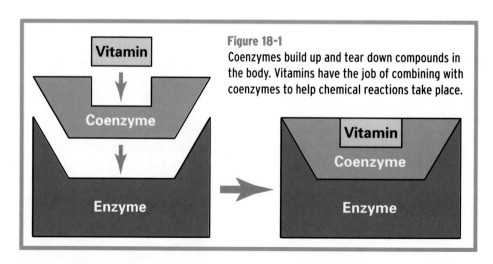

Figure 18-1
Coenzymes build up and tear down compounds in the body. Vitamins have the job of combining with coenzymes to help chemical reactions take place.

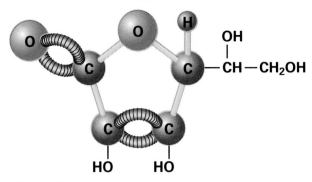

Vitamin C - $H_2C_6H_6O_6$

Figure 18-2
The structure of vitamin C has similarities to other vitamin structures. What are they?

- Attached hydrogen atoms. One or two hydrogens may form single bonds with the carbons in the ring.
- Attached *hydrocarbons,* which are compounds of hydrogen and carbon atoms. These compounds include one carbon and up to three hydrogen atoms. They, too, may form single or double bonds.

The Vitamins You Need

As organic compounds, natural vitamins are found only in living things. The human body cannot synthesize most of these substances; they must be supplied in foods or dietary supplements. As with other nutrients, different foods are good sources of different vitamins. Also, recommended intakes vary according to a person's age and gender. **Figure 18-3** gives the Dietary Reference Intakes for different groups.

Vitamins fall into two main groups: water-soluble and fat-soluble.

Water-Soluble Vitamins

Water-soluble vitamins *dissolve in water.* As you might expect, these are found in the water that's present in foods. They circulate freely in the blood and cell fluids. Water-soluble vitamins are quickly absorbed from the small intestine, and excess amounts are excreted in urine. Since the body stores these vitamins only in small amounts, if at all, they are best consumed daily.

Figure 18-3

Dietary Reference Intakes for Selected Vitamins						
Vitamin	Children 1-3	Children 4-8	Males 14-18	Females 14-18	Males 19-50	Females 19-50
Vitamin A (μg RE)*	400	700	1000	800	1000	800
Vitamin D (μg)	5	5	5	5	5	5
Vitamin E (mg TE)*	6	7	10	8	10	8
Vitamin C (mg)*	40	45	60	60	60	60
Thiamin (Vitamin B_1) (mg)	0.5	0.6	1.2	1.0	1.2	1.1
Riboflavin (Vitamin B_2) (mg)	0.5	0.6	1.3	1.0	1.3	1.1
Niacin (mg)	6	8	16	14	16	14
Folate (mg)	150	200	300	400	400	400

μg = microgram   RE = retinol equivalent   TE = tocopheral equivalent

*DRIs for this nutrient have not been released by the National Academy of Sciences; figures shown here represent the RDAs.

Bioavailability

Whether a food is a good source of a vitamin or mineral depends on *bioavailability,* or how efficiently your body absorbs and uses a nutrient. Bioavailability varies with the following circumstances:

- **Nutrient interaction.** Nutrients can interfere with each other. The presence of vitamin C decreases copper absorption but increases iron absorption. Fiber and some proteins are known to bind certain vitamins and minerals, limiting bioavailability.
- **Need.** The body can regulate its absorption of some vitamins and minerals based on need. A body that's well-supplied with zinc might take in only one-fifth of the amount a food contains.
- **Form.** Supplements in particular vary in nutrient form. Calcium pills, for instance,

may be bone meal, oyster shell, or dolomite (a component of limestone). Nutrients also have different forms in foods. While milk provides more calcium, you absorb a greater percentage of the calcium found in mustard greens.

- **Food preparation and storage.** Vitamins and minerals that are vulnerable to light, heat, or certain pH levels are diminished if foods are prepared or stored in such conditions.

Bioavailability is a concern mostly when it involves long-term or limited eating habits. If you always boil vegetables, for instance, your diet may not be as rich in vitamins and minerals as you thought. The advice to eat a variety of nutritious foods, healthfully prepared, holds true again.

Applying Your Knowledge

1. At a farmers market, produce is sold from open crates on a hot day. How might this affect bioavailability? What buying tips could help you get the most nutrition for your dollar?
2. Why would nutritionists be interested in bioavailability?

Nine water-soluble vitamins—the major B vitamins and vitamin C—are essential for humans.

Thiamin

Thiamin (THIGH-uh-min), or vitamin B_1, is needed to metabolize carbohydrates. It is also involved in transmitting high-speed impulses in the nervous system. Thiamin is present in many foods, notably peas, pork, peanuts, and wheat germ. Thiamin-rich foods are not always available in eastern and southern Asia, which have the highest rates of beriberi.

Thiamin is easily destroyed by oxidation. It is degraded by high temperatures unless the pH is kept below 5.0. Long, slow cooking can reduce thiamin levels by over one-third.

Riboflavin

Riboflavin (RY-bo-flay-vin), also called vitamin B_2, is important for growth. It's needed to metabolize protein and for tissue repair. Riboflavin is found in both animal and vegetable foods, including broccoli, beef liver, asparagus, and milk.

Riboflavin is stable when heated to temperatures above 100°C. Therefore, milk can be used in cooking without the loss of this vitamin. Can riboflavin be lost in any other way?

As a partner of several enzymes, riboflavin affects many chemical reactions in the body. It helps form red blood cells and release energy from glucose. It also helps make another B vitamin, niacin, from the amino acid tryptophan.

Riboflavin is more stable than thiamin. It is unaffected by acids, heat, or oxidation. However, it is vulnerable to high pH and breaks down under light. After four hours in the light, milk in a clear container loses 75 percent of its riboflavin—an advantage of having the light go out when you close the refrigerator door.

Niacin

A need for niacin (NY-uh-sin) is one thing you share with every living organism. This vitamin is crucial for the release of energy from carbohydrates, fats, and protein. It's also needed for forming deoxyribonucleic acid (dee-ahk-see-ry-bo-noo-KLAY-ik), better known as DNA, the compound that carries genetic instructions for each living cell.

Niacin is another stable vitamin, resisting light, heat, oxidation, acids, and bases. As a result, very little is lost during food processing. Niacin is found in dry peas and beans, peanut butter, liver, chicken, and fish.

Other B Vitamins

A number of other B vitamins are closely related to the vitamins discussed here. They are the following:

- **Vitamin B$_6$.** This vitamin helps your body use carbohydrates and proteins. It helps keep your nervous system healthy and helps your body make nonessential amino acids. Poultry, whole grains, and legumes are good sources.
- **Vitamin B$_{12}$.** With this vitamin, your body can use carbohydrates, fats, and proteins better. It teams with folate to help build red blood cells and form genetic material. It's also needed for a healthy nervous system. Only animal products supply this vitamin.
- **Folacin (FAWL-uh-sin).** Folacin, also called folate or folic acid, has earned recent attention for helping prevent birth defects of the brain and spinal cord. Functions also include teaming with vitamin B$_{12}$ and helping your body use proteins.
- **Pantothenic acid (pan-tuh-THEH-nik).** This vitamin helps the body release energy from carbohydrates, fats, and proteins and helps produce cholesterol. It promotes a healthy nervous system as well as growth. Meat, poultry, fish, eggs, legumes, milk, and whole-grain foods are good sources.
- **Biotin (BY-uh-tin).** Biotin helps your body use carbohydrates, fats, and proteins. It's found in green, leafy vegetables, whole-grain products, liver, and egg yolks.

Vitamin C

Oranges are a food often associated with vitamin C. This nutrient is found there and in all citrus fruits, as well as broccoli, bell peppers, potatoes, tomatoes, and cabbage. For a single vitamin, vitamin C does a lot of work, including the following:

The need for folate rises considerably for a pregnant woman. She must be sure to get the recommended amount in order to prevent birth defects.

- Helps prevent cell damage. Vitamin C belongs to a class of vitamins called antioxidants. Antioxidants "save" cells from oxidation during metabolism by being oxidized instead. They also protect cells from the harmful effects of some chemicals.
- Assists in forming the protein collagen in connective tissue.
- Promotes iron absorption.
- Aids in the formation of dentin, the hard, dense layer of the teeth.
- Helps the body heal wounds and resist infection.

You may have heard that vitamin C can cure colds if taken in **megadoses**, or *excessively large amounts.* No evidence conclusively proves this claim. In fact, no vitamin in any dose has been proven to cure any illness except deficiency diseases. As you've seen with other nutrients, megadoses of vitamins, even water-soluble ones, can be harmful. Massive doses of vitamin C, for instance, cause kidney stones in some people and diarrhea in others.

Vitamin C is stable in the presence of acids but is destroyed by oxidation, light, heat, and

The food service industry is concerned about nutrition. The longer a cooked vegetable sits, the greater the loss of nutrients. How does this principle impact food service?

If you were mixing orange juice from frozen concentrate, what temperature should the water you use be? Why?

uses this nutrient mostly to maintain healthy skin and mucous membranes. These membrane cells secrete protective mucus in the walls of the respiratory and digestive systems. However, vitamin A is also essential to normal vision. Children and teens need adequate vitamin A to ensure full bone development. Reproduction and the manufacture of red blood cells and hormones are other areas where vitamin A plays a role.

The liver can store up to a year's supply of vitamin A, yet it's hard to get too much from foods. Taking megadoses of vitamin A pills, on the other hand, is apt to cause problems. Painful joints, poor blood clotting, vomiting, diarrhea, irritability, and fatigue are some possible ill effects.

Vitamin D

Growing, repairing, and strengthening bones is the main job of vitamin D, also called cholecalciferol (ko-luh-kal-SIFF-uh-rawl). Vitamin D works with the bone-building minerals calcium and phosphorus, making them available in the blood so they can be deposited in bone tissue.

bases. Because it is so easily broken down, vitamin C is added to many foods after processing. Humans, along with guinea pigs and trout, are among the few creatures that don't produce vitamin C internally.

Fat-Soluble Vitamins

Fat-soluble vitamins *dissolve in lipids,* rather than water. Since they occur in the fats and oils of foods, fat-soluble vitamins are usually carried in the blood in lipoproteins. Like fat, they are stored in the body until needed. Thus, they accumulate more easily than water-soluble vitamins. This is an advantage in preventing deficiency diseases, but is also potentially dangerous.

Vitamin A

Vitamin A, or retinol (RET-un-awl), is found in animal products, such as liver, butter, egg yolk, and cheese. The body

Night blindness is one of the first noticeable signs of vitamin A deficiency. It's caused by a deficiency in the retina. Total blindness can be caused by a deficiency in the cornea. Vitamin A deficiency is the major cause of childhood blindness in the world.

Can you substitute oil for butter in a packaged cookie mix?
Ask a . . . Consumer Representative

As a consumer representative for a food manufacturer, Luisa Gutierrez believes all questions are important. "People deserve respect for their concerns. It may not be an emergency, but if someone cares enough to call me, I take the question seriously.

"People want to know where they can buy something they saw on TV or how a food fits a diabetic diet. Nutrition information is a real source of confusion. After a 30-second news story on antioxidants, I get requests for the whole story. People trust me for the facts. Of course, some people have complaints. They're already angry and disappointed in a product. Giving them a friendly reply is just part of good customer relations."

KEY SKILLS
- Communication
- Organization
- Food preparation
- Record keeping
- Problem solving
- Curiosity

Life on the Job: Consumer representatives work for food companies, responding to customers' inquiries by telephone, mail, and e-mail. They often work with marketing, legal, and packaging experts to identify trends in consumer concerns. Since they are often the public voice of large companies, they need to be clear, helpful, and positive to maintain good public relations.

Education Needed: Consumer representatives in the food industry need an understanding of food preparation and nutrition, which a bachelor's degree in family and consumer sciences, food science, or nutrition provides. Courses in communication are useful for relating to people from diverse backgrounds.

Your Challenge

Suppose as a consumer representative, you take a call from a young person who wants to gain weight and build muscle by taking megadoses of supplements.

1. Would large doses of vitamins likely be helpful?
2. What would your advice be?

Just a small amount of time in the sun each week is enough to enable your body to produce the vitamin D it needs. What precautions do you need to take to prevent skin damage from too much sun?

Until the 1930s, many children developed **rickets**, *a vitamin D deficiency disease that causes soft, weak bones.* At that time, processors began adding the vitamin to milk, which is rich in calcium and phosphorus. Today rickets is relatively unknown in the United States.

Except for eggs, liver, and fish oils, few foods naturally contain large amounts of vitamin D. Most vitamin D in the body is synthesized in the skin with exposure to sunlight. For fair-skinned people, fifteen minutes of clear sunshine on the arms and face, about three times a week, is usually enough. The process takes progressively longer for darker complexions.

Adult rickets is called **osteomalacia** (ahs-tee-oh-muh-LAY-shuh). It most often strikes women who spend little time in the sun or who have had numerous pregnancies. During pregnancy, vitamin D, calcium, and phosphorus in the woman's body are used to form the baby's bones. Repeated pregnancies can drain the woman's own store of vitamin D. Supplements are often needed.

On the other hand, too much of this vitamin can cause problems with excess calcium deposits, including kidney stones. You cannot build up too much vitamin D from exposure to sun, for the body makes only the amount it needs.

Vitamin E

Vitamin E, scientifically known as tocopherol (toe-KAHF-uh-rawl), is an umbrella term for at least eight different chemicals. Vitamin E is gaining appreciation for its role as an antioxidant. It helps stabilize cell membranes and protects vitamin A and lipids from damage due to oxidation.

Vitamin E is primarily found in vegetable oils—soybean oil and wheat germ oil are good sources—but small amounts are present in grains, fruits, and vegetables. Because it occurs in so many foods, deficiency is rare. Megadoses can cause nausea and intestinal distress.

Vitamin K

Would you believe that without vitamin K, a nosebleed could be fatal? This vitamin is necessary to make several proteins responsible for clotting blood. It also helps produce a protein that allows bones to absorb minerals needed for growth.

Vitamin K is found in milk, liver, and many dark green vegetables. However, about half of the vitamin K you need is made by bacteria that live in the intestines. Thus, vitamin K is produced in, but not by, the body.

These bacteria can be killed by antibiotics. To avoid problems with poor blood clotting, people on such medication should be sure their diet includes plenty of foods rich in vitamin K.

Vitamin Precursors

While the human body cannot make most vitamins, it sometimes assembles them from raw materials. A **precursor** (pri-KUR-sur) is *a compound that can be changed into a vitamin in the body.* Vitamin precursors are also called provitamins.

Precursors are not vitamins themselves. **Beta carotene** (BAY-tuh KAIR-uh-teen), for example, is *an orange plant pigment that can be changed into vitamin A in the intestines and liver.* This pigment is what gives

some deep-yellow and orange fruits and vegetables their showy colors. It occurs in green vegetables, though hidden by the pigment chlorophyll. Beta carotene supplies up to 50 percent of the vitamin A used in the body. Carrots, spinach, broccoli, sweet potatoes, and apricots are all excellent sources.

A vitamin D precursor is made in the liver. The precursor circulates in the blood until exposed to the ultraviolet rays of the sun through the skin. The rays trigger a chemical process that converts the precursor into vitamin D. This gives Vitamin D its nickname, the sunshine vitamin.

The amino acid tryptophan can be a precursor of the B vitamin niacin. The liver can produce one milligram of niacin from 60 milligrams of tryptophan. The conversion of tryptophan to niacin helps illustrate two important concerns in nutrition and food science: deficiency disease and nutrient interaction.

A lack of niacin can cause **pellagra** (puh-LA-gruh), *a disease that causes skin eruption, digestive and nervous disturbances, and eventual mental decline.* It was a major health problem in the late 1800s, especially in the South, where the diet was corn-based. Corn is low in both tryptophan and niacin. Moreover, any niacin present is bound to the corn protein. Both conditions limit the food's value as a niacin source.

Pellagra still occurs where corn is a dietary staple, including part of Egypt and India. Corn is a mainstay in the Central American diet, yet pellagra is not common in that region. Corn tortillas, which are used in numerous dishes, are traditionally made with soda lime water. Soda lime is a powdery, basic mixture of sodium hydroxide and calcium oxide. By raising the pH of the corn flour, it chemically frees the niacin for use in the body.

You can get too much of a good thing if you take megadoses of vitamins. An excess of water-soluble vitamins will just leave your system, although some can cause health problems. What happens if you take megadoses of fat-soluble vitamins like A and D?

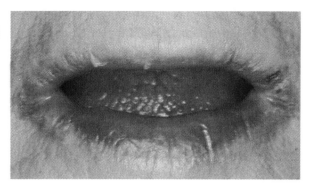

Pellagra can cause a skin condition that looks like a sunburn. The tongue may also become bright red and inflamed. *What causes this disease?*

The Function of Minerals

Minerals are *inorganic elements needed by the body.* Combined, they make up only about five percent of your body weight, yet they are critical to your mental and physical health. Like vitamins, minerals further chemical reactions in the body. The two often work as a team. For example, vitamin C boosts iron absorption. Zinc helps the liver release vitamin A. Minerals may be cofactors of enzymes, teaming with enzymes as vitamins do.

Minerals are sturdy nutrients. They withstand most food preparation methods, although some may leach into water used for cooking.

The Minerals You Need

About 17 minerals are considered essential to humans. Some of these, and their DRIs, are listed in **Figure 18-4**.

Major Minerals

Major minerals are *those needed in amounts of 0.1 g or more daily.* Some of these are described here.

Sodium and Potassium

Sodium and potassium work together in several ways. Both are found in the body as positive ions. With this electrical charge, they help balance the flow of water in and out of cells. They are also needed in the functioning of nerves and muscles, including the heart.

Potassium-rich foods include potatoes, bananas, dried fruits, and orange juice. Sodium occurs naturally in meats and vegetables.

Figure 18-4

Dietary Reference Intakes for Selected Minerals						
Mineral	Children 1-3	Children 4-8	Males 14-18	Females 14-18	Males 19-50	Females 19-50
Calcium (mg)	500	800	1300	1200	700	1000
Phosphorus (mg)	460	500	1250	1250	700	700
Magnesium (mg)	80	130	410	360	400	310
Iron (mg)*	10	10	12	15	10	15
Zinc (mg)*	10	10	15	12	15	12
Iodine (μg)*	70	90	150	150	150	150

μg = microgram

*DRIs for this nutrient have not been released by the National Academy of Sciences; figures shown here represent the RDAs.

Added sodium, especially as table salt, may increase blood pressure, which can cause problems for people with high blood pressure or heart disease. Preparing food with salt substitutes, herbs, and flavored vinegars is a healthful alternative.

Processing foods often adds sodium, while decreasing their natural potassium content. This can draw water out of cells, upsetting their chemical balance. Low-sodium varieties of some processed foods are available.

Calcium and Phosphorus

Your body contains more calcium than any other mineral, ninety-nine percent of it in the bones and teeth. There it combines with phosphorus to form the compound calcium phosphate, providing strength and structure.

Calcium is also critical in the blood. Calcium ions aid blood clotting, muscle action, regular heartbeat, and the fluid balance on both sides of body cell membranes. Its role in the blood is so crucial, in fact, that calcium may be drawn from the bones if not supplied in the diet. The result over time may be a condition called **osteoporosis** (ahs-tee-oh-pore-OH-sus), *a loss in bone density due to a prolonged deficiency of calcium.* The bones thin and become more fragile, increasing the risk of fractures.

All milk products are high in calcium. Milk itself is a particularly good source. It contains equal amounts of calcium and phosphorus as well as lactose, and usually vitamin D. All of these factors aid calcium absorption.

Phosphorus has added functions in the body. You may recognize it as part of ATP, the molecules that carry and release energy during metabolism. Phosphorus combines with lipids and proteins, transporting those nutrients to cells. It is a component of DNA. This mineral is plentiful in muscle tissue. Any type of animal protein is an excellent source.

Other Major Minerals

Other minerals, though less known, also make important contributions.

- **Chloride.** This essential nutrient is the negative ion of chlorine, a highly poisonous gas. Often ingested as table salt, chloride ions help control blood pH and form hydrochloric acid, a compound used to digest food in the stomach.
- **Sulfur.** As part of several amino acids, sulfur lends strength to the skin, nails, and hair. Sulfur is found in all protein foods.
- **Magnesium.** The roles of this mineral include tacking on the final phosphate group on the ATP molecule. By inhibiting blood

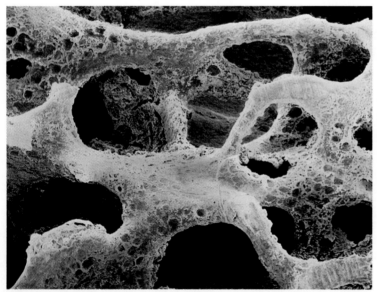

Peak bone mass is reached around age thirty. A few years later, gradual bone loss begins. This photo shows unhealthy bone that could easily break. You can prevent the bone loss of osteoporosis by building dense bones now. How can you do that?

Hard water contains minerals like calcium and magnesium. Some studies indicate that people who live in areas with hard water have low rates of heart disease. Since magnesium is critical to heart function, there could be a connection.

clotting and muscle action, it balances the effects of calcium to help regulate circulation and respiration. Magnesium also helps the teeth retain calcium. Good sources include legumes, nuts, whole grains, and dark green vegetables.

Trace Minerals

Minerals the body needs in daily amounts of 0.01 g or less are known as **trace minerals**. The tiny amounts you need contradict their great importance to your health. Eating a variety of fresh, unprocessed foods is the best way to get the proper amounts of these needed minerals.

- **Iron.** Forms part of the hemoglobin molecule, which carries oxygen in the blood. With an adequate supply, the body can store, release, and recycle iron to maintain needed levels.
- **Iodine.** Changes to iodide ions in the intestines. It's a component of thyroid hormones that regulate many areas of metabolism, including growth rate and cell oxidation.
- **Zinc.** Needed to form proteins, including many enzymes. Zinc, thus, plays a role in making DNA, metabolizing carbohydrates, and digestion. It also aids in healing wounds.
- **Copper.** Part of various enzymes involved in oxygen-related processes. These include releasing energy and making collagen. Copper helps metabolize iron to make red blood cells.

- **Manganese.** Aids enzymes that metabolize protein and release energy. It is needed for bone growth and nervous system functions.
- **Fluoride.** Important for the mineralization of bones and teeth. In this process, fluoride, as fluoride ions, acts with phosphorus and calcium to form large crystals that strengthen bones and help teeth resist cavities.
- **Chromium.** Plays related roles in the body's use of glucose. It is a cofactor of enzymes that break down carbohydrates. It also enhances the action of insulin.
- **Selenium (sih-LEE-nee-um).** Teams with several enzymes. It's an antioxidant in one combination and an aid to producing a thyroid hormone in another.
- **Molybdenum (muh-LIB-duh-num).** Partners with certain enzymes to help release iron stored in the liver and metabolize lipids.

As with vitamins, scientists continue to study whether other minerals are important to the functioning of a healthy body.

Phytochemicals

Recent research shows that the nutrients described aren't the only substances in foods that provide health benefits. Many foods also contain **phytochemicals**, *compounds that come from plant sources and may help improve health and reduce the risk of some diseases.* Current estimates suggest that every plant has at least 50 to 100 different phytochemicals.

Figure 18-5 shows just a few of the many phytochemicals under study. Every day brings new information about how these and many other compounds in foods impact health. While no single food is the complete answer to good health, eating a variety of foods is. A diet that includes plenty of fruits and vegetables just makes good sense.

Pumpkin is one source of beta carotene. What are others?

Figure 18-5

Phytochemicals in Foods

Food Sources	Phytochemical	Possible Links to Health
Garlic, onions, leeks, chives	Allyl sulfides (A-lul)	May trigger enzyme production to help eliminate carcinogens (cancer causers)
Fruits and vegetables with deep pigments, such as carrots, pumpkin, and spinach	Carotenoids (kuh-RAH-tun-oids), such as beta carotene and lycopene	Antioxidants; may reduce cancer risk
Grapes	Ellagic acid (uh-LA-jik)	May reduce risk of heart disease
Broccoli, cauliflower, cabbage, and other cruciferous vegetables	Indoles (IN-dohlz) Sulforaphane (sul-FOR-uh-fane)	May reduce risk of breast cancer May play a role in cancer prevention
Soy and legumes	Isoflavones (i-soh-FLAY-vonz) Phytosterols (fy-TAHS-tuh-rahl)	May reduce risk of some cancers May lower cholesterol
Flaxseed	Lignans (LIG-nuns)	May reduce risk of breast and ovarian cancers

Iron as an Additive in Cereals

SAFETY FIRST

Review these safety guidelines before you begin this experiment.

Many cereals sold in the United States today have iron added to them. This is intended to help consumers meet the minimum daily value recommended for this important mineral. In this experiment you'll determine the amount of iron present in a serving of cereal.

Equipment and Materials

plastic bag	magnetic stirrer	400-mL beaker	crucible tongs or forceps
cereal sample	magnetic stirring bar	250-mL beaker	

Procedure

1. In the plastic bag, crush enough of the cereal assigned to you by your teacher to produce a 150-mL sample.
2. Carefully examine the magnetic stirring bar. Describe its appearance in your data table.
3. Place the magnetic stirring bar in a 400-mL beaker. Carefully pour the crushed cereal into the beaker; avoid spilling any. Add 250 mL of water.
4. Place the beaker on a magnetic stirrer and stir for 15 minutes.
5. Using crucible tongs or forceps, remove the stirring bar from the mixture and put it on a paper towel to dry.
6. Observe the appearance of the stirring bar and again record your observations in your data table. Add results of others in the class to your data table.

Analyzing Results

1. Did all of the cereals tested produce a coating of iron on the magnetic stirring bar?
2. Which brand of cereal appeared to have the largest quantity of iron in a 250- mL sample?
3. How much iron does someone your age and gender need on a daily basis?
4. Based on the information in the Nutrition Facts panel on the cereal box, would a bowl of any of these cereals meet your daily requirement?

SAMPLE DATA TABLE

Brand of Cereal	Appearance of Bar Before Stirring	Appearance of Bar After Stirring

Chapter Summary

- Historically, vitamins are a relatively new discovery.
- Vitamins and minerals often function with enzymes to sustain cell life.
- The basic vitamin molecule includes a ring of carbon and nitrogen atoms with attached hydrogen atoms.
- Vitamins and minerals perform specific, vital functions in the body.
- A lack of vitamins or minerals can result in various deficiency diseases.
- Different foods are good sources of different vitamins and minerals.
- Because they often interact, an imbalance of one vitamin or mineral can affect many body processes.

Using Your Knowledge

1. How do vitamins and minerals affect enzymatic action?
2. How are vitamin molecules similar to and different from molecules of other nutrients you have studied?
3. Contrast water-soluble with fat-soluble vitamins.
4. Describe the functions of riboflavin.
5. Based on the example of pellagra, explain why eating a nutrient-rich food may not prevent a deficiency of that nutrient.
6. How does vitamin C help extend the life of cells?
7. How can a vitamin D deficiency lead to rickets?
8. What health problems might arise from a lack of vitamin A?
9. What are the main sources of vitamin D?
10. Explain whether this statement is true about vitamins: If a little is good, a lot is better.
11. How might a diet that is heavy in processed foods upset the body's mineral balance?

Real-World Impact

Vitamin A Deficiency

Vitamin A deficiency in developing countries is the leading cause of preventable severe visual impairment and blindness in children. Estimates say that children in 118 countries have this deficiency. According to the World Health Organization's Micronutrient Deficiency Information System, approximately 2.8 million children under the age of five years have visual problems from vitamin A deficiency. Many countries are implementing food fortification programs to combat this problem.

Thinking It Through
1. Why do you think it might take many months for a vitamin A deficiency to show up?
2. Why might it be difficult for children in some countries to get enough vitamin A?

Using Your Knowledge *(continued)*

12. What contributions does calcium make in the body?
13. What are phytochemicals and why are they useful?

Skill-Building Activities

1. **Language.** Research the ways that various vitamins and minerals were named. What languages are the most common sources of names? What properties do the names reflect?
2. **Critical Thinking.** In northern regions that get little sun, children were commonly given cod-liver oil to prevent rickets. Why do you think this treatment was especially effective?
3. **Teamwork.** In a group of four, compare six vitamins or minerals on these points: functions in the body; good sources; symptoms of deficiency or excesses; and interaction with other nutrients. Present your findings in chart form.
4. **Critical Thinking.** Can supplements effectively replace foods as a source of vitamins and minerals? Explain your reasoning.
5. **Problem Solving.** In Emma's family, mealtime is often rushed or nonexistent. On many days the family doesn't sit down to a meal together. Frequently, individual family members either skip meals or grab fast-food. What implications does this lifestyle have for vitamin and mineral intake? What would you suggest to Emma?
6. **Communication.** Locate an article related to a vitamin or mineral in a magazine, newspaper, or on the Internet. Present a brief oral report, summarizing the article's key points.

Thinking Lab

Non-Vitamin Vitamins

The actions and interactions of vitamins and minerals are complex and varied. Some are still being researched.

Analysis

People are taking more responsibility for their own health and nutrition. Some are turning to "dietary compounds" such as PABA and choline. These substances are sometimes promoted as vitamins, though they are not true vitamins.

Thinking Critically

1. Why might some substances be mistaken for vitamins?
2. What might be some risks of taking supplements advertised as "new vitamins"?

Understanding Vocabulary

On a slip of paper, write a question that can be answered by one of the "Terms to Remember." Repeat with another term. As one person reads all the questions written by the class, take turns providing answers.

TechWatch

GENETICALLY ALTERED PLANTS

Genetically altered plants are becoming more common as scientists change them slightly to improve crop yields and make heartier, more nutritious plants. The tomato was the first genetically engineered, whole product approved by the FDA. The use of recombinant DNA processes produced tomatoes that could be picked ripe, shipped, and sold without rotting first. One company hopes to use a new gene that regulates plant aging (deoxyhypusine synthase) to increase the shelf life of vegetables and fruits. By transplanting three different genes in rice, scientists recently enabled rice to produce beta carotene, which the body converts to vitamin A. In many countries a serious vitamin A deficiency may be lessened by this development in genetic research.

UNIT 5 The Chemistry of Food

TechWatch Activity
Since some people worry about potential negative effects, controversy surrounds the genetic alteration of plants. Investigate this controversy and debate the issues with your class.

Objectives

- Explain the function of enzymes as catalysts in chemical reactions.
- Describe the relationship between an enzyme and a substrate.
- Compare the functions and activities of enzymes and coenzymes.
- Explain how enzymes are used in digestion.
- Describe how food scientists manage enzyme activity.
- Identify influences on enzyme activity.
- Explain how enzyme reactions are involved in food preparation.

Terms to remember

activation energy
active site
blanch
catalyst
coenzymes
enzymatic browning
papain
substrate

Why do some fruits turn brown after you peel them? Do you know what helps the center of chocolate-covered candies remain soft? Enzymes are the special proteins responsible for these and many other chemical reactions.

Of the thousands of different enzymes that exist, each has a specific job to do. Some enzyme reactions take place in your body, where food is broken down into compounds the body can use. For example, you digest food efficiently because of enzyme activity. In the food industry, scientists apply their knowledge of enzymes in order to control enzyme activity in food. They also use enzyme technology to create and improve food products. Enzymes are clearly important pieces in the food science puzzle.

Enzymes Influence Reactions

When a chemical reaction occurs between two substances, chemical bonds are broken and new ones form. Energy causes particles of the two substances involved to collide with enough force to break the bonds. *The energy that supplies the force needed for a reaction* is known as **activation energy**. You've produced this kind of energy when striking a match. Friction from the two surfaces rubbing together gives the substance on the head of the match enough energy to burst into flame.

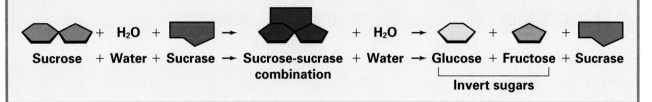

Figure 19-1
The enzyme sucrase, named for the substrate sucrose, acts as a catalyst in the hydrolysis of sucrose. What happens to the sucrase during this chemical reaction?

If it weren't for something called a catalyst, the heat energy required to start most cell reactions would kill the cell. A **catalyst** (KAT-ul-ist) is *a substance that speeds up the rate of a reaction without being permanently changed or used up itself.* When a catalyst is added to a reaction, it lowers the amount of activation energy needed to start the reaction. As a result, the reaction can take place faster at lower temperatures. Since less activation energy is required, the cell is able to survive the reaction.

Although a catalyst participates in a reaction, it isn't the *reactant* (the starting substance in a chemical reaction) or the *product* (substance formed in a chemical reaction). In fact, the catalyst is neither used up nor destroyed during a reaction. It remains available for use again and again. See **Figure 19-1**.

Because enzymes can act as catalysts, scientists call them "protein catalysts" or "biological catalysts." These powerful catalysts found in nature speed up the reactions necessary for a cell to function efficiently. Without them, for example, protein from food would not be digested quickly enough to supply amino acids for the body's needs.

How Enzyme Reactions Work

Enzymes participate in chemical reactions through a unique process. By comparing the process to a lock and key, you can understand it more clearly.

In a reaction, *the substance on which the enzyme works (the reactant)* is called the **substrate**. One or more substrate molecules of different shapes may be part of a reaction. Each enzyme has a three-dimensional structure that includes pocket or groove shapes. The

Enzymes perform many essential functions in the body. When you eat an apple, enzymes convert the sugar into a form the body can use.

Substrate

Enzyme

When picturing how enzymes and substrates work, think of a lock and key. The enzyme (or key) cannot do its job unless it combines with a substrate (or lock) that it fits.

area of the enzyme that fits into the shape of a substrate is called the **active site**. When the enzyme "key" fits into the right substrate "lock," the reaction takes place. **Figure 19-2** shows how these chemical reactions occur.

Identifying Enzymes by Name

To identify enzymes by name, look for the ending "ase." Typically, an enzyme name begins the same as its matching substrate. What would the enzyme that breaks down the starch amylose be called? That enzyme is amylase. In the same way, maltase is the enzyme that attacks the sugar maltose.

The Coenzyme Connection

In certain chemical reactions, enzymes need a partner in order to work. These enzymes temporarily join with other molecules known as coenzymes. **Coenzymes** are *heat-stable, organic molecules that must be loosely associated with an enzyme for the enzyme to function*. Coenzymes are usually smaller than

Figure 19-2
The molecules known as substrates undergo a chemical change to form new substances called products. Enzymes help this process.

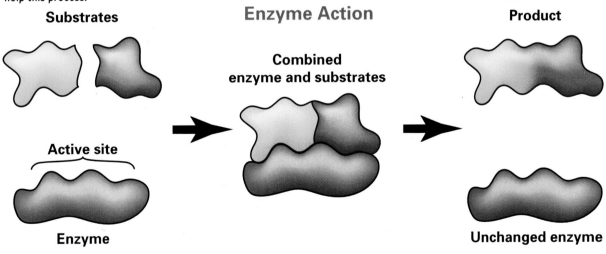

Enzyme Action

Substrates

Active site

Enzyme

Combined enzyme and substrates

Product

Unchanged enzyme

A	B	C
Each enzyme acts on a specific molecule or set of molecules called substrates. Each substrate fits into an area of the enzyme called the active site.	The enzyme holds the substrate or subtrates in a position that allows the chemical reaction to occur easily.	After the reaction, the enzyme releases the product and can go on to carry out the same reaction repeatedly.

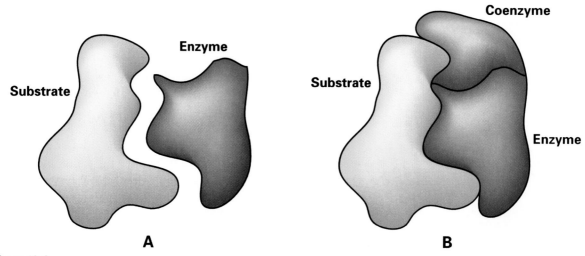

Figure 19-3
A coenzyme become an enzyme's "buddy." Why won't a reaction occur in A above? Why will a reaction occur in B?

enzymes themselves. Like enzymes, they are not used up during a reaction, but can be reused.

Coenzymes work with enzymes in the following ways:

- **Influencing the active site.** Coenzymes often alter the shape of the enzyme's active site so the enzyme "fits" the substrate better and the reaction can take place. **See Figure 19-3**.
- **Serving as transfer agents.** Sometimes coenzymes act as transfer agents for atoms,

electrons, or groups of atoms during a reaction. For example, an enzyme might remove a hydrogen atom from a molecule during a reaction. The coenzyme accepts this hydrogen atom until donating it to another substance. If the coenzyme were not available, the reaction couldn't take place.

Some very important coenzymes are made from vitamins or fragments of vitamins. If a person doesn't consume enough of the needed vitamin, the coenzyme cannot be made. Then

Glucose

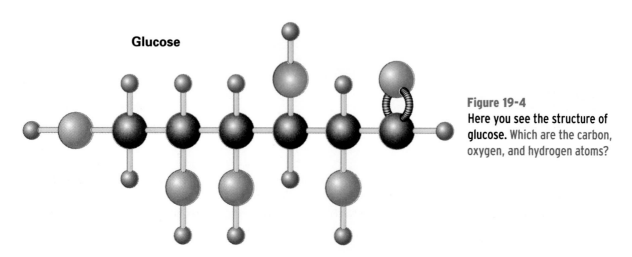

Figure 19-4
Here you see the structure of glucose. Which are the carbon, oxygen, and hydrogen atoms?

Nutrition *Link*

Enzymes Affect Digestion

Enzymes are needed for digestion. Without them, your body couldn't use the nutrients in food.

- **Digestion begins in the mouth.** The next time you eat a slice of bread, chew a bite slowly. Do you notice a change in taste? The bread should taste sweeter. Why would that happen? The taste change you notice occurs as digestion begins. Saliva contains amylase, an enzyme that breaks down the starch amylose, found in bread. Amylase converts amylose to the sugar maltose in the mouth, creating the sweet taste. The breakdown of amylose to maltose explains why starches cause dental cavities as easily as sugars.

- **Digestion continues in the stomach.** Some enzymes aid digestion by working in the acidic environment of the stomach. The enzyme renin curdles milk protein, preparing it for pepsin. Pepsin denatures protein by attacking the peptide bonds that hold the amino acids together. Stomach enzymes are the only proteins in the body protected from acid.

- **Digestion is completed in the small intestine.** When food moves from the stomach to the small intestine, other enzymes go to work. The proteases work on protein, the lipases break down lipids, and the carbohydrases work on carbohydrates. The enzyme lactase, which breaks down lactose, is an example of carbohydrase. Phospholipase is a lipase enzyme that acts on phospholipids.

Applying Your Knowledge

1. How does the enzyme amylase help digestion?
2. Why do starches cause dental cavities as easily as sugars?

---

certain cellular reactions can't take place. For example, thiamin functions as a coenzyme. It acts as a coenzyme in reactions that help the body store energy.

Your Body Uses Enzymes

Enzyme activity is vital to good health. When your body uses enzymes to metabolize food, nutrients become available in usable forms.

How you digest and use food depends upon the enzymes in your system.

Your body's treatment of glucose is one example of how enzymes change compounds for use in living cells. As you know, glucose is a six-carbon sugar with the formula $C_6H_{12}O_6$. The structure of glucose is shown in **Figure 19-4**.

During metabolism, enzymes help change glucose to glycine, an amino acid. The process takes a series of steps.

Careers

Product Development Scientist

How are new food products invented? Ask a . . . Product Development Scientist

T he frozen foods enjoyed today aren't anything like what was in the stores years ago," says Alex Hamden, senior product development scientist for the frozen foods division of a large food manufacturer. "Busy schedules mean people want food that can be fixed quickly but still tastes good. At the same time, they don't want to sacrifice nutrition. We have employees who study the market to find out what customers will buy. My team works with them to turn ideas into real products."

Key Skills
- Scientific method
- Leadership
- Data analysis
- Understanding of market
- Creativity
- Communication

Education Needed: A senior product development scientist is an upper level job that requires at least a bachelor's degree. Most employers, however, prefer a graduate degree in food science or a related field along with experience in the particular industry. Supervisory skills and sales experience help these scientists lead teams and work with partners.

Life on the Job: Working with people in sales, research and development, and manufacturing is part of a typical day in this career. Product development scientists spend plenty of time in the lab developing prototypes. After they analyze opinion data from flavorists and tasting panels, it may take many more tries to get a product right. In large corporations, these scientists may visit plants and stores in other cities and countries.

Your Challenge

Suppose your team has been working unsuccessfully on a new frozen food product. Half the team want to scrap the project, but half want to make changes and try again. You're already over budget.

1. What would you do?
2. Explain how four key skills above would help in this situation.

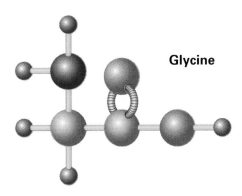

Glycine

Figure 19-5
Glycine is the simplest amino acid. Where is nitogen in this structure?

After the body breaks down carbohydrates into glucose, enzymes work on the glucose molecule. They add a phosphate group, alter the arrangement of atoms, and break the compound in two. In addition, atoms are removed, an amine group is added, and other changes are made.

Enzymes bring about each step in the metabolism of glucose. Finally, the enzymes produce the amino acid glycine. The body can then use this compound, shown in **Figure 19-5**, to produce protein.

Enzymes and Food Science

In addition to their use in the body, enzymes are also used in the food science industry. What basic knowledge about enzymes would food scientists need?

For one thing, they must learn which enzymes control particular reactions. Only then can they look for ways to manage those reactions.

Food scientists must also learn how to denature, or inactivate, enzymes. For example, they may want to stop a chemical reaction that causes food to overripen or spoil.

Learning how to slow down an enzymatic reaction is also fundamental in food science. Scientists might look for a way to eliminate either the enzyme or the partnering substrate in order to slow or stop a reaction. For instance, sugar cane is harvested for its sucrose. The su-

crose will break down, however, if it contacts the enzyme sucrase, which is contained in the tassel. During harvesting, therefore, the tassels are carefully separated from the cane so the sucrase won't hydrolyze the sucrose.

In some cases it's an advantage for enzyme activity to continue. For example, when rice grains are stored for long periods of time, they are less sticky after cooking than fresher grains are. This is apparently because the enzyme amylase acts on the starch amylose, breaking down the amylose chains. The broken chains are less sticky than the longer chains. Knowledge of enzyme activity enables food scientists to explain occurrences such as this.

What Affects Enzyme Activity?

A food scientist's knowledge about enzymes must include how certain conditions affect their activity. Temperature, pH, and water all affect enzymes.

Temperature
Temperature has a strong impact on enzyme activity. Usually, enzymes function very slowly at temperatures below freezing. Enzyme activity increases as temperature increases, reaching a peak between 30°C and 40°C. Once the temperature rises past 45°C, most enzymes are inactivated. This occurs because their protein is denatured (deactivated). See **Figure 19-6**.

Figure 19-6
Heat can denature an enzyme. Why won't the enzyme be able to work with its substrate?

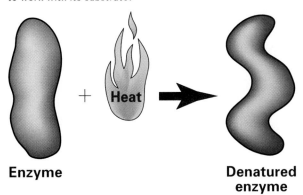

Enzyme   **+** **Heat** → **Denatured enzyme**

When foods are dried, enzyme activity slows down but still continues. What could be done to stop the enzyme activity completely?

If enzymes continue to react in frozen food, the food develops a bitter flavor. To deactivate enzymes, fruits and vegetables are blanched before freezing. **Blanching** means *briefly immersing food in boiling water*. The high temperature of the water or steam denatures the enzymes.

pH

Another influence on enzyme activity is pH. Most enzymes function best at a pH of about 8.5, although they can react when the pH ranges from 7-10. If the pH falls or rises beyond this range, however, activity decreases.

Water

Water affects enzyme activity. It provides a medium in which enzymes and their substrates can interact. If water is removed, the enzyme and substrate have more difficulty reacting. Although a low water level tends to limit enzyme activity, the activity doesn't stop completely.

Even though the water level is low in dried food, enzyme activity continues until the food

Figure 19-7
No one wants to eat bananas that have turned brown. How can lemon juice make a difference chemically?

Preventing Enzymatic Browning

When oxygen reacts with the enzymes in freshly cut banana, the bananas turn brown.

has an unpleasant aroma. Therefore, vegetables and fruits to be dried also need to be blanched to stop unwanted enzyme activity.

Enzyme Activity in Food

Enzymes affect food preparation—in both positive and negative ways. You may have seen what enzymes do without even realizing they were at work.

Enzymatic Browning

Have you ever noticed how a peeled peach turns brown after sitting for a while? Pears, apples, figs, bananas, plums, avocados, and potatoes also change color when bruised or cut. Their normal color turns to gray or brown when the cell tissues are injured. *When enzymes produce discoloration in fruits and vegetables*, this is called **enzymatic browning**. See **Figure 19-7**.

Three substances are needed for enzymatic browning to occur—a substrate, an enzyme, and oxygen. To stop enzymatic browning, one of these three substances must be limited or eliminated. The food itself is the substrate, so that can't be eliminated. Denaturing the enzyme, however, can be done by:

- Using heat, as in blanching prior to freezing.
- Lowering the pH.
- Using sulfur dioxide, a gas that affects enzymes.

Lowering the temperature doesn't denature the enzyme, but it will slow down the process.

Excluding oxygen is another way to slow enzymatic browning. This can be accomplished by placing vegetables in salt water, and fruits in sugar water. The solutions prevent oxygen from contacting the food. Ascorbic acid (vitamin C) is an antioxidant that is available commercially to prevent enzymatic browning. The powder form can be mixed with water and sugar before serving or freezing fruits.

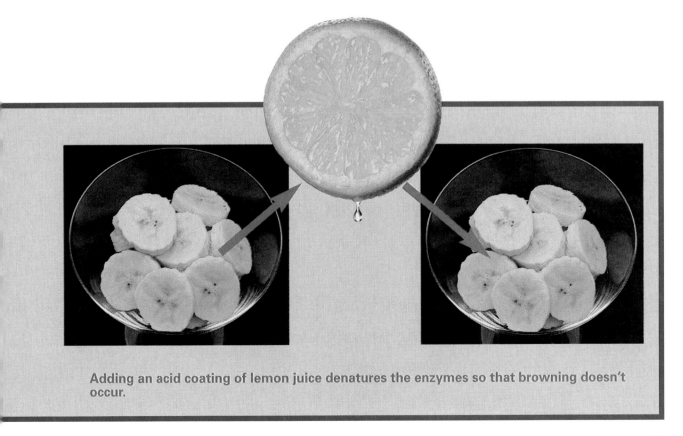

Adding an acid coating of lemon juice denatures the enzymes so that browning doesn't occur.

The enzyme maltase helps make yeast breads rise. What would happen if the dough were not baked?

Baking Yeast Bread

Has anyone in your family ever made a yeast bread dough that failed to rise? If so, something went wrong with an enzyme reaction.

The enzyme maltase helps in baking yeast bread. Yeast cells contain maltase, which acts as a catalyst for the breakdown of maltose to simple sugars. As the simple sugars break down, they produce carbon dioxide. That's what causes the dough to rise. The process stops when the bread bakes, because the rising temperature eventually kills the yeast cells.

Tenderizing Meat

Another use of enzymes is in meat tenderizers. *Three enzymes, which come from the papaya fruit, are diluted with salt to make a dry powder* called **papain** (puh-PAY-in).

The enzymes in papain are used to attack the connective tissue in muscle fiber when sprinkled on meat. By breaking down the muscle fiber, they make your steak more tender to eat. Other fruits that contain enzymes that break down muscle fiber include pineapple and kiwi.

The Impact of Enzymes

Without a doubt, the ability to manage enzyme activity has wide-reaching impact on society. A tomato at the supermarket has better flavor and a longer shelf life because of enzymes. Enzyme reactions help make apple and grape juices clear rather than cloudy. The bread you buy stays fresh longer and has a nicely browned crust and softer texture because of certain added enzymes. Even the center of some chocolate candies stays soft due to enzyme activity. The more you learn about what enzymes do, the more you can appreciate their many contributions to everyday life.

Food science has brought about many benefits. Less tender cuts of meat are typically also less expensive. Consumers can buy and use these economical cuts because of the tenderizing products available. How do these products work?

Enzymes in Foods

     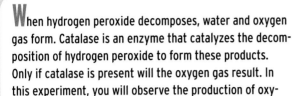

When hydrogen peroxide decomposes, water and oxygen gas form. Catalase is an enzyme that catalyzes the decomposition of hydrogen peroxide to form these products. Only if catalase is present will the oxygen gas result. In this experiment, you will observe the production of oxygen gas when catalase is present. This is the same reaction you've seen if you've ever put hydrogen peroxide on a cut or open wound. In those situations catalase in your blood reacted with the hydrogen peroxide.

Equipment and Materials

3 percent hydrogen peroxide   6, 100-mL beakers   burner or hotplate
three food samples   10-mL graduated cylinder

Procedure

1. Obtain two samples each of three different foods, as directed by your teacher; place each sample in a 100-mL beaker, covering the bottom.
2. To only one beaker of each food, add enough water to cover the sample. Place the beakers with water on a burner and heat until the water boils. Continue heating until the food sample is tender.
3. Remove the cooked samples from the burner and carefully drain off the water.
4. Label the beakers, or set them on paper marked with the contents of the beaker. Use your 10-mL graduated cylinder to add 10 mL of hydrogen peroxide to each of the six beakers containing the food samples.

5. In each case record whether or not bubbling occurred, and if so, to what extent.
6. Dispose of materials as directed by your teacher.

Analyzing Results

1. Which raw sample produced the greatest degree of bubbling?
2. Which cooked sample bubbled the most?
3. Based on your observations, what effect does cooking have on enzymes?

SAMPLE DATA TABLE

Food Sample	Observations: Cooked	Observations: Uncooked

Chapter Summary

- Enzymes are critical participants in chemical reactions.
- Enzymes perform specific functions on specific substrates.
- In many chemical reactions, small molecules called coenzymes participate in the reaction.
- Certain enzymes are active during digestion, breaking down food so nutrients can be used by the body.
- Food scientists need to know what conditions affect enzyme activity in order to control reactions.
- Enzymes play a desired role in baking yeast bread and tenderizing meat.
- The enzyme activity that causes enzymatic browning and food spoilage can be managed in several ways.

Using Your Knowledge

1. How does activation energy affect chemical reactions?
2. Why are catalysts needed in chemical reactions?
3. What happens to catalysts during chemical reactions?
4. What basic function do enzymes have?
5. Why are substrates and enzymes like a lock and key?
6. Why are coenzymes useful in enzyme reactions?
7. How do enzymes help the digestive process?
8. In what ways do food scientists control enzyme activity?
9. What conditions control enzyme activity in food?
10. Why is food blanched prior to freezing or drying?
11. What are four methods for denaturing enzymes?
12. What role does the enzyme maltase play in bread making?

Real-World Impact

Cutting Food Waste

Since feeding the world's people is a serious concern, finding ways to cut down on food waste is critical. Getting food to those who need it can be a challenge. What will a fruit or vegetable be like after it travels for many miles? Newly harvested food must still be edible by the time it reaches a destination. The more food that is wasted, the less there is for consumption.

Thinking It Through

1. What role do you think food scientists can play in preventing food waste?
2. What chemical actions can cause problems when transporting food supplies? What chemical actions might help?
3. Check the Internet and other sources for information on gene alteration that extends the freshness of food. Report your findings.

Skill-Building Activities

1. **Problem Solving.** Suppose you are making gelatin for a family dinner. You add fresh, diced pineapple (which contains protease enzymes that break down protein molecules). Several hours later the gelatin is still liquid. What happened? What would have happened with canned pineapple? Why?

2. **Communication.** Work with a partner to create a one-page magazine advertisement that educates people on the value of enzymes. Possible topics are: enzymes as catalysts; enzymes and temperature; enzymes and food processing.

3. **Problem Solving.** Cut three apples of different varieties into slices. Set them out at room temperature. After 30 minutes check the color of the apples. Which variety of apple would be best for use in making fruit leather (a snack food made from dried, cooked fruit)? Why?

4. **Leadership.** An estimated one million people in the United States are allergic to sulfiting agents, which can cause severe reactions, even death. Locate information on sulfites and report to the class on why you believe they should or shouldn't be used in food preservation. Explain the relationship between enzymes and sulfites in food preservation.

Understanding Vocabulary

Choose three "Terms to Remember" from this chapter. How could you demonstrate what each of these terms means to someone who isn't familiar with them? Write a description of each demonstration.

Thinking Lab

How Do Meat Tenderizers Work?

Enzymes are used in meat tenderizers. As you read, three enzymes are present in the tenderizer called papain. These enzymes attack the connective tissue in muscle fiber.

Analysis

When using a meat tenderizer, you must pierce the meat many times with a fork to carry the enzyme into the muscle of the meat. If this isn't done, the enzyme's ability to tenderize will be reduced. Temperature is also important. Papain has very little effect until the temperature of meat reaches at least 55°C. Tenderizing works best at 80°C but becomes inactive at 85°C.

Thinking Critically

1. What effect might cooking with pineapple have when preparing a less tender cut of beef, such as round steak? Why?
2. Some people marinate a steak for flavor and tenderizing. Why does this work?
3. If you used twice the amount of meat tenderizer recommended on a label, what would you expect to happen to the meat?

CHAPTER 20

Solutions and Colloidal Dispersions

Objectives

- Describe the properties of solutions.
- Calculate the concentration of a solution, using mass percent.
- Describe the properties of colloidal dispersions.
- Identify and describe three types of colloidal dispersions.
- Use examples to explain how solutions and colloidal dispersions exist as foods.

Terms to remember

aggregate
continuous phase
dispersed phase
homogenization
mass percent
phospholipids
saturated
Tyndall effect
unsaturated

What do you think makes grape jelly and grape juice different? Although both are clear, homogeneous mixtures, one is a solution and one is a colloidal dispersion. Which one is the dispersion? If you said the jelly, you're right. This chapter will help you see why, as you learn more about how solutions and colloidal dispersions relate to foods.

Solutions

In thinking about solutions, you may recall that they are homogeneous mixtures of at least two substances—a *solute* that is dissolved in a *solvent*.

Solutes can be solids, liquids, or gases. Examples of solid solutes in foods are salt, sugar, and vitamin C. A fruit syrup, for instance, is a solution of sugar and water. The acetic acid found in vinegar is a liquid solute. Gases can also be solutes. Every time you open a carbonated beverage, you hear the sizzling sound of the carbon dioxide (CO_2) that was dissolved in the liquid.

Although solvents can be solids, liquids, or gases, the solvent in foods is almost always a liquid—usually water.

Both nutritionists and food scientists are interested in how solutions function. They study the many reactions in the body that take place in water solutions. For example, glucose and water-soluble vitamins are in solution when they are carried in the blood. After studying such solutions as fruit juices and other beverages such as coffee and tea, nutritionists and food scientists gather new information, create new methods of production and preservation, and pass along nutrition advice to people.

Most beverages are solutions. What might be the solvents and solutes in these beverages?

To understand solutions, you need to investigate their properties. What happens when different solutes are added to a liquid? The amount of solute and the temperature make a difference.

Comparing Solutions

In a solution, every part of a given sample is the same. You can observe this by taking 1-mL samples from the bottom, middle, and top of a 100-mL sample of salt water. After you evaporate the 1-mL samples and mass the remaining salt, the same amount of salt is found in each sample.

Although each sample of a specific solution has the same composition, samples from another solution may be different. If the procedure just described were repeated with a different salt-water solution, equal amounts of salt would be found in the three samples; however, this amount might not be the same as in the first solution. Thus, there is no set formula for a solution.

Concentration of Solutions

Food scientists often need to know how much solute a solution contains. In other words, what is the concentration of the solution? *To calculate and describe the concentration of a solute in a solution*, food scientists commonly use the term **mass percent**. The equation below shows how to calculate mass percent, referring to the solute as solute A.

$$\text{Mass percent of solute A} = \frac{\text{mass of solute A} \times 100}{\text{total mass of the solution}}$$

If you dissolve 15 g of sodium chloride (NaCl) in 85 g of water, the calculation would be as follows:

$$\text{Mass percent of NaCl} = \frac{15 \text{ g} \times 100}{85 \text{ g} + 15 \text{ g} = 100 \text{ g}} = 15\%$$

Saturation

Suppose someone with a strong sweet tooth adds sugar to a glass of iced tea. The person continues adding sugar until a layer just sits in the bottom of the glass. What has happened?

When first added to the tea, the sugar dissolved. As long as *a solution contains less solute than can be dissolved in it at a given temperature*, it is called **unsaturated**. For any given amount of solvent, however, a point is eventually reached when no more solute will dissolve. *A solution that contains all the solute that can be dissolved at a given temperature* is **saturated**. If the sugar will no longer dissolve in the tea, the tea is saturated. See **Figure 20-1**.

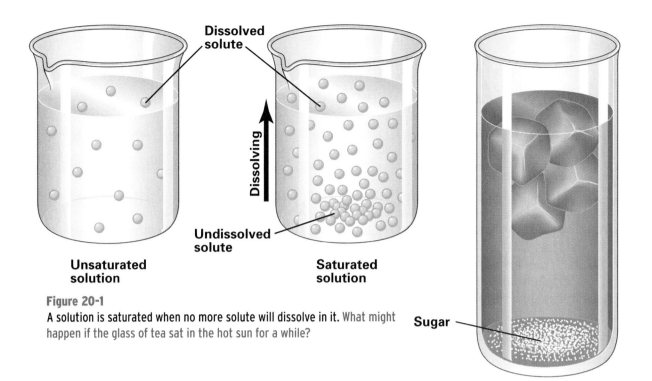

Dissolved solute

Dissolving

Undissolved solute

Unsaturated solution

Saturated solution

Sugar

Iced tea

Figure 20-1

A solution is saturated when no more solute will dissolve in it. What might happen if the glass of tea sat in the hot sun for a while?

Solubility

The term you learned earlier to describe the maximum amount of solute that can be dissolved in a definite amount of solvent at a specific temperature is solubility. The temperature of a solution affects solubility.

If you open a carbonated beverage, will it go flat faster if the can is cold or at room temperature? With higher temperatures, the gases dissolved in a liquid become less soluble. Thus, a warm can of soda will go flat faster than a cold can will. The gases escape more quickly at the higher temperature. This principle also explains why water loses flavor after boiling. The gases that were dissolved in the water are driven off in the heating process.

In contrast, the solubility of most solids dissolved in a liquid increases as the temperature increases. If you have more solid substance in a beaker than the amount of water present will dissolve, you could heat the solution. As the temperature increases, more of the solid will dissolve.

Heating a mixture to dissolve additional solid solute, then cooling it undisturbed, can make a supersaturated solution. A supersaturated solution is one that contains more dissolved solute than it would normally have at that temperature. As you saw in Chapter 15, the crystallization from a supersaturated sugar solution is a step used in candy making.

Heat can change a saturated solution to an unsaturated one. If you dissolve more solute in a heated solution and then cool it, what does it become?

The Solution Is Fluoride

Dental caries (tooth decay) used to be common. During the twentieth century, however, dental caries declined significantly. Fluoride has contributed to this decline.

Fluoride is an ion of the element fluorine. Fluoride helps form enamel, the outer layer of a tooth. Numerous studies show that fluoride helps make enamel more resistant to decay. Bacteria are prevented from dissolving tooth surfaces and penetrating soft tissues below, which reduces dental decay, disease, and tooth loss. Also, recent studies show that fluoride may slow the loss of bone in osteoporosis.

Fluoride ions are often found naturally in water. When the concentration of fluoride in drinking water is greater than one part per million, a noticeable decline in the rate of tooth decay occurs.

Adding fluoride to water supplies. Many communities add fluoride to water. Since tooth decay is considered a major public health problem, the American Medical Association, the National Institute of Dental Health, and the National Cancer Institute all endorse fluoridation of drinking water.

Not every community, however, fluoridates water. Some people believe that adding fluoride ions to the water supply is unnatural; others worry about harmful effects from too much fluoride. If fluoride ions exceed 2.5 parts per million, it is true that the excess can cause mottled, or discolored, enamel. The mottled teeth, however, are not harmed. In fact, they are very resistant to decay.

Other fluoride options. If your water supply isn't fluoridated and you want extra protection for your teeth, you have several options. You can use a fluoride toothpaste, take fluoride tablets, or have your dentist treat your teeth with a fluoridated substance. Dental fluoride treatments are generally given only to those between ages three and sixteen.

Applying Your Knowledge

1. If your drinking water comes from a municipal water system, find out whether it is fluoridated.
2. Do you think water supplies should be fluoridated? Why or why not?

Solutes and Phase Changes

The amount of solute in a solution affects the temperature at which the solution boils or freezes. When a solid is dissolved in a liquid, the solution boils at higher temperatures and freezes at lower temperatures than the pure solvent does. The more solute present in a solution, the higher the boiling point and the lower the freezing point will be.

When making homemade ice cream, you add salt to the ice during the process. By lowering the freezing point, the salt makes the ice cream freeze faster.

Vitamins and Minerals in Solutions

If you were going to boil potatoes, what size would you cut the pieces? Your knowledge of what happens during cooking will affect what you do.

Whenever food is cooked in water, some of the water-soluble vitamins and minerals dissolve. The amount of nutrients lost in the water depends on certain factors. The following are two influences:

• **Surface area.** The greater the surface area, the greater the vitamin and mineral loss. Surface area, of course, depends on the size of the cooking pieces. Thus, a thinly sliced potato will lose more nutrients than if the potato is boiled whole.

Surface area affects the loss of vitamins and minerals when cooking vegetables. Which potatoes, the diced, pieces, or whole, will lose more nutrients during cooking?

• **Cooking time.** The length of cooking time also affects the amount of vitamins and minerals that dissolve in water. Longer cooking times remove more of the vitamins and minerals. Fewer nutrients dissolve during shorter cooking times.

If vitamins and minerals are not deactivated by heat, they remain in the water. Throwing the water away throws away the vitamins and minerals too. If the water is used, however, nutrients can be saved and used. Some recipes suggest mashing potatoes with part of the potato water. Liquids from cooked vegetables can be used to make soup. Both actions save nutrients that would otherwise be lost.

Colloidal Dispersions

Not every mixture allows particles to dissolve as they do in solutions. In the mixtures you know as colloidal dispersions, the particles of one substance are distributed, or dispersed, in another substance without dissolving.

At first thought, a colloidal dispersion can seem much like a solution. What makes it different, however, is mainly the particle size.

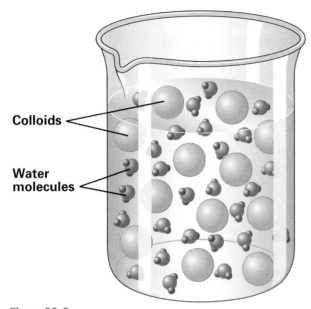

Figure 20-2
The particles in a colloidal dispersion can be large molecules. They can also be groups of ions or molecules, called aggregates. What are they here?

Colloids

Water molecules

In solutions, the solute is present as individual ions or molecules. In colloidal dispersions, the solids are much larger—either very large molecules, such as protein, or aggregates of ions or molecules. See **Figure 20-2**. Because of the large size of protein molecules, any protein in liquid forms a colloidal dispersion. An **aggregate** (AG–ruh–git) is *a group or dense cluster of ions or molecules.* Vitamin B_{12} molecules form aggregates in a colloidal dispersion.

Phases in Colloidal Dispersions

In a colloidal dispersion, the two substances may be a combination of any of these phases: solid, liquid, or gas. For example, an Italian salad dressing is a liquid dispersed in another liquid (vinegar and oil). In a beaten egg white, gas is dispersed in liquid. Examples of other colloidal systems in food are solid in liquid (gravy) and liquid in solid (baked custard).

In all of these systems, *the substance that is dispersed within another* is called the **dispersed phase**. *The substance that extends throughout the system and surrounds the dispersed phase* is called the **continuous phase**.

Properties of Colloidal Systems

Colloidal dispersions have certain properties that help identify them. Some are described here.

- **Particle dispersion.** If you let a colloidal dispersion sit, will any particles settle to the bottom? That won't happen under stable conditions. Even though the particles in a colloidal dispersion are larger than those in a true solution, they are still small enough to remain indefinitely suspended in the liquid. The calcium and magnesium phosphates that are colloidally dispersed in milk, for example, do not settle after the milk has been standing for a while. In this way, dispersions and true solutions are similar.

- **Tyndall effect.** *When a light beam passes through a colloidal dispersion, you can see the path of light. The dispersed particles are large enough to scatter, or deflect, visible light.* This is known as the **Tyndall effect**, as shown in **Figure 20-3**. If you've ever noticed a visible sunbeam shining through a window, you've seen the Tyndall effect. The dust particles suspended in the air scatter the light, making the beam visible. Can you see a beam of light passing through a glass jar of gelatin? You can because gelatin is a colloidal dispersion. Since the individual particles present in a true so-

Figure 20-3
Since light passes through a true solution, you can't see the beam. The light is visible, however, in a colloidal dispersion. Why is that true?

The food industry uses many colloids that have the ability to thicken or gel liquid foods. Pectin is a natural substance that is found in fruits and can be used to produce the gel structure in jelly.

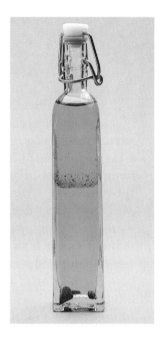

When vinegar and oil are mixed, the two liquids eventually separate. Why does this happen?

lution are too small to scatter the light waves, true solutions don't exhibit the Tyndall effect.

- **Concentration.** In contrast to solutions, the concentration of solids in a colloidal dispersion has almost no effect on boiling or freezing point.
- **Stability.** Even though the particles in some colloidal dispersions don't settle, many colloidal dispersions are unstable. That is, the particles will settle out under certain circumstances. This is because of the size of the particles involved. The curdling of milk, for example, is due to the instability of the protein casein.

In the food industry different kinds of colloidal dispersions exist. The main types you will read about here are emulsions and foams.

Emulsions

Emulsions are one type of colloidal dispersion. As you know, an emulsion is a mixture of two liquids that do not normally blend with each other. Because the two liquids can't combine, they are said to be immiscible.

The two basic types of emulsion are oil-in-water and water-in-oil. When oil is dispersed in water as very tiny droplets, this is known as an oil-in-water emulsion. Many emulsions in food science are of this type.

Emulsions are common in the food industry. Milk, cream, ice cream, butter, processed cheese, and mayonnaise are all examples. Why do you think all of these have an opaque, milky appearance? Because the particles are large enough to deflect light rays from their normal path through the surrounding liquid, they have this characteristic look.

Temporary Emulsions

Vinegar-and-oil salad dressing is a temporary emulsion. By shaking or stirring, you can get the two liquids to mix, but only for a while. Soon the oil is floating on the vinegar.

Why won't these two liquids stay mixed together? The reason is that the molecules of each liquid attract the like molecules of the same liquid, while repelling the dissimilar molecules of the other liquid. As the mixture sits,

How is food kept fresh and safe to eat? Ask a . . . Microbiologist

"The *E. coli* scare in the mid-1990s really focused public attention on safe food processing techniques," says Owen Cramer, a microbiologist for a large food supplier. "Many people became sick and some even died from tainted food. Even though we knew our plant safety and sanitation procedures were excellent, we still took precautions. I spent many extra hours in the lab testing and analyzing food samples for bacteria. We can't have people sick because of mistakes. Peoples' health and our company's reputation are always at stake."

Education Needed: Microbiologists like Owen need a bachelor's degree in microbiology or another applied life science, such as food science. Research and laboratory experience are critical. A background in chemistry helps microbiologists design and carry out experiments and perform calculations. Because they must often explain and interpret their results for people with much less scientific knowledge, microbiologists need strong communication skills.

KEY SKILLS
- Laboratory skills
- Accurate record keeper
- Detail-oriented
- Organizational skills
- Able to work independently
- Communication

Life on the Job: People in this career may work in private industry or for a government agency, such as the USDA. They sometimes work with quality assurance teams to ensure safe and sanitary food production processes. On a typical day microbiologists use and maintain lab equipment as they conduct experiments. Microbiologists are instrumental in avoiding public scares like the one Owen Cramer mentioned.

Your Challenge

Suppose you were a microbiologist for a meat processing company. You evaluate safety in food production.

1. Identify three sanitation concerns.
2. What would you do if your boss told you to disregard the results of an experiment that indicate a possible problem?

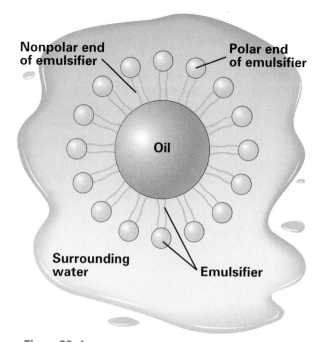

Nonpolar end of emulsifier

Polar end of emulsifier

Oil

Surrounding water

Emulsifier

Figure 20-4

In a sense, an emulsifier acts like glue in an oil-and-water mixture. The polar end of the emulsifier hangs on to the water by dissolving in it. The nonpolar end hangs on to the oil by dissolving in that. Thus, the liquids are held together.

the molecules of each liquid come together. At the same time, the lighter liquid rises, leaving the oil layer floating on top.

Permanent Emulsions

For practical reasons, certain foods need to stand or be stored without the components separating. The substances that you've learned can help with this need are emulsifying agents, or emulsifiers. **Figure 20-4** is a drawing of an emulsifier. Adding an emulsifier to a mixture can keep one liquid permanently dispersed in the other. In other words, emulsifiers keep immiscible liquids mixed.

How does an emulsifier work? An emulsifier is an active compound that absorbs or sticks to the surface of a substance to keep it dispersed. The molecules of an emulsifier usually have a polar end that dissolves well in water and a nonpolar end that dissolves in fat. When these two effects take place, the emulsifier holds the emulsion together.

Oil and water stay together when an emulsifier works in another way. As the mixture is ag-

itated, the oil breaks into tiny droplets. When an emulsifier is added, it coats the droplets so they cannot rejoin.

Emulsifiers in Foods

Many different substances can be used as emulsifiers in food. Egg yolk and starch are two examples.

Egg yolk is a good emulsifying agent for fats or oil in water. Egg yolk contains 10 percent phospholipids (fahs-foh-LIH-puds), which act as natural emulsifiers. A **phospholipid** is *a compound similar to a triglyceride but with a phosphorus-containing acid in place of one of the fatty acids*. A phospholipid called lecithin (LESS-uh-thin) is an emulsifier used to make mayonnaise.

Starch is an emulsifier in such foods as gravies and sauces. Starch molecules, which are long and thin, move between droplets of the dispersed substance (the fat or oil) and prevent them from coming together.

When starch is the emulsifier, evaporation of water can be a problem. These emulsions are often created by heating for long periods of time. If too much water evaporates, the fat concentration becomes so high that the starch molecules can't keep the fat droplets separated, which causes the fat to separate from the emulsion. When this happens, the water that has evaporated must be replaced to remake the emulsion. For example, if fat begins to separate from gravy, adding water while heating and stirring the mixture should cause the gravy emulsion to re-form.

When too much water evaporates from gravy, the fat separates from the emulsion. What could you do to fix this situation?

Common Food Emulsions

Although you may not have realized, hardly a day goes by that you don't consume some type of food emulsion. Here are a few common ones:

- **Milk.** Milk is an emulsion, but without processing, it is unstable. The fat globules can clump together and rise to the top of the milk. Through the process of **homogenization** (ho–mahj–un–ih–ZAY–shun), *fat globules are reduced to a smaller, more equal, size and distributed evenly*. Homogenization increases the stability of milk by preventing the fat from separating. The droplets of fat in homogenized milk are stabilized by what is called the fat globule membrane. This membrane, which surrounds the fat globules, helps prevent the fat globules from joining together. See **Figure 20-5**.

- **Mayonnaise.** Mayonnaise is a familiar example of an oil-in-water emulsion, shown in **Figure 20-6**. Although the ingredients in mayonnaise are liquids, mayonnaise is a solid. It can be made stiff enough to cut with a knife. In making mayonnaise, the oil is dispersed in vinegar, with egg yolk as the emulsifier. When the yolk surrounds the oil droplets, the droplets are immobilized. They can no longer flow as a liquid would.

Likewise, the water present in the vinegar cannot flow because of the immobilized oil droplets scattered throughout.

- **Butter.** Butter is a water-in-oil food emulsion. Butter is made by agitating cream. When the cream is shaken, the fat globules collide with each other and stick together. Eventually the globules of fat become large enough to separate and form butter. Most of the water from the cream is left behind, but some water remains in the butter. According to the legal description, butter must be 80 percent butterfat by weight. Margarine, usually made from vegetable oil, is also a water-in-oil emulsion.

Food Foams

Foams are another type of colloidal dispersion. These dispersions of gas in liquid are actually very complex colloidal systems that often include emulsifiers.

In an egg foam, the egg white is the continuous phase. The air beaten into the white is the dispersed phase. The emulsifier, which keeps the air dispersed in the egg white, is denatured protein.

Homogenization of Butterfat Globules in Homogenized Milk

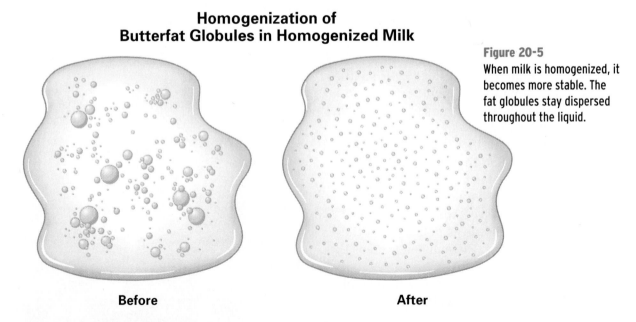

Before

After

Figure 20-5
When milk is homogenized, it becomes more stable. The fat globules stay dispersed throughout the liquid.

With whipped cream, air is beaten into the cream, while protein in the cream serves as an emulsifier.

The more viscous a liquid is, the better the foam. This is why cream creates a better foam than milk.

Not all liquids can make a foam. A liquid must have a low surface tension to form a foam. With high surface tension, liquid molecules are more attracted to each other than to the gas molecules in the air, and the foam won't form. A low surface tension, on the other hand, allows the liquid to surround the gas bubbles. Then the stabilizing agent in the liquid, such as protein in cream, can keep the gas bubbles apart.

Whipped cream is a gas-in-liquid emulsion. If air is the gas and cream is the liquid, what is the emulsifier?

Figure 20-6
In mayonnaise, the egg yolk is an emulsifier. The oil content is at least 65 percent. Why would this much oil be needed?

Oil

Egg yolk emulsifier

Liquid (vinegar)

Temperature and Solubility

When molecules or ions are pulled apart by the molecules of a solvent, the solvent is said to have dissolved the solute. While the rate at which a solute dissolves is always increased by raising the temperature, the amount of a substance that ultimately dissolves—the solubility of the substance—varies from one solute to another. In this experiment you'll compare the effect of temperature on the solubility of two solids, sodium chloride and sucrose, as well as on carbon dioxide gas.

Equipment and Materials

sodium chloride	sucrose	club soda	water
balance	3 test tubes	2, #2 rubber stoppers (optional)	10-mL graduated cylinder
400-mL beaker	burner or hotplate	thermometer	safety goggles

Procedure

1. Mass a 10-g sample of sodium chloride and pour it into a clean test tube.

2. Mass a 10-g sample of sucrose and pour it into another clean test tube.

3. Using your 10-mL graduated cylinder, add 10 mL of room-temperature water (approximately 20°C) to each test tube. Stopper the test tubes or cover with your thumb and shake each tube for 2 minutes to dissolve as much solid as possible.

4. Allow the undissolved solid to settle to the bottom of the test tubes until the liquid above the solid is clear. Measure the height of the remaining solid. Record this information in your data table.

5. Fill a 400-mL beaker half full of water. Place both test tubes in the beaker. Remove stoppers, if used. Place a thermometer in the water in the beaker.

6. Measure 10 mL of club soda using your 10-mL graduated cylinder and add it to a third clean test tube. Record the rate of bubbling in your data table. Place this test tube in the 400-mL beaker with the other test tubes.

7. After putting on safety goggles, heat the beaker on a burner set on high until the water temperature reaches 80°C. Note any changes in appearance as the test tubes and their contents are heated.

8. Remove the test tubes from the beaker and place them in a test tube rack. Use your ruler to measure the height of the solid remaining in any of the tubes.

1. Describe any difference in the behavior of the solutes as the solutions were heated.
2. Which substance(s) seemed to increase in solubility as the temperature rose? Which substance(s) decreased in solubility? Did the solubility of any of the solutes seem to remain the same as the temperature increased?
3. Why is it easier to dissolve sugar in hot tea than in iced tea?
4. Why is it best to store open containers of carbonated beverages in the refrigerator?

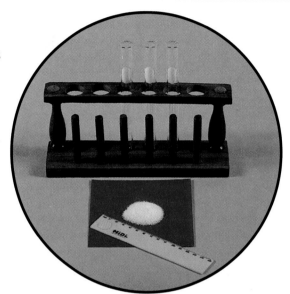

SAMPLE DATA TABLES

Substance	Height of Solid in 20°C Water	Height of Solid in 80°C Water	Observations
Sodium chloride			
Sucrose			

Substance	Rate of Bubbling	Rate of Bubbling	Observations
Club soda*			

*Should contain carbon dioxide

Chapter Summary

- Solutions are homogeneous mixtures in which a solute is dissolved in a solvent.
- The concentration of a solute affects the solution's boiling and freezing points.
- The concentration of a solution can be measured by calculating mass percent.
- Gases become less soluble in solutions when heated, while solids generally become more soluble in solutions when heated.
- Some vitamins and minerals easily become part of a solution.
- Protein molecules are so large that they readily form colloidal dispersions in liquids.
- The Tyndall effect distinguishes colloidal dispersions from solutions.
- Oil-in-water emulsions include salad dressings and mayonnaise; butter and margarine are water-in-oil emulsions; egg foams and whipped cream are gas-in-liquid colloidal dispersions.

Using Your Knowledge

1. Using terms from the chapter, explain what it means when a carton of orange drink says it contains 10-percent orange juice.
2. Compare unsaturated, saturated, and supersaturated solutions.
3. Can more sugar be added to iced tea or hot tea before the solution is saturated? Why?
4. How do the freezing and boiling points of water solutions compare with those of pure water?

Real-World Impact

Hollandaise Sauce

Eggs Benedict is a favorite dish for many special occasions. This dish layers Canadian bacon, a poached egg, and hollandaise sauce over an English muffin. A well-made hollandaise sauce is thick and rich, with a slight lemon flavor. Making hollandaise is very tricky. One reason is that the recipe requires not one, but two, emulsions. The basic ingredients in a hollandaise sauce are: egg yolks, water, butter, lemon, and salt and pepper to taste.

Thinking It Through

1. In making hollandaise, the first emulsion occurs when egg yolks and water are whisked together. Then butter is added to create a second emulsion. During this process, the mixture is slowly cooked. What problems could occur? What precautions should be taken?
2. Suppose you're planning to serve Eggs Benedict for a holiday brunch. If one of your relatives has heart trouble, should you change your plan? Why or why not?
3. Hollandaise can be kept warm for only an hour before using. What do you think happens to the sauce if held longer?

Review and Activities CHAPTER 20

Using Your Knowledge *(continued)*

5. Should the carrots that you're serving as a side dish be cut differently from those you're putting in soup? Explain.
6. Compare solutions and colloidal dispersions.
7. Are water and vinegar immiscible? Why or why not?
8. What is the function of an emulsifying agent?
9. Why do you think eggs are an ingredient in so many recipes?
10. How can you remake gravy after some of the fat has separated?
11. Why is mayonnaise an emulsion?
12. Why does whipped cream remain in the foam state?

Skill Building Activities

1. **Mathematics.** Calculate the mass percent of sugar if 40 g of sugar is dissolved in 50 g of water. Use the formula in the chapter.
2. **Science.** With a partner, plan a demonstration of the Tyndall effect for the class. Be sure your demonstration compares colloidal dispersions with solutions.
3. **Communication.** Interview an adult you know who is an experienced cook. Collect tips on making sauces and gravies. Discuss ingredients, techniques, timing, and temperature. Ask about mistakes and solutions. Report to the class, including scientific principles that support what you learned.
4. **Science.** Soap is an emulsifier. Write an explanation of how the soap used to wash dishes gets them clean.

Thinking Lab

Preparing Solutions

Food science instructors, as well as quality control technicians, must frequently prepare large quantities of solutions of a particular concentration.

Analysis

In evaluating the quality of canned vegetables, food scientists often make use of brine solutions that are 10-20 percent sodium chloride. In preparing these solutions, they know that the density of water is 1.0 gram per mL, which means the volume of water in a mL has the same numerical value as its mass in grams. In other words, 100 mL of water has a mass of 100 g.

Practicing Scientific Processes

Describe how you would prepare enough 10-percent, sodium chloride solution for six lab groups to each use 200 mL of the brine.

5. **Problem Solving.** Suppose you're making a sugar syrup for fruit, but when you combine and stir the sugar and water, the sugar forms a thin layer on the bottom of the container. How can you get the sugar to dissolve?

Understanding Vocabulary

Choose one or two "Terms to Remember" from this chapter. Draw a poster that demonstrates the meaning of the term or terms.

CHAPTER 21 Leavening Agents

Objectives

- Explain the purpose of leavening agents in baked goods.
- Identify natural leavening agents and describe how they work.
- Explain the chemical process by which baking soda and baking powder leaven baked goods.
- Describe the role of yeast in leavening.
- Explain how quick breads are different from other baked products.
- Compare the leavening agents used in different types of cakes.
- Evaluate the nutritional value of specific wheat products.

Terms to remember

baking powder
baking soda
double-acting baking powder
fermentation
quick breads
single-acting baking powder

Nearly every beginning cook can recall at least one cooking effort that turned into a disaster. One new "chef" stared in dismay at the sunken center and outer edges of his freshly baked cake. With good humor, he decided to frost the oddly shaped cake anyway, dubbing it a "volcano cake." What might have caused this mishap?

Proper leavening is the key to creating light, fluffy baked goods—including cakes. To leaven means to lighten. When heat is applied to a batter or dough, bubbles of gas cause the mixture to "inflate." The main gases that leaven baked goods are air, steam, and carbon dioxide. Although one gas may be the leaven in a product, often a combination of gases is at work. Air and steam occur naturally. Carbon dioxide, however, is produced by a chemical or biological process.

Using Air and Steam to Leaven

Air and water are as natural to baked goods as they are to you. Air and the steam produced by water are common gases used to leaven many products.

Air

Air is not the main leavening agent in most baked goods, but it does give many products a "lift." Air is mixed in through many means. When you sift flour, beat fat with sugar, and whip batter, you are adding air.

Air is the principal leavening agent in meringues and angel food cakes. These products are based on egg foams, where beating egg whites introduces air into the batter. The denaturation of protein is crucial here, because the protein traps the air bubbles in the foam. The entrapped air creates a light and fluffy texture.

A cake recipe tells you to beat the batter for a certain length of time. Why should you follow that direction?

Using Carbon Dioxide to Leaven

Carbon dioxide gas (CO_2) is a common leaven. Anyone who has done much baking has used one of these two basic methods for creating carbon dioxide:

- Using common baking products, typically baking soda or baking powder, to create a chemical reaction that yields the gas.
- Using yeast in a biological process to create the gas.

Producing CO_2 Chemically

In most cake and cookie recipes, you'll see a chemical leavening agent in the ingredient list. In the United States, only chemical leavening agents that produce carbon dioxide are used. The two most widely used agents are products you may have seen and used—baking soda and baking powder. These products participate in a chemical reaction that produces carbon dioxide for leavening. See **Figure 21-1**.

Baking Soda

Baking soda is *the chemical compound sodium bicarbonate*. Sodium bicarbonate is a salt with the formula $NaHCO_3$. It is formed from sodium hydroxide (a strong base) and carbonic acid (a weak acid). Because sodium hydroxide is stronger than carbonic acid, the salt formed by their interaction is basic.

The equation for the breakdown of sodium bicarbonate by heat is shown here.

$$2NaHCO_3 \xrightarrow{\text{heat}} CO_2 + Na_2CO_3 + H_2O$$

| Sodium bicarbonate | Carbon dioxide | Sodium carbonate | Water vapor |

Steam

All recipes for baked goods include some liquid. The water in the liquid produces steam when heated. Recipes that use steam as a primary leaven require a very hot oven. The temperature must usually be at least 204°C to convert the water to steam.

As in other baked goods, eggs and gluten provide the structure of steam-leavened products. Steam forms, and the batter expands around it. Baking coagulates the protein, setting the structure. You may have noticed the large pockets that form inside a popover or cream puff. These were created by air and steam.

Cream puffs must be baked until they are well browned and crusty or they will collapse. What would cause a collapse to occur?

How a Cake Is Leavened

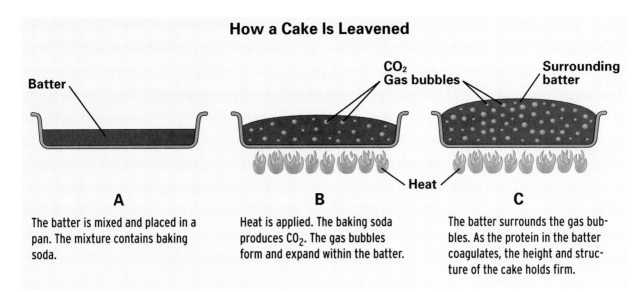

A

The batter is mixed and placed in a pan. The mixture contains baking soda.

B

Heat is applied. The baking soda produces CO_2. The gas bubbles form and expand within the batter.

C

The batter surrounds the gas bubbles. As the protein in the batter coagulates, the height and structure of the cake holds firm.

Figure 21-1

As you can see, baking soda releases sodium carbonate as well as carbon dioxide when heated. Sodium carbonate is not a welcome addition to your favorite baked items. It gives a bad taste and a yellowish color.

Why, then, doesn't baking soda ruin a recipe? Baking soda is always used with an acid, which alters the chemical reaction to prevent sodium carbonate from forming. Some acid ingredients that can be used with baking soda are buttermilk, vinegar, lemon juice, molasses, honey, fruits, fruit juices, and cream of tartar.

As an example, suppose a recipe that uses baking soda also asks for cream of tartar. Cream of tartar is an acidic compound called potassium bitartrate. Its formula is $KHC_4H_4O_6$. When cream of tartar is added, the breakdown of the baking soda becomes a two-step reaction, as shown in **Figure 21-2**.

Many recipes call for ingredients such as these. What purpose do they serve?

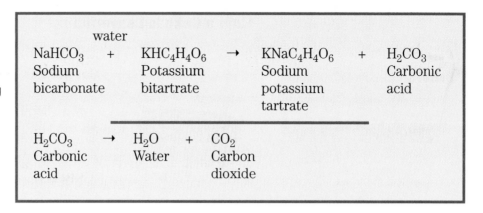

Figure 21-2
Some recipes combine baking soda with cream of tartar. What is the chemical reason for this combination?

$$\text{NaHCO}_3 \ + \ \overset{\text{water}}{\text{KHC}_4\text{H}_4\text{O}_6} \ \rightarrow \ \text{KNaC}_4\text{H}_4\text{O}_6 \ + \ \text{H}_2\text{CO}_3$$

Sodium bicarbonate   Potassium bitartrate   Sodium potassium tartrate   Carbonic acid

$$\text{H}_2\text{CO}_3 \ \rightarrow \ \text{H}_2\text{O} \ + \ \text{CO}_2$$

Carbonic acid   Water   Carbon dioxide

In the first step, the baking soda reacts with the cream of tartar to produce a salt, sodium potassium tartrate, as well as carbonic acid. In the second step, the carbonic acid breaks down into carbon dioxide and water.

Thus, combining baking soda and cream of tartar produces the carbon dioxide needed for leavening the baked product. The sodium potassium tartrate and other remaining products don't affect flavor or cause the other undesirable side effects of sodium carbonate.

Baking Powder

Baking powder is *a leavening compound that contains baking soda, dry acids, and starch or some other filler.* The filler in baking powder, usually cornstarch or calcium carbonate, absorbs moisture in the air, which helps prevent a chemical reaction from taking place too soon. You can make baking powder by blending baking soda, cream of tartar, and cornstarch, but most people buy it at the store.

Baking powder comes in two types: single-acting and double-acting.

- **Single-acting baking powder.** *As soon as liquid is added to* **single-acting baking powder***, carbon dioxide starts to be released.* Water, milk, potato water, orange juice, or other liquids may begin this process. This quick reaction occurs because the acid in baking powder is soluble in a cold liquid. Homemade baking powder is single-acting, but this type is rarely sold in the United States.

- **Double-acting baking powder.** This baking powder is the type preferred in the United States. **Double-acting baking powder** *contains two acids—one that reacts with cold liquid and one that reacts with heat.* Some carbon dioxide is released as soon as liquid is added, but most is produced as the batter heats in the oven. The double-acting baking powder you find in stores is often a combination of sodium bicarbonate and phosphate compounds. One common version contains sodium bicarbonate, monocalcium phosphate, or $\text{CaH}_4(\text{PO}_4)_2$, and sodium aluminum sulfate, chemically known as $\text{NaAl}(\text{SO}_4)_2$.

Some beginning cooks don't realize that baking soda and baking powder are two different products. Do baking powder recipes require an acid ingredient? Why or why not?

FOOD SCIENCE
Careers
Chef

Why has dining out become so popular? Ask a . . . Chef

A s the executive chef in a big-city restaurant, Jeremy Grover does more than take care of business; he's a public relations person too. "The restaurant business is very competitive," Jeremy says, "so we work hard to keep our customers coming back. Today's diners are very sophisticated, so first of all, I have to make sure we serve first-rate food. But good restaurants are everywhere these days. That's why chefs with personality are so valuable. I get to know my customers. I also do public cooking classes and donate my time to charity events by preparing food and doing cooking demonstrations. It's a great way to draw customers and make our restaurant stand out."

KEY SKILLS
- **Eye for detail**
- **Creativity**
- **Organization**
- **Oral communication**
- **Supervisory skills**

under the direction of an established executive chef.

Life on the Job: Chefs work long hours in a fast-paced environment. In addition to overseeing food preparation, the executive chef implements safe food handling procedures and trains and supervises the kitchen staff. At smaller restaurants, the executive chef may do menu development, purchasing, and even market research.

Education Needed: Chefs generally hold a two-year or four-year degree in culinary arts. Some chefs specialize in baking and pastry making, while others pursue a more general culinary degree. A chef must also take courses in flavor evaluation, marketing, finance, and even foreign language. Chefs also complete an apprenticeship at a restaurant

Your Challenge

Suppose you have been hired as the executive chef for a new restaurant in your town.

1. What are five principles you've learned in food science that would help you set up and manage your kitchen?
2. What would you do to distinguish your restaurant from others?

The Strength of Baking Powder

Baking powders vary in the amount of carbon dioxide they release. By federal law, baking powder must yield at least 12 g of carbon dioxide for every 100 g of powder (12 percent). Most powders for home use are 14 percent, while the food industry has higher levels available. Since the carbon dioxide is formed from the breakdown of the baking soda, baking powder must be at least 25 percent baking soda by volume to yield the carbon dioxide needed.

When baking, not just any amount of baking powder will do. Too much can cause the walls of the flour mixture to stretch too far, break, and collapse. Too little will mean a compact product. If not stored with a tight cover, baking powder can absorb moisture and start reacting early. The resulting loss of some carbon dioxide can make the powder less effective.

Ammonium Bicarbonate

Baking soda and baking powder are the most common chemicals used to produce carbon dioxide during baking. However, other substances are used for this purpose too. One is ammonium bicarbonate, known as baker's ammonia. As it decomposes, ammonium bicarbonate produces carbon dioxide according to the following equation:

$$\overset{heat}{NH_4HCO_3 \rightarrow NH_3 + CO_2 + H_2O}$$

| Ammonium bicarbonate | Ammonia | Carbon dioxide | Water vapor |

Although it effectively produces carbon dioxide, ammonium bicarbonate is used mostly in crackers and certain types of cookies. Why do you think its use is so limited? The ammonia gas produced may affect the taste of the baked good. In thin products that have a large surface area, the unpleasant tasting gas can escape completely.

Producing CO$_2$ with Yeast

Yeast is a microscopic organism that produces carbon dioxide through **fermentation** (fur-men-TAY-shun). Fermentation is *a biological reaction that slowly splits complex organic compounds into simpler substances.* You will learn more about this process in Chapter 22.

During fermentation, yeast converts sugar, usually glucose, into ethyl alcohol and carbon dioxide. The chemical equation for the reaction is given below.

$$\overset{yeast}{C_6H_{12}O_6 \rightarrow 2C_2H_5OH + 2CO_2}$$

| Glucose | Ethyl alcohol | Carbon dioxide |

The ethyl alcohol evaporates during baking, while the carbon dioxide causes the product to rise. The yeast is killed by the high baking temperature, which stops fermentation.

The Origins of Yeast

Yeast is found throughout the environment. Thousands of years ago, perhaps a little yeast landed on some bread dough and started to ferment. The dough would have begun to rise, forming a lighter bread than had been known before. "Yeast power" was soon harnessed.

Most of the carbon dioxide from the baking powder in these biscuits will be released by the oven heat.

Once the discovery of yeast was made, bread products started to become more varied. By experimenting with recipes and the addition of organisms besides yeast, new flavors were developed for breads.

The ancient Egyptians at first leavened bread by using a yeasty piece of leftover dough. Eventually, they achieved the same result using beer froth, which also contains yeast. The word "yeast" originally meant the froth or sediment of a fermenting liquid.

These two techniques were continued and sometimes modified by later civilizations. At harvest time in ancient Rome, grape juice was mixed with wheat bran or with the grain millet. These mixtures were fermented to develop yeast growth, then sun-dried in cakes. The cakes were soaked in water as needed for leavening. The Romans valued raised bread as a particularly nutritious, body-building food.

Likewise, beer froth remained the preferred leavening in England in the 1700s. The froth was dried slowly. When needed, it was mixed with water, flour, and sugar and left to stand for a day.

Leavened bread made from leftover dough was more common, but less popular, among English seafarers. The yeast and other microbes that grew in the stored bread produced a sour taste. Less than a century later, however, bread made with this technique was becoming a tradition of the American West. Gold miners made sourdough bread from a starter, keeping the yeast alive from one batch of bread in order to start the next one.

Today, sourdough bread contains other organisms besides yeast, including the bacteria *Lactobacillus* (LAK-toh-buh-SI-lus). These organisms give sourdough bread its distinctive flavor.

Making Leavened Products

Meals and snacks are routinely made more appealing with leavened products. Using leavens to create these products is a skill that builds with practice and understanding.

Batters and Doughs

When baking, leavens are used to make many different products. Recipes produce both doughs and batters that use all of the leavening methods you've read about. Batters contain more liquid than doughs.

- **Pour batters.** Pour batters are made with a nearly equal ratio of flour to liquid, usually water. These batters range from thin to hard-to-pour. Salt, if it's added, generally

Nutrition*Link*

Why Eat Wheat?

As early as the fifth century B.C., the Greek historian Herodotus described Egyptian bread baking in his writings. Today nearly one-third of the world's population depends on wheat for nourishment, making it a highly valued grain. Many baked products, including most breads, are made with wheat flour.

Types of wheat. Ninety percent of all wheat consumed today belongs to one of three main groups—common, club, and durum. Common wheat is best for making bread. Club wheat is processed into flour for general baking purposes. Durum wheat is used to make pasta.

You may hear wheat-based foods called "starchy." That's because wheat contains mostly starch grains surrounded by protein. Although wheat has high protein value, the protein is incomplete. To provide the body with all the amino acids needed, wheat proteins must be combined with those in legumes or seeds.

The value of whole wheat. A wheat kernel is made of three parts, the endosperm, the bran, and the germ, which you can see in **Figure 21-3**. The *endosperm* is high in complex carbohydrates and proteins, with just a small amount of vitamins and minerals. The *germ*, which is the sprouting section of the kernel, contains B vitamins, vitamin E, iron, zinc, and some protein and trace minerals. *Bran*, the outer layer of cells, contains most of the water-soluble B vitamins and iron found in wheat. Bran cells have thick walls of cellulose and hemicellulose, which humans can't digest. This makes bran an excellent source of fiber in the diet. Scientists continue to explore the link between fiber-rich diets and lowered risk of certain diseases, including some types of cancer.

Typical white breads are made from wheat that has the germ and bran removed. Only products made from wheat with the bran and germ intact can use the term "whole wheat" on the package label. Otherwise the label will just say "wheat." When buying bread, look for the word "whole" to be sure your choice is a nutritious one.

Nutrient density. Bread is sometimes called a nutrient dense food. It provides a relatively high quantity of nutrients compared to the number of kilocalories. Still, breads vary greatly in nutritional value, depending on their ingredients. Even the most nutrient-packed bread should make up only part of your diet, however. Eating moderate amounts of a variety of healthful foods is the cornerstone of good nutrition.

Applying Your Knowledge

1. What type of nutrient are starches? How do they contribute to bread's nutritional value?
2. What qualities besides nutrition might food scientists and recipe developers consider when choosing a type of wheat for a product?

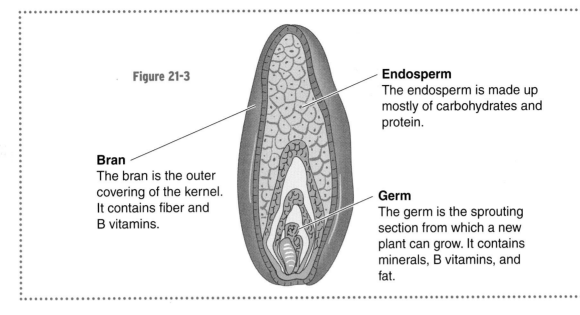

Figure 21-3

Endosperm
The endosperm is made up mostly of carbohydrates and protein.

Bran
The bran is the outer covering of the kernel. It contains fiber and B vitamins.

Germ
The germ is the sprouting section from which a new plant can grow. It contains minerals, B vitamins, and fat.

equals 2.5 mL for every 240 mL of flour. Steam is the main leavening agent in some pour batters, such as those used to make popovers and cream puffs. The batter for these products is the only one that has enough water to make the steam needed to puff up the final product. Waffles and pancakes are also made from pour batters.

- **Drop batters.** To make muffins, quick bread loaves, and some cookies, you use a drop batter, mixing about two parts flour to one part liquid. Baking powder and steam are both leavening agents in these recipes.

- **Soft doughs.** Mixing three parts flour to one part liquid gives a soft dough. Soft doughs need more mixing to develop gluten. Some must be kneaded to develop the gluten sufficiently. Yeast breads, pizza crusts, and baking powder biscuits are made with soft doughs.

- **Stiff doughs.** Stiff doughs use six to eight times as much flour as liquid. The drier consistency is needed for certain products.

Pizza crusts are made from soft doughs. Why is kneading necessary?

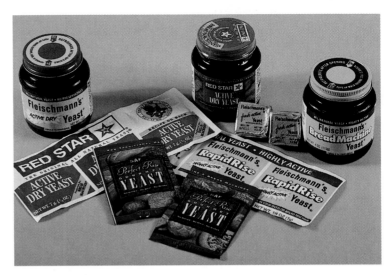

Yeast comes in different forms. Active dry yeast can usually be stored for about a year. Typically, a date is provided on the package.

Because all flour mixtures contain some water and are usually heated so that the water vaporizes, steam leavens all flour mixtures to a certain degree. Even the stiff dough of pie crusts and some cookies will leaven slightly with steam.

Making Yeast Products

Yeast bread is made from soft dough. The yeast used is *Saccharomyces cerevisiae*. (SA-kuh-roh-MY-seez ser-uh-VI-see-eye). It's sold in both granular and compressed forms. The granular or pellet form is stable for long periods at room temperature due to its low moisture. It's usually less than 9 percent water by weight. The compressed yeast contains considerably more water—about 70 percent water by weight—and should be refrigerated or frozen.

Besides the yeast, the basic ingredients in these breads are flour, liquid, and salt. Each ingredient has specific purposes.

- **Flour.** The flour in bread provides starch as well as the proteins that produce gluten. Flour varies in its ability to make gluten. For yeast bread, flour with a strong gluten quality produces bread with high volume and fine texture. This is because the gluten traps the starch grains in the flour when the dough is mixed. Then as the bread bakes, these starch grains surround the gas cells and form the solid structure of the bread.

- **Liquid.** The liquid in bread making is usually water or milk. Milk is often chosen because it provides nutrients. Milk also helps bread stay fresh longer. In bread making, liquid provides a medium that dissolves other ingredients and transports them to the yeast cells. During baking, the steam from the liquid combines with the carbon

Yeast produces carbon dioxide, which leavens many breads. If the gas production is poor for some reason, the result will be disappointing.

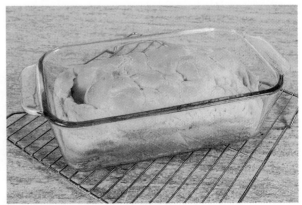

You may have eaten muffins that were full of tunnels inside. What caused this problem?

dioxide produced by the yeast, which helps the bread expand and rise.

- **Salt.** Salt not only adds flavor but also prevents enzymes from breaking down protein. A yeast dough made without salt would be very sticky and difficult to work with. As the amount of salt is increased, carbon dioxide production decreases, creating a more compact product.
- **Sugar, fat, and eggs.** Additional ingredients can be added to yeast bread. Sugar helps the crust brown and can be added for flavor. Fat makes bread tender. Adding eggs makes the bread richer in texture and flavor.

Baking successfully with yeast can be a challenge. Those who have mastered the art are admired, both for their knowledge and the delicious products they create. In the next chapter, you'll learn more about how managing the fermentation process leads to greater success in baking yeast breads.

Making Quick Breads

Watching yeast bread transform from a mass of dough into an airy puff can be impressive. The process takes time, however. As an alternative to yeast breads, **quick breads** *need no time to rise. These products are made with a leavening agent other than yeast, usually steam or carbon dioxide that is produced with baking powder or baking soda.* Muffins, coffee cakes, cake-like breads, popovers, and biscuits are all examples of quick breads.

All quick breads contain flour, liquid, and salt. In addition to the chemical leavening agent, they may also contain fat, sugar, eggs, and other ingredients.

Different products require specific mixing methods. Proper mixing becomes an art in making light and delicate quick breads. With muffins, for example, a quick mix is the secret. These drop-batter recipes usually say "Mix ingredients only until moistened" or "Batter will be lumpy." Stirring past this point overdevelops the gluten protein in the flour, producing a tough product with tunnels.

Making Leavened Cakes

What makes a chocolate layer cake moist and tender, while an angel food cake is tall, light, and airy? A skilled baker gets some of the credit. The leavening used also affects a cake's texture and other properties. Cakes may use any of the leavening agents you've read about.

Sheet cakes and layer cakes, traditional favorites at birthday parties, are most often *shortened* cakes. They contain fats (familiarly called shortening) and usually a chemical agent as a primary leaven. Air is a minor leavening agent in shortened cakes. It is added when the fat and sugar are beaten together and when the ingredients are blended. Air is also added if the recipe calls for eggs to be beaten separately. Fat and eggs tenderize the batter so it expands easily.

Shortened and unshortened cakes are not the same—in ingredients and leavening. What are some differences?

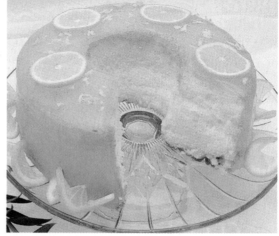

Unshortened cakes are also called foam cakes. A basic angel food cake is cholesterol- and fat-free. Sponge cakes, however, contain egg yolks and sometimes butter. Chiffon cakes are low in cholesterol but have oil.

EXPERIMENT 21

Using Baking Powders to Produce Carbon Dioxide

SAFETY FIRST

Review these safety guidelines before you begin this experiment.

Baking powder releases carbon dioxide gas. In this experiment, you will identify sources of carbon dioxide and compare the approximate amounts of CO_2 produced by each.

Equipment and Materials

2 different brands of baking powder	baking soda	cream of tartar
pH indicator paper with 1-11 range	albumin solution	2, 50-mL or 100-mL graduated cylinders
3, 250-mL beakers	electronic balance	shallow saucepan or waterbath
plastic ruler		

Procedure

1. Place 15 mL of albumin solution in each of three, 250-mL beakers.

2. On squares of paper, mass 3.5 g of two different brands of baking powder.

3. Mass 2 g of sodium bicarbonate and 3.9 g of cream of tartar. Mix them to form baking powder.

4. Place tap water in a shallow container to a depth of about 2.5 cm. Bring water to a simmer. Wear safety goggles during heating.

5. Pour the baking powders into the three, 250-mL beakers at the same time (one powder in each beaker, marked for identification). Stir quickly but only enough to disperse the baking powder. Check pH of foam.

6. At 1-minute intervals for 5 minutes, measure the height of the foam in each beaker. After the final height measurement, determine the liquids' pH with pH paper. Record this information.

7. After 5 minutes, place the three beakers in the shallow container of simmering water.

8. Heat the beakers for 5 minutes. Measure the height of the foam each minute, and observe changes during that time. Again, test the liquids with pH paper after the final height measurement to determine pH. Record this information in your data table.

Pound cakes are primarily leavened with air and steam. Air is incorporated when the batter is mixed. Heat changes the liquid in the batter to steam. The steam enlarges the air bubbles created by mixing and causes the cake to rise.

Unshortened cakes, including sponge, chiffon, and angel food cakes, depend on the air beaten into an egg foam for leavening. Steam is an additional leaven. An angel food cake expands more than other cakes because of the large number of egg whites in the recipe, as many as twelve. Besides providing protein, the egg whites contain a large percentage of water that converts to steam, creating a tall, light cake.

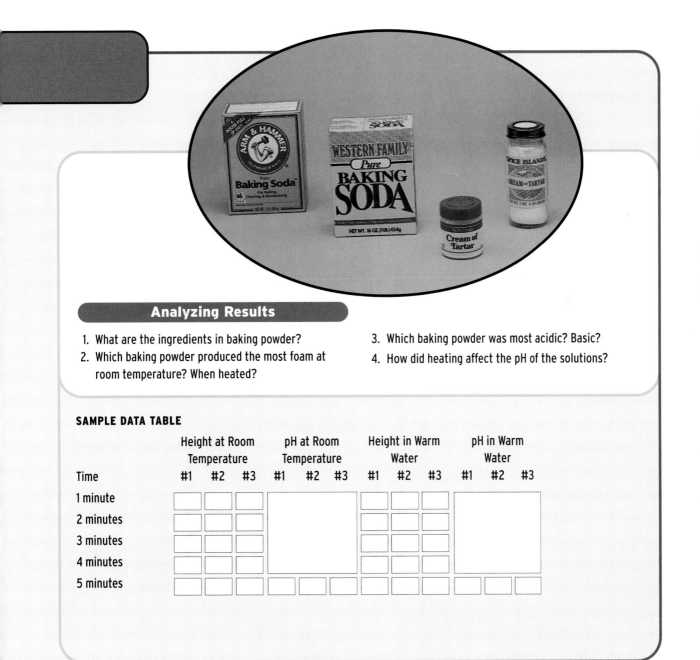

Analyzing Results

1. What are the ingredients in baking powder?
2. Which baking powder produced the most foam at room temperature? When heated?
3. Which baking powder was most acidic? Basic?
4. How did heating affect the pH of the solutions?

SAMPLE DATA TABLE

Time	Height at Room Temperature			pH at Room Temperature			Height in Warm Water			pH in Warm Water		
	#1	#2	#3	#1	#2	#3	#1	#2	#3	#1	#2	#3
1 minute												
2 minutes												
3 minutes												
4 minutes												
5 minutes												

CHAPTER 21 Review and Activities

Chapter Summary

- Leavening uses the production of gas to cause baked goods to rise.
- Leavening agents can occur naturally or be produced by chemical and biological processes.
- Baking soda is combined with acid to produce carbon dioxide in a two-step process.
- Baking powder is made from baking soda, dry acid, and a filler.
- Depending on composition, baking powders may be single- or double-acting.
- Yeast ferments sugar to produce the carbon dioxide used in leavening.
- Correct temperatures and ingredients are keys to using yeast successfully.
- Quick breads offer a faster baking alternative than yeast breads.
- A shortened cake is made with a fat, such as shortening.

Using Your Knowledge

1. What are the three main leavens in baked goods?
2. How do each of the two natural leavens work?
3. What are two methods for producing carbon dioxide for leavening?
4. How does baking soda work as a leaven?
5. What would happen if you left out the lemon juice in a recipe that called for lemon juice and baking soda?
6. Compare single-acting baking powder with double-acting.
7. How much baking soda is in a recipe that calls for 2 teaspoons of baking powder? How do you know this?
8. Why does the amount of baking powder used in a recipe matter?
9. Describe how yeast produces carbon dioxide for leavening.

Real-World Impact

Bread Machines

More and more families are coming home to the taste and aroma of fresh, home-made, yeast bread. These families haven't hired a baker; they've bought a bread machine. Affordable technology lets them enjoy homemade loaves without hours of preparation. The machine has a well for the ingredients and dials that indicate time and desired browning. After you choose the setting, the bread machine mixes the ingredients, kneads the dough, monitors its rising, and bakes the bread. You return a few hours later to find a warm, freshly baked product.

Thinking It Through

1. What are the possible advantages and disadvantages of a bread machine? Would this be a wise purchase for your family?
2. Using library resources, learn about the technology used in bread machines. What role do computers play?
3. Compare a conventionally prepared yeast bread recipe with one to be used with a bread machine. What might account for any differences you note?

Using Your Knowledge *(continued)*

10. What is the scientific basis for the ancient Romans' choice of grape juice and a grain product for growing yeast?
11. Should yeast be refrigerated? Why or why not?
12. Why do batters need to be thin when steam is the main leavening agent?
13. Are cakes leavened with one or more leavening agents? Explain.

Skill-Building Activities

1. **Problem Solving.** Suppose you and Audrey are making a quick bread. The recipe says to mix dry ingredients and liquid ingredients separately before combining the wet with the dry. Audrey says that's a waste of time and starts to put the baking powder in the milk first. What would you point out?
2. **History.** Yeast was used to leaven long before chemical agents came on the scene in the mid-1800s. What kind of impact do you think chemical leavening agents had on bread baking?
3. **Problem Solving.** At high altitudes the carbon dioxide gas generated by yeast and baking powder encounters less resistance from surrounding thin air, giving the CO_2 more leavening power. What could you do to compensate for this?
4. **Critical Thinking.** While making popovers, Ben opened the oven a couple times to see how they were doing. At the end of the baking time, he discovered that the popovers had collapsed. What do you think happened?
5. **Mathematics.** About how much flour by volume would you need to prepare the following recipes? a) a drop batter that uses 125 mL of water; b) a soft dough made with 175 mL of water; c) a pour batter that contains 250 mL of water.

Thinking Lab

Making a Better Batter

Have you ever made a recipe and thought: This recipe would be better if . . .? Most recipes that you come across were developed, intentionally or by trial-and-error, to produce the best possible product. Just the same, a little tinkering with the ingredients or their amounts may better results.

Analysis

Breads, of course, vary greatly in flavor, texture, and density. If you, as a food scientist, found that a particular yeast bread recipe yielded a product with a delicious flavor but a heavy consistency, you could take steps to make an improvement.

Thinking Critically

1. What do you think is the problem with the recipe described?
2. How would you determine whether your theory is right?
3. What changes would you make to try to correct the problem?

6. **Problem Solving.** Suppose you are making a cake and the recipe calls for baking soda and vinegar. You are out of vinegar. What could you do?

Understanding Vocabulary

With a team of students, gather different types of leavening products. Include those listed in "Terms to Remember" for this chapter. Show these products to the class, explaining each one and its uses.

CHAPTER 22

The Fermentation of Food

Objectives

- Describe the history of fermentation.
- Explain why food is fermented.
- Compare respiration in human metabolism to anaerobic respiration in food science.
- Explain what causes fermentation.
- Summarize information on bacterial fermentation.
- Describe how yeast fermentation works in bread making.
- Explain the value of molds and enzymes in food production.
- Describe how various fermented beverages are made.

Terms to remember

aerobic
anaerobic
anaerobic respiration
brine
brine pickling
cell respiration
fresh-pack pickling
indigenous
microbes
microorganisms
pasteurization

What food or beverage first comes to mind when you hear the word fermentation? Many people know that pickles are fermented, but did you realize that cheese, yogurt, and even chocolate are all fermented foods?

Fermentation in History

Fermentation has been used in food processing since before recorded history. Thousands of years ago, tribes of nomads noticed that under certain conditions milk would change to solid cheese or semisolid yogurt. Although the nomads didn't know it, fermentation was the cause. The odds are this happened first by accident, but that's exactly how many important scientific discoveries have been made.

For centuries people have made alcoholic beverages by using yeast to ferment fruit juices. In the 1850s, the French wine industry was in serious trouble because their wine was spoiling. Emperor Napoleon III called in scientist Louis Pasteur for help.

Pasteur knew that yeast caused the fermentation used in wine making. He discovered that certain bacteria in the wine, however, were also fermenting and spoiling the wine. While yeast created alcohol, these bacteria produced vinegar instead. Pasteur suggested that wine makers heat the wine for a short time to destroy the bacteria.

At first, wine producers were horrified by Pasteur's idea. They soon found, however, that it worked. Known as **pasteurization**, *the process of using heat treatment to destroy bacteria* is still used today, especially for milk.

As you can see, fermentation has a long history in human civilization. It remains a topic well worth exploring today.

The delicious taste of chocolate is a result of fermentation.

- Fermentation makes some food more usable. Few people would like the taste of chocolate or coffee if they weren't fermented and processed before eating.

Making Use of Fermentation

Have you ever stored a container of leftover fruit in the refrigerator for too long? If so, you probably noticed a strong aroma after opening the container. The fruit had begun to ferment. In this case, fermentation caused a problem; in other situations fermentation is very useful.

Through fermentation, certain foods you eat are safely preserved and prepared. Food is fermented for the following basic reasons:

- Fermentation extends the time food can be stored without spoiling.
- Some food is more enjoyable to eat when fermented. Wild rice, for example, is easier to chew.

The Fermentation Process

When the cells of your body need energy, they get it through the metabolic process of **cell respiration**. This *multistage process releases energy from glucose when the glucose molecule in the cell is taken apart*. Oxygen is required during the last stage of respiration. A chemical reaction such as respiration, which *must take place in the presence of air or oxygen*, is called **aerobic**. See Figure **22-1**.

Figure 22-1
An aerobic reaction requires oxygen, but an anaerobic reaction doesn't. Which of these applies to fermentation? Why?

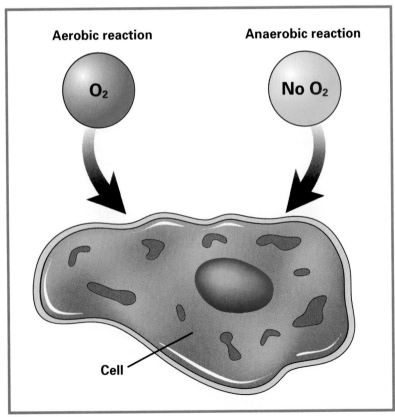

Aerobic reaction

Anaerobic reaction

O_2

No O_2

Cell

Some organisms, called microorganisms, are so small they can be seen only through a microscope. What role do microorganisms play in fermentation?

Just as you need energy, other living systems do too. Sometimes in food science, however, no oxygen is available or oxygen is withheld. If cells are deprived of oxygen, the process of respiration can't be completed. Fermentation can produce energy without air or oxygen. A reaction such as fermentation, which *can occur in the absence of oxygen,* is called **anaerobic**. This is why fermentation is sometimes called **anaerobic respiration**, meaning *respiration that occurs without oxygen*. Since the early stages of respiration can occur without oxygen, respiration can begin in an oxygen-deprived cell. Without oxygen, however, respiration cannot continue, so fermentation reactions take over to generate needed energy.

As you know, fermentation splits complex organic compounds into simpler substances. During this chemical reaction, sugar (usually glucose) is changed to carbon dioxide and alcohol or various organic acids.

What causes this process to occur in foods and beverages? Fermentation is the result of the anaerobic respiration of microorganisms. **Microorganisms** *are single cells of microscopic size that cannot be seen by the human eye but can be studied through a microscope. Microorganisms are also called* **microbes** (MY-krohbs).

The microorganisms that ferment food can be bacteria, yeast, and mold. Some foods are produced by a combination of these microorganisms. Soy sauce, coffee, Camembert cheese, and many sausages are in this category.

Bacterial Fermentation

Because of bacteria's common link to illness, not everyone thinks kindly of them. Without bacteria, however, such foods as pickles, sauerkraut, and yogurt wouldn't exist.

Of the many types of bacteria that can cause fermentation, the following four types are discussed in this text:
- *Lactic-acid bacteria* are commonly used to ferment food in order to produce pickles, sauerkraut, and other foods.
- *Acetic-acid bacteria* are also commonly used for fermentation in the vinegar-making process.
- *Carbon-dioxide bacteria*, discussed in the next chapter, are used in making Edam, Gouda, and Swiss cheeses.

Yogurt is a milk product. What kind of bacteria is used in the fermentation process that produces yogurt?

sample of milk are the best ones for making yogurt. Therefore the milk is heated to kill all bacteria. Then the producer chooses a particular microbe to add to the milk.

In some food products, lactic-acid bacteria are added. The scientific names for the bacteria added to yogurt are *streptococcus thermophilus* (strep-toh-KAH-kus thur-MAH-fuh-lus) and *lactobacillus bulgaricus* (LAK-toh-buh-SI-lus bul-GAR-ih-cus). Adding these bacteria allows yogurt producers to have more control over what the end product is like.

The microbes that are added to make yogurt and other fermented products are produced commercially. They have been preserved through drying, freeze-drying, or concentrating and freezing. The uniform quality of microbes that are commercially prepared helps produce quality foods.

- Less-common bacteria used in food fermentation are *proteolytic bacteria* (pro-tee-uh-LI-tik). These differ from other fermenting bacteria in that they break down protein rather than carbohydrate. Proteolytic bacteria are used in the production of cocoa and chocolate for candy and beverages.

Sources of Bacteria

Some bacteria occur naturally on plants used as food. For example, cucumbers and cabbage have bacteria on them when they are harvested. The microorganisms found on foods such as these are called native, or indigenous (in-DIJ-uh-nus), microorganisms. **Indigenous** means *found naturally in a particular environment*. Indigenous bacteria are used in making pickles from cucumbers, and sauerkraut from cabbage.

In other foods, indigenous bacteria may not be present or useful. For instance, it's impossible to know whether the bacteria in a given

Lactic-Acid Bacteria

Lactic-acid bacteria are used in the production of other foods besides yogurt. Sour cream, cottage cheese, cheddar cheese, dill pickles, olives, sauerkraut, and vanilla are all examples. Lactic-acid bacteria carry out the same reaction that takes place in muscle cells. In both cases, glucose is converted into lactic acid and energy. The equation for the reaction is shown here.

$$\overset{\text{bacteria}}{G_6H_{12}O_6 \quad \rightarrow \quad 2HC_3H_5O_3 \quad + \quad \text{Energy}}$$
$$\text{Glucose} \qquad\qquad \text{Lactic acid}$$

Careers

Research Chemist

Are fermented products safe?
Ask a . . . Research Chemist

As a research chemist for a privately funded foundation, Omar Ahmed works with a team of chemists who deal primarily with fermentation. "Certainly, the public doesn't need to worry about the safety of the fermented foods they buy," says Omar, "but when you're a chemist, you do have a high level of respect for the chemical process of fermentation. Fermentation can spoil food, yet we carefully harness the positive side of fermentation to create new food flavors and textures, and even a longer shelf life. Right now my team is researching fermented cereal products. Koko, for example, is a sour cereal porridge that's a staple food in parts of Africa. From our research we hope to develop new products that are safe and nutritious."

Education Needed: Jobs for chemists with bachelor's degrees to Ph.D.s are growing rapidly. Chemistry students typically take additional coursework in mathematics, physics, and biology. A background in computer science is useful for modeling and simulation tasks

KEY SKILLS
- Mathematical skills
- Curiosity
- Mechanical ability
- Ability to work alone
- Attention to detail
- Communication

and for operating computerized lab equipment.

Life on the Job: The work of a chemist can be very solitary. Chemists spend much of their time collecting samples, carrying out experiments, and reporting results. Chemists often work with manufacturing as they troubleshoot processes and perform quality control.

Your Challenge

Suppose you were a chemist working with a research team. You make a key discovery that contributes greatly to your team's progress. Another team member, however, takes credit for the discovery.

1. What would you do?
2. What qualities do you think are needed for people to work well together as a team?

Lactic-Acid Fermentation

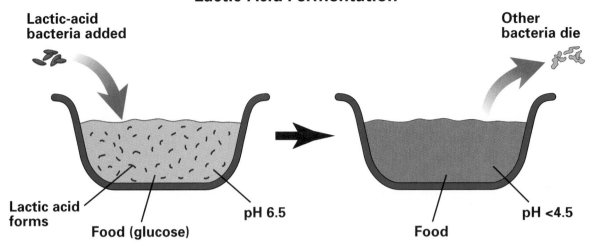

Figure 22-2
When food is fermented with lactic-acid fermentation, the acid that forms lowers the pH of the food and kills unwanted bacteria. How low does the pH level usually need to go?

During lactic-acid fermentation, the pH of the solution lowers, as shown in **Figure 22-2**. This continues until no more glucose is available for reaction; then fermentation stops. When the pH goes below 4.5, other bacteria, which could spoil the food, normally die. If the fermentation doesn't go on long enough to lower the pH below 4.5, the bacteria remain alive and can make the food inedible.

Pickles

Pickles are usually made from cucumbers. However, watermelon, cauliflower, onions, okra, and other foods can also be pickled. Pickle making uses **brine**, *a water-and-salt mixture that contains large amounts of salt.* Pickling is currently carried out in two ways.

The process that uses brine for pickle making is known as **brine pickling**. *The vegetable remains in the brine for several weeks. During this time, fermentation by lactic-acid bacteria takes place until an acceptably low pH is reached.* The food changes in appearance and texture. Green vegetables change to an olive or yellow-green color and become translucent, or partly transparent. The texture of the food becomes firm and crisp, but tender. Characteristic flavors develop.

The other process for making pickles is **fresh-pack pickling**. *With this process, the food is not fermented as in brine pickling. Instead, it is placed in brine for only a few hours or overnight, if at all. The food is then drained and immersed in a boiling mixture of vinegar (acetic acid) and spices.* The high temperature of the solution, combined with the low pH of the vinegar, kills undesirable bacteria, while the vinegar and spices flavor the pickles.

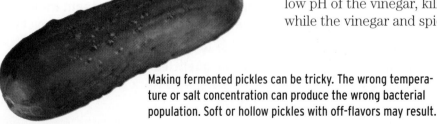

Making fermented pickles can be tricky. The wrong temperature or salt concentration can produce the wrong bacterial population. Soft or hollow pickles with off-flavors may result.

Sauerkraut

Sauerkraut is another food fermented by lactic-acid bacteria. When salt is mixed with cabbage, the bacteria on the cabbage cause fermentation, which creates sauerkraut. Here's what happens.

- **Bacteria growth is promoted.** Fermentation couldn't occur without the soil-borne, lactic-acid bacteria on the surface of cabbage. To help make this bacteria grow, the cabbage is washed before shredding. Washing increases the growth of lactic-acid bacteria, while decreasing the percentage of unwanted bacteria. Since temperature also affects growth of the lactic-acid bacteria, an ideal temperature of 21°C is used to make sauerkraut.

- **Salt helps produce lactic acid.** Salt has two main purposes when making sauerkraut. One, it releases fluids that provide a growth medium for the friendly bacteria. Two, it inhibits the growth of many other bacteria that could spoil the sauerkraut. As the salt pulls water and sugar from the cabbage, bacteria turn the sugar into lactic acid, the main product of the fermentation. Lactic acid is what puts the "sour" in sauerkraut.

- **The pH lowers during fermentation.** Ordinarily, the pH of fresh vegetables like cabbage is 5.5–6.5. This pH is perfect for growing such microorganisms as yeast, mold, and bacteria, including both spoilage bacteria and bacteria that produce lactic acid. As the cabbage ferments, bacteria produce enough acid to lower the pH to between 3.0–4.6. Once the pH is below 4.6, the solution is too acid for anything except fermenting bacteria to grow. The other microorganisms die, while the fermenting bacteria continue to multiply. The result is a flavorful sauerkraut that is safe to eat.

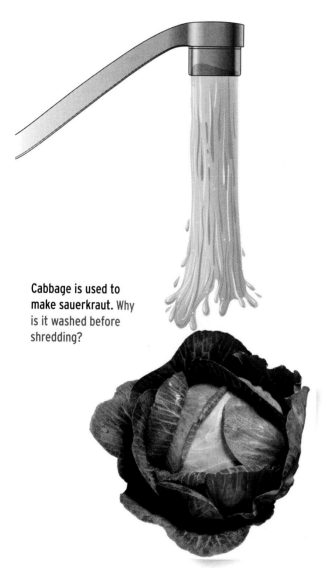

Cabbage is used to make sauerkraut. Why is it washed before shredding?

Acetic-Acid Bacteria

If you've heard of raspberry and apple cider vinegars, you may realize that vinegar can be made from fruit. The raw material chosen affects the final flavor of the vinegar. The process takes two steps, one to produce the alcohol and one to convert the alcohol to vinegar.

In the first step, sugar from the fruit is fermented to ethyl alcohol (C_2H_5OH). Yeast makes this happen.

During the second step, acetic-acid bacteria attack the alcohol that was produced. When the ethyl alcohol combines with oxygen, oxidation occurs. This leads to the production of acetic acid. Vinegar is a dilute solution of acetic acid.

Vinegar comes in many flavors. Each one is chosen to give the dish a distinctive taste. Why is vinegar created in two steps?

Yeast Fermentation

When you read about leavening breads, you learned how yeast participates in fermentation. The process converts glucose into ethyl alcohol and carbon dioxide.

During fermentation, yeast grows and produces carbon dioxide at different rates. When making yeast breads, the challenge is to manage the rate and length of fermentation to create the best food product. Expert bakers have learned these principles about yeast fermentation:

- Temperature is critical. Yeast ferments best at 27°C. A temperature that's too cool won't allow fermentation to take place. One that's too hot can kill the yeast. A sudden temperature change, either up or down, may hinder gas production.
- Yeast needs sugar to produce carbon dioxide. In addition to glucose, yeast can also ferment maltose, sucrose, and fructose. However, it can't ferment lactose, the sugar found in milk. Too much sugar will slow down the fermentation process.
- Too much salt can cause fermentation to slow down. Too little salt may weaken the gluten and cause cells to break. Following a reliable recipe carefully will yield the best results when using yeast fermentation in baking.

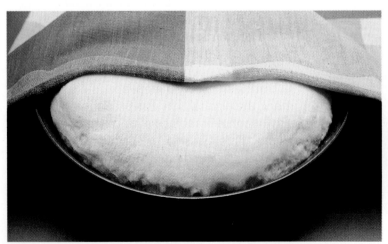

Yeast dough is placed in a warm, draft-free place to rise. A warm temperature increases the rate of fermentation. A temperature that's too warm, however, allows microorganisms that cause a sour, yeasty taste to flourish.

Nutrition and Fermented Food

Will a food be more or less nutritious after fermentation? Both are possible.

When water-soluble vitamins dissolve in brine, they can be lost during pickling. The amount of energy (kilocalories) present in a food may decrease during fermentation. This occurs because the food releases energy as it ferments. The bacteria that cause fermentation use some of this energy.

Some foods are more nutritious after fermentation than before. During fermentation, certain microorganisms produce vitamin B_{12} and riboflavin. Enzymes in mold can split the fiber coatings of grains, nuts, and seeds, making the nutrients in these foods available.

More commonly, fermented foods are equally nutritious to the foods used to make them. Fat-soluble vitamins, protein, and carbohydrates are not usually changed much by fermentation. With the chart in **Figure 22-3**, you can compare the nutritional value of fermented foods with similar sizes of their unfermented sources.

Applying Your Knowledge

1. Using information from the text and the chart, explain whether vitamin C is water-soluble or fat-soluble and why.
2. If energy is released in fermentation, why do most of the fermented foods in the chart contain more kilocalories than their unfermented sources?

Figure 22-3

Nutrient Comparison of Fermented Foods and Their Source					
Food	Kcal	Riboflavin	Vitamin A	Vitamin C	Calcium
Cucumber	5	0.01 mg	70 I.U.	3 mg	8 mg
Pickle	15	trace	30 I.U.	1 mg	8 mg
2 Percent Milk	120	0.42 mg	450 I.U.	2 mg	297 mg
2 Percent Yogurt	145	0.44 mg	170 I.U.	2 mg	415 mg
Cabbage	20	0.04 mg	120 I.U.	42 mg	34 mg
Sauerkraut	40	0.09 mg	120 I.U.	33 mg	85 mg

Mold and Enzyme Fermentation

Usually mold growing on food leaves makes them inedible. When used in fermenting certain food products, however, molds are helpful.

Since the human body lacks the enzymes needed to metabolize cellulose, molds can help. They contain enzymes that split cellulose molecules. Thus, mold fermentation of cellulose lets you eat foods that would otherwise be indigestible. For example, many grains have a cellulose or hemicellulose coating. Mold fermentation breaks down these coatings so the body can use the grains' nutrients. Nuts and seeds can also be made more digestible with mold fermentation.

Molds often alter the texture of foods and make them more pleasing to eat. They are used to age many cheeses, such as blue cheese. Mold fermentation also helps produce soy sauce.

Enzymes participate in fermentation too. As mentioned, they play a role in mold fermentation. During yeast fermentation, the enzyme sucrase breaks down sucrose molecules and speeds up the process as well.

Fermented Beverages

The beverage industry has used fermentation for thousands of years. Depending on the beverage, different processes are used.

Wine

Most alcoholic beverages are produced through yeast fermentation. Fermentation of barley produces beer and ale. Other alcoholic drinks rely on the fermentation of corn and rye. Most wine, however, is made from fermented grapes, although apples, blackberries, and other fruit can also be used.

Records indicate wine was produced in the Middle East as early as 3000 B.C. In the beginning, it is quite possible that grapes fermented accidentally before they were eaten. This could happen because the yeast was present naturally on the surface of the grapes.

Since then, wine making has become a highly scientific process. While it is an art, technology is making it a more controllable art. Wine makers, or vintners, employ elaborate systems with stainless steel tanks, electronic monitors, and computerized temperature con-

Many nuts, seeds, and grains have a hard coating that can't be digested. Through mold fermentation, these coatings can be broken down.

Wine producers must select from many varieties of grapes to choose those that make the best wine. Some wines are allowed to age for many years in order to improve the quality.

trols. They carefully monitor and test the grapes as they grow. In deciding when to harvest, wine producers look at the sugar levels in the growing grapes. They know that variations in sugar content cause different amounts of alcohol to be produced during fermentation.

Even the most careful scientific testing, however, cannot remove all the mystery from wine making. No two yields of wine are ever exactly the same. Even when the fermentation process is consistent, such factors as temperature, fermentation time, and nutrients in the grape juice all affect the final product. The grapes, too, always vary from one year to the next. They are affected by the amount of rainfall, whether the summer is long and hot or short and cool, and dozens of other variables. These influences make it difficult for a wine maker to predict whether the grape harvest will produce a mediocre wine or one of premium quality.

Coffee

Coffee trees 6 to 20 feet tall grow in tropical countries. They produce cherry-like berries from which the beans inside are removed. The curing processes that produce the beans include fermentation by bacteria and enzymes. A wet or dry method can be used.

- **Wet method.** With this method, the coffee cherries are washed and placed in machines that remove the outer pulp. The cherries are soaked in tanks of water and left to ferment for 12–24 hours. Then they are sprayed with water until the sticky coating formed on them has been removed. Finally, they are allowed to dry in the sun.
- **Dry method.** The dry method is the older and more natural method. The cherries are washed and then spread in thin layers to dry for two or three weeks. During this time, fermentation occurs. The cherries are raked several times each day to ensure uniform drying. About 60 percent of all the world's coffee is produced with this method.

Coffee trees produce berries that contain the beans needed to produce the beverage. Each berry contains two beans. What are the two methods that can be used to extract the beans from the berries?

After the beans are produced with either the wet or dry method, they are roasted. The roasting method determines the resulting flavor and aroma. During roasting, the beans become drier, and their oils become water-soluble. Finally, the beans are ground for brewing.

Many people prefer decaffeinated coffee, especially in the evening. As a stimulant, caffeine in the system can prevent a person from falling asleep. Indeed, an overdose of caffeine can produce an irregular heartbeat and muscle trembling.

One method for producing decaffeinated coffee is to steam the unroasted beans and then extract the caffeine with a solvent. This solvent must then be thoroughly washed off before the beans are dried and roasted. Another decaffeinating method also begins with steaming the beans, followed by scraping off the beans' outer layers to remove the caffeine. People who are concerned about solvent residues in coffee prefer the "decaf" prepared through this process.

Tea

The earliest records of tea cultivation come from China and date from the fourth century A.D. By 800 A.D. tea had become a popular drink in China and was becoming so in Japan. In spite of its long history in that part of the world, the largest consumers of tea worldwide are the English. In 1700, England imported 20,000 pounds of tea. One hundred years later the amount had risen to 20 million pounds. Today the British still consume more than twice as much tea per person as the Japanese do.

The tea plant, which is part of the camilla family, grows in a wide range of climates. Once leaves have been plucked from a tea bush, they are dried until they become wilted and structurally weak. The leaves are then rolled repeatedly for an hour or two. This used to be done by hand and was very time-consuming. The process is now done by machine, making it much less labor-intensive.

After rolling, fermentation begins. The leaves stand at 27°C for one to three hours, during which time they turn a coppery brown and develop their flavor. When the tea maker decides fermentation has gone as far as desirable, the leaves are "fired," or dried, at temperatures close to 93°C until the moisture content is about five percent. This fermentation process doesn't involve yeast or bacteria but is caused instead by the enzyme oxidation of the tea leaves. By controlling fermentation, tea can be made with color variations and subtle flavor differences.

Chocolate

If you're a chocolate lover, you may not be surprised to learn that 1.5 million tons of cocoa are produced worldwide each year. While Brazil is the largest producer of cocoa in the Americas, three-quarters of the world's supply comes from West Africa.

Making chocolate is a complicated process. First, *cacao* (kuh-KAH-oh) beans, which are actually seeds, are removed from the pod, piled on the ground, and left to sit in the sun for several days. Fermentation and enzyme reactions occur. A variety of microorganisms, including proteolytic bacteria, multiply in the juicy pulp and raise the temperature. This kills the seeds' embryos while causing a number of other biochemical changes. The cell walls in the beans break down, allowing various substances to mix together. Also, a number of bitter-tasting compounds in the beans react with each other, giving them more agreeable flavor.

After the beans are cleaned of pulp and dried, they are shipped to countries around the world. Chocolate makers then roast the beans for about an hour to develop the rich, characteristic, chocolate flavor. The process involves over 300 different chemical compounds, produced in part by enzymatic browning. By using different procedures and ingredients, producers can create either cocoa or chocolate.

The best tea is made from the small young shoots and unopened leaf buds on a bush. These contain the most enzymes and caffeine and are more easily crushed than older leaves. With regular pruning, a bush produces more of these leaves and buds.

After roasting, cacao beans are cracked open. The kernels that emerge are more than half cocoa butter. When ground up, the kernels form a thick liquid that can then be refined to create the chocolate a manufacturer uses to produce different products.

In earlier times chocolate may have been consumed for its nutritional value. One ounce of chocolate contains about 150 calories and two to three grams of protein. Like many seeds, the cacao bean contains significant amounts of B vitamins and vitamin E. Moreover, the Aztecs, Mayans, and Totecs also considered cocoa and chocolate a mild stimulant. In chocolate candy, however, the vitamins in the cocoa beans are extremely diluted by fat and sugar, as is any potential nutritional value. Chocolate does contain caffeine, but only a fraction of that found in coffee. You would have to eat five ounces of dark chocolate candy or drink 15 cups of hot chocolate to get the caffeine found in one cup of coffee.

Fermented Food from Around the World

Most cultures of the world enjoy some fermented foods. In some Asian countries, people eat paw tsay, a dish consisting of fermented turnips and radishes. Soy sauce, saki (rice wine), and kimichi (fermented cabbage) are also traditional parts of some Asian meals. Fermented fish has been popular Japanese fare for centuries.

In Hawaii, the famous dish called poi, which is served at luaus, is prepared from the fermented taro root. This root grows in the moist valleys found throughout the Hawaiian Islands.

Fermented foods of European origin include sausages, salami, bologna, and hundreds of cheeses. Buttermilk is a favorite in Bulgaria. Each of these cultures has developed its distinctive cuisine and identity with the help of the humble bacteria and yeast that produce fermentation.

In many cultures, people enjoy the flavors of various fermented meat products.

Yeast Growth

In this experiment, you will grow yeast in a variety of environments to determine how each affects yeast growth.

Equipment and Materials

dry yeast	sucrose	sodium chloride	100-mL graduated cylinder
250-mL beaker	laboratory thermometer	saucepan	metric ruler

Procedure

1. Add 100 mL of water to a 250-mL beaker. Heat the water to 35ºC over medium heat.
2. Put a small saucepan half full of water on a second burner over low heat.
3. When the temperature of the water in the beaker has reached 35ºC, remove it from the heat. Add 3.2 g dry yeast to the beaker and mix thoroughly so all the yeast dissolves.
4. Follow the variation assigned by your teacher.
   Variation 1: Add nothing to the yeast and water mixture.
   Variation 2: Add 4.3 g sugar to the yeast and water; mix well.
   Variation 3: Add 6.6 g of salt to the yeast and water; mix well.
   Variation 4: Add 4.3 g sugar and 6.6 g salt to the yeast and water; mix well.
5. Remove the saucepan from the stove. Determine the temperature of the water in the saucepan.
6. Add cool water if necessary to adjust the temperature of the water to 30ºC. Place the beaker containing the yeast mixture in the warm water in the saucepan for 15 minutes.
7. After 15 minutes, remove the beaker from the saucepan. Measure the height of the yeast mixture. In your data table, record the height of the mixture for the variation.
8. Observe the odor and consistency of your mixture, and record this information.
9. Write the information for your variation on the chalk-board. In your data table, copy the information for the other variations.

Analyzing Results

1. In which environment did the yeast grow best? Worst? Why?
2. Why wasn't the beaker placed directly on the heating element during yeast growth?
3. Check at least three different bread recipes. Do the ingredients used and directions given agree with the results of this experiment?

SAMPLE DATA TABLE

Variation	Height	Odor	Consistency
1			
2			
3			
4			

Chapter Summary

- Fermentation has a long history of use in civilization.
- Cell respiration is a multistep process that involves the breakdown of glucose and the release of energy.
- Anaerobic respiration is another name for fermentation.
- Fermentation in food can be caused by bacteria, yeast, mold, and enzymes.
- Food can be fermented by several different types of bacteria.
- Yeast works with other ingredients in bread to ferment and cause the bread dough to rise.

- Some foods are created through fermentation caused by molds and enzymes.
- Yeast, bacteria, and enzyme fermentation can all be used to produce certain beverages.

Using Your Knowledge

1. Why did Pasteur's contribution to wine making have long-term value?
2. What are three reasons for fermenting food?
3. Why is fermentation called "anaerobic respiration"?
4. Explain what causes fermentation to begin.
5. What are four types of fermenting bacteria and their products?
6. If indigenous bacteria aren't present, can fermentation take place? Explain.

Real-World Impact

Chocolate

Chocolate, which grows in a belt 20° north to 20° south of the equator, can be traced to ancient Mayan and Aztec civilizations. These cultures brewed a rich, pepper-spiced drink called "chocolate." The first Europeans to see cacao beans were probably the crew of Columbus' fourth voyage in 1502. When Cortez brought the bean from Mexico in 1528, the drink became popular in Spain. By 1650, many Europeans were drinking chocolate, and fifty years later it was the rage of London society. English importers renamed the bean "cocoa," probably due to a spelling error. Today, chocolate factories are found in almost every country.

Thinking It Through

1. Describe the ways people make use of chocolate today.
2. If a major chocolate producer raised prices on chocolate, what would the economic effect be?
3. Why do some forms of chocolate candy taste better than others do?
4. Check the Internet to discover more about chocolate.

Using Your Knowledge *(continued)*

7. What role does pH play in lactic-acid fermentation?
8. How do brine pickling and fresh-pack pickling compare?
9. How is the growth of spoilage bacteria, yeast, and mold stopped in making sauerkraut?
10. What role does gluten play when yeast dough forms a foam?
11. What are two positive effects of mold fermentation in food?
12. Describe two methods used to process coffee.
13. How is fermentation used to produce chocolate?

Skill-Building Activities

1. **Critical Thinking.** In fermentation why would microorganisms that can survive high levels of acidity be needed?
2. **Social Studies.** Select a country and describe the fermented foods that are part of its cuisine. Check the Internet or other resources for information. Why do you think these foods became popular in this country?
3. **Teamwork.** In groups, design an advertising campaign for a new coffee that includes a description of the coffee-making process described in the chapter.
4. **Communication.** Suppose you had to explain fermentation to a group of sixth-grade students. Write what you would tell them.

Thinking Lab

Pickling Cucumbers

To deactivate enzymes, fruits and vegetables are blanched, or briefly immersed in boiling water, before freezing.

Analysis

In brine pickling, cucumbers are placed in a brine solution for several weeks. During this time, lactic-acid fermentation takes place. Although the cucumbers are generally washed before being placed in the brine solution, they are not blanched.

Thinking Critically

1. Would blanching affect the pickling process? In what way or ways?
2. Why do you think cucumbers shrink when they become pickles?

Understanding Vocabulary

Categorize the "Terms to Remember" for this chapter into two lists: those you can define and those you can't. Look up definitions for the terms in the second list and write them in your own words.

Objectives

- Identify the components of milk and describe how they are dispersed in milk.
- Explain what happens when milk protein is coagulated.
- Describe how milk is processed and the effects of pasteurizing, homogenizing, and fortifying milk.
- Distinguish the characteristics of various milk products.
- Describe how cultured milk products are produced and give examples.
- Explain how milk and milk products should be stored.
- Relate certain factors to cream's ability to foam.
- Describe reactions that may occur when milk is heated.

Terms to remember

carrageenin
casein
cream
creaming
culture
curds
fortification
incubation period
inoculation
lactose intolerance
micelles
milk solids
precipitate
shelf life
whey

For most people, milk is a pillar of nutrition for life. It's their first food as infants. As they grow, children are encouraged to make milk a basic beverage at mealtimes. Milk supplies the calcium for building bones and teeth, especially during adolescence and the young adult years. In older adults, milk on the menu combats the bone loss caused by aging. People of all ages value milk as the perfect accompaniment to the ever-popular cookie. Whether from cows, goats, or llamas, milk and milk products are a tasty, healthful part of the diet.

The Complex Nature of Milk

What's in that plain-looking glass of milk that makes it so vital to health and versatile for cooking? Actually, milk is 87 percent water. That water, however, is a solvent for over 250 chemical compounds, the milk sugar lactose, water-soluble vitamins, and many trace minerals and mineral salts. As you will see, milk is not only a solution, but it's also a colloidal dispersion as well as an emulsion. All of these characteristics make milk a pretty amazing substance.

The Composition of Milk

Around the world, people use milk from different kinds of animals. Cow's milk, however, is the most common type consumed. This chapter focuses on the composition and properties of cow's milk and the products made from it.

Although milk is mostly water, it contains all the major nutrients. Milk is a source of fat (about 3.5 percent), protein (about 3.5 percent), carbohydrate (about 5 percent), vitamins, and minerals. The exact composition of these nutrients depends on the cow's breed, health, and environment.

Protein

Two main proteins are dispersed as colloids in milk. *The protein* **casein** *makes up about 80 percent of milk.* Twenty percent of milk is whey.

In milk, casein molecules associate with each other and with certain minerals in tiny, raspberry-like structures called **micelles** (my–SELLS). A micelle is *an aggregation, or cluster, of molecules, often found in colloidal dispersions.* The micelles in milk are more or less sphere-shaped. The light reflecting from micelles makes milk white.

As long as milk keeps its normal acidity level, about pH 6.6, casein remains stable. However, if an acid is added, lowering the pH to about 4.6, the casein breaks down and forms lumps. In other words, the milk coagulates. *When milk coagulates, the casein clumps that separate from the liquid* are called **curds** or clots. If you've prepared a recipe that calls for milk and vinegar or lemon juice to stand together for a time, you've seen the curdling effect of acidic food on milk.

When cooking, you can prevent curdling by using starch to thicken the milk or the acidic food. The starch surrounds the casein and prevents it from coagulating when the foods combine.

Another way to coagulate milk is with the enzyme rennin. Rennin is used to clot milk in the first stage of cheese making.

Whey protein is also suspended colloidally in milk. **Whey** is the *protein found in the liquid that remains after fat and casein have been removed from milk.* Whey is also called *serum protein.* Although heat has little effect on coagulating casein, it will easily coagulate whey.

In addition to the proteins, casein and whey, milk contains a number of enzymes that are also protein molecules. Most of these are denatured during pasteurization.

To ensure that milk is high quality, the Food and Drug Administration (FDA) outlined sanitary practices that must be followed as milk goes from the dairy farm to you. These practices include farm, dairy herd, and plant inspections as well as monitoring the stages of manufacturing.

Fat

Milk is an emulsion because small globules of fat (lipids) are dispersed throughout the water. The globules vary in size, depending on the animal that produced the milk. Around each fat globule is a thin membrane that keeps the globules apart. This membrane contains proteins, phospholipids, and bound water. **Figure 23-1** shows a fat globule.

Although the fat droplets in fresh milk are suspended throughout, some cluster loosely together. If you watch freshly drawn milk stand for a while, you'll see a process called **creaming**, in which *some of the fat droplets come together in larger clusters that rise and float to the top of the milk.* These clusters rise to the surface because fat is less dense than the watery portion of milk. **Cream**, then, is simply *milk that is extra rich in emulsified fat droplets.*

Casein curds are the basis for many cheeses. The whey, a by-product of cheese making, has many food applications. In concentrated form, it is used in high-protein beverages, baked goods, prepared mixes, soups, and margarines.

Figure 23-1
In milk, fat globules are dispersed in water. The surrounding membrane, which contains proteins, phospholipids, and bound water, is what keeps one globule separate from another.

A Fat Globule in Milk

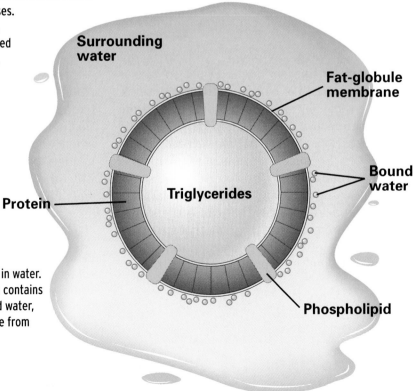

Surrounding water

Fat-globule membrane

Bound water

Triglycerides

Protein

Phospholipid

The fats in cow's milk are the most complex lipids known. These lipids include over 400 different fatty acids, although only 20 account for most of the fat. Milk fat contains very little cholesterol compared to egg yolk. The cholesterol that is present is in the membrane that surrounds the fat globules. The fats in milk influence its flavor, texture, and price. Typically, the greater the percentage of milk fat, the more costly the milk.

Carbohydrate

Lactose, or milk sugar, is the main carbohydrate found in all varieties of milk. Lactose is made only by cells of the lactating mammary gland. Lactose provides food energy. It also adds body to milk as well as a delicately sweet flavor.

When milk is heated, lactose reacts with amino acids in the protein. This chemical reaction is what gives cooked milk products a golden brown color and slightly caramel flavor.

When milk products are consumed, lactose becomes available to the body because of the enzyme lactase. This enzyme breaks lactose down into galactose and glucose, which can be used by the body for fermentation during digestion.

Perhaps you know people who can't consume milk or milk products. Doing so causes them to complain of bloating, stomach cramps, and diarrhea. These are symptoms of **lactose intolerance**, *the inability to digest milk due to the absence of the enzyme lactase in the intestines*. Because lactase is missing, lactose remains undigested. Then certain intestinal bacteria can attack and ferment the lactose, producing acids and gas that cause the symptoms. Sufferers can use these methods to manage their symptoms: limit consumption of milk products; try cultured products such as butter-milk and yogurt; buy special, lactose-reduced products; or use lactase supplements.

Minerals and Vitamins

Although you may not taste it, milk has salt that occurs naturally. Salts in milk include chlorides and phosphates, as well as citrates of potassium, calcium, sodium, and magnesium. In fact, most of the minerals present in milk are in the form of salts. Of course, they are present in very small amounts, usually equaling less than one percent of the milk. Milk is no threat to a low-sodium diet.

One function of salts is to prevent milk from curdling. Calcium and magnesium ions in milk, both salts, help keep casein micelles stable, which prevents curdling. If an acid is added to milk, however, the pH of milk lowers, which removes the calcium ions from the casein micelles. As a result, the milk curdles, allowing curds and whey to form. Likewise, when the enzyme rennin is added to milk in cheese making, it coagulates the milk more quickly when fewer salts are present.

Among the trace elements present in milk are cobalt, copper, iodine, iron, magnesium, nickel, and molybdenum. The exact amount of each mineral depends on the soil conditions where the cows' feed was grown.

Of the four vitamins found naturally in milk, riboflavin is present in the largest amount. It has a greenish tint that can be seen only in milk that contains no fat. Riboflavin breaks down when exposed to light, so careful storage is needed to preserve this important nutrient. The other vitamins in milk are thiamin, niacin, and vitamin A.

Because of lactose intolerance, some people cannot eat regular dairy products. What effect could this have on health? What can they do?

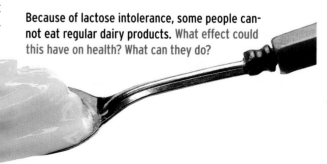

The Nutritional Value of Milk

"Drink your milk so you'll grow strong" may be a favorite line of parents to children, but it's true for everyone. Milk is a storehouse of good nutrition.

Milk provides quality protein—complete protein, in fact, containing all the essential amino acids. You can consume milk as the only source of protein at a meal or use it to supplement incomplete-protein foods. You might enjoy milk on cereal or in a cream soup.

On the mineral front, milk is the most important source of dietary calcium—in a form the body can easily use. Milk also contains significant amounts of phosphorus. This is a convenient arrangement, since phosphorus is needed for the body to absorb calcium.

Several important vitamins also appear in milk's nutritional profile. You'll find the water-soluble B vitamins: riboflavin, thiamin, and niacin. Whole milk also provides the fat-soluble vitamin A.

Finally, milk is a good source of energy. One cup of milk provides 12 g of carbohydrate in the form of the sugar lactose. Lactose also increases the absorption of calcium, phosphorus, magnesium, and zinc.

No single food supplies all your nutrient needs, of course. Milk isn't a good source of vitamin C, iron, or copper. This is why vitamin C and egg yolk are added to a baby's diet early in life.

Likewise, not all milk products are equally nutritious. Cheese is a nutritious food because of protein. Similar weights of cheese and ground beef have equal amounts of protein. In contrast, cream and butter contain few of the nutrients found in fluid milk and are high in fat.

GRADE A PASTEURIZED HOMOGENIZED FAT FREE MILK VITAMIN A & D ADDED

Nutrition Facts

Serving Size 1 Cup (240 mL)
Servings Per Container about 8

Amount Per Serving

Calories 90	Calories from Fat 0

	% Daily Value*
Total Fat 0g	0%
Saturated Fat 0g	0%
Cholesterol <5mg	1%
Sodium 125mg	5%
Total Carbohydrate 13g	4%
Dietary Fiber 0g	0%
Sugars 12g	
Protein 9g	17%

Vitamin A 10%	•	Vitamin C 2%	
Calcium 30%	•Iron 0%	•Vitamin D 25%	

* Percent Daily Values are based on a 2,000 calorie diet. Your daily values may be higher or lower depending on your calorie needs:

	Calories	2,000	2,500
Total Fat	Less than	65g	80g
Sat Fat	Less than	20g	25g
Cholesterol	Less than	300mg	300mg
Sodium	Less than	2,400mg	2,400mg
Total Carbohydrate		300g	375g
Dietary Fiber		25g	30g
Protein		50g	65g

Calories per gram: Fat 9 • Carbohydrate 4 • Protein 4

INGREDIENTS: FAT FREE MILK, VITAMIN A PALMITATE, VITAMIN D3.

DIST. & SOLD EXCLUSIVELY BY:
TRADER JOE'S, PORTLAND, OR 97202
PROCESSED AND PACKAGED AT

Applying Your Knowledge

1. Suppose you were in charge of preparing food for a child who resists drinking milk. What would you do?
2. Evaluate your own consumption of milk and milk products. What changes, if any, do you need to make?

Milk production takes place under rigid sanitary conditions. What signs of sanitation are apparent in this manufacturing situation?

Processing Milk

Consumers expect foods to have good flavor, high nutritive value, and a satisfactory **shelf life**, *the time a food product can be stored before deteriorating.* When properly processed—that is, pasteurized, homogenized, and fortified—milk meets all of these conditions. Local health departments carefully regulate the processing and sale of milk and milk products.

Pasteurization

In pasteurization, milk is heated to high temperatures for a short time to destroy harmful bacteria. This process has been used for over 100 years to help ensure safety of the milk supply.

Even though pasteurization eliminates bacteria that cause diseases, milk will still spoil over time. The remaining bacteria (that don't cause diseases) will multiply enough to cause spoilage.

Pasteurization denatures enzymes that cause milk to spoil, which helps delay spoilage.

In fact, one way to check milk for pasteurization is to test for the specific enzyme alkaline phosphatase. After milk has been pasteurized, this enzyme is no longer present. A check for this enzyme is known as the phosphatase test.

Pasteurization has a small effect on milk's nutritional value, creating a slight loss of thiamin, vitamin B_{12}, and vitamin C. Some of the calcium becomes insoluble and thus not usable by the body.

You may have heard that unpasteurized (raw) milk is more nutritious, but scientific research hasn't given support to this claim. The safety benefits of pasteurization far outweigh any slight decrease in nutritional value it may cause. In many states, selling raw milk is illegal.

Homogenization

Most milk sold in the United States is homogenized. This process is used to eliminate creaming. Under pressure of 2,000 to 2,500 pounds per square inch, milk is forced through small openings in a machine called a homogenizer. The fat particles break down and are sur-

The milk that reaches you has been through certain processes since leaving the dairy farm. What three processes make milk safer and more wholesome?

rounded by an emulsifier that keeps the tiny particles permanently separated. As a result, the milk you buy doesn't have cream on the top.

Fortification

Milk is fortified during processing as well. **Fortification** is *the addition of a nutrient to a food*. Most milk producers voluntarily fortify milk with vitamin D. Without this vitamin, your body can't absorb and use calcium and phosphorus very well, and milk contains both of these nutrients. You can get vitamin D from sunlight but not many other sources. That makes the addition of vitamin D to milk a good idea.

To fortify with vitamin D, milk may be exposed to ultraviolet light, which converts some of the milk fat components to vitamin D. Vitamin D concentrate can be added before pasteurization.

When fat is removed to make low-fat milk products, most of the vitamin A is removed as well. By law, any vitamin A removed during processing must be replaced by producers.

Most milk is fortified with vitamin D. Why is this necessary?

Types of Milk Products

Milk products come in many forms. Federal law sets standards for manufacturing and labeling the different products.

Fluid Milk

Fresh fluid milk is categorized by fat content. The level of fat in each type of milk is set by law. These amounts are shown in **Figure 23-2**.

Ultrahigh-Temperature Milk (UHT)

With special processing, milk can be stored without refrigeration for three months or more if unopened. This is called UHT milk because it's prepared by heating for a short time at ultrahigh temperatures. These temperatures are higher than those used for pasteurization.

UHT milk is placed in specially sealed packaging that keeps spoilage bacteria out. The flavor of UHT milk is not very good right after processing, but it improves in time. If storage time is too long, however, flavor problems develop again.

Concentrated Milk

Evaporated milk is aptly named. This is whole milk that has been heated at low pressure, causing up to 60 percent of the water present to evaporate at temperatures well below milk's

If UHT milk is stored at lower than ordinary room temperatures, chemical and enzyme activity is less likely to affect the flavor. What uses might be made of UHT milk?

normal boiling point. The result is a concentrated form of milk that is usually homogenized, sterilized in processing, and canned. **Carrageenin** (kair-uh-GEE-nun), *a vegetable gum, is often added to evaporated milk before processing to stabilize the casein proteins*. Due to the temperatures needed for sterilization, evaporated milk has an off-white color caused by the chemical reaction of lactose with amino acids.

Some dessert and candy recipes call for *sweetened condensed milk*. This is a canned product in which 50 percent of the water has been removed and sugar added. Sweetened condensed milk, in fact, is 44 percent sugar. Because the high sugar content acts as a preservative, this milk does not need sterilization. The sugar produces a high osmotic pressure that draws the remaining water in the milk

Figure 23-2

Whole milk contains considerably more fat than the other types and has a creamier texture. How might someone get used to the thinner consistency of milk that is lower in fat?

Fat Content of Milk	
Type of Milk	**Amount of Fat***
Whole milk	8 g fat
Reduced-fat milk	5 g fat
Low-fat milk	2.5 g fat
Fat-free milk	Trace of fat

*Amount of fat is per 8-ounce (250-mL) serving.

Although these two products are both concentrated milk and come in cans, they are different. If a potato soup recipe called for evaporated milk, why wouldn't you want to use condensed milk instead?

away from the nutrients needed for bacterial growth. Therefore, harmful bacteria aren't able to grow and contaminate the milk.

Cream

Cream is another product classed according to the amount of fat it contains.

- *Half-and-half* is 10.5 to 18 percent fat.
- *Light cream, or coffee cream,* contains 18 to 30 percent fat.
- *Light whipping cream* has 30 to 36 percent fat.
- *Heavy whipping cream* contains at least 36 percent fat.

Whipping cream often bears the label UHT. This product has been pasteurized at an ultra-high temperature to lengthen its shelf life.

Dry Milk

Milk can be dried into powder form. The milk used to make dried milk is pasteurized. Then the water is removed, leaving a dry, solid material. These remaining solids, called **milk**

solids, are *the protein, carbohydrate, fat, minerals, and vitamins that were dissolved in the liquid portion of the milk.*

Nonfat milk that has been dried has a longer shelf life than whole milk or low-fat dried milk. This is because no fat is present to oxidize and spoil the product.

Dry milk can be returned to its liquid form after purchase by mixing with water. Some people enrich certain dishes, such as casseroles, by adding dry milk.

Cultured Milk Products

For thousands of years, people have known that milk would ferment into different forms if not consumed quickly. These new products

The thickness of cream depends on fat content. Generally, the higher the fat content, the thicker the cream. Chilling makes cream more viscous because the fat globules become firmer.

could be stored without spoiling for longer periods of time than milk itself could. As a result, people all around the world developed a taste for the unique flavors of such fermented milk products as buttermilk, yogurt, and cheese.

Cultured is the term used for any milk product that has been fermented to produce a distinctive flavor or texture. This name comes from the substance that begins fermentation in milk—the culture, also known as a *starter*. A **culture** is *a controlled bacterial population that is added to milk*. The starter bacteria produce acid and flavors that are characteristic of each fermented milk product.

In preparing a cultured milk product, the milk is generally pasteurized. The producer selects and prepares the specific culture to be used. Commercial producers use lactic-acid bacteria to break down lactose, creating a low pH that inhibits the growth of other, undesirable bacteria. Next **inoculation** takes place as *the starter is added to the milk*. During the **incubation period**, *the bacteria is given time to grow and ferment the milk*. The product is sometimes agitated. Finally, it is cooled to stop or slow bacterial growth.

Several hundred different cultured milk products are enjoyed around the world. Each region has developed a particular system for preparing them. A few cultured foods that you may be familiar with are described here.

Buttermilk

Originally, buttermilk was the fluid left over from the production of butter. This "leftover" beverage was popular on farms. Today, buttermilk is a cultured product prepared from skim or low-fat milk. It is also sold in a powder, similar to dried milk, for use in baking.

Sour Cream

A culture of *Streptococcus lactis* (strep-toh-KAH-kus LAK-tis) organisms is added to cream to make sour cream. It is held at a certain temperature until the proper acidity level is reached.

Despite its name, sour cream doesn't have a sour taste. Indeed, sour cream is popular served as is or used in dips and other dishes. Although it contains fewer calories and fats

Sour cream is often the base in dips that are eaten with raw vegetables and chips. What ingredients could be added to flavor the sour cream?

Today most cheese that is consumed is produced in factories. The process takes place in three basic steps. First, the casein is precipitated into curds through bacterial action. Second, the curds are concentrated. Third, the cheese is ripened.

than mayonnaise or salad oils, regular sour cream is not a low-calorie food. Low-fat varieties, however, are available.

Yogurt

Yogurt is an eastern European food that has become popular in the United States. Its thickness ranges from beverage-like to a gel. Yogurt contains lactose, lactic acid, B vitamins, and high concentrations of protein. It's usually low in calories.

Yogurt is prepared by adding a bacterial culture to milk that has been heated. Heating the milk not only kills unwanted bacteria but also denatures the whey proteins. Denaturation increases the capacity of the proteins to bind water and to promote the growth of the fermenting bacteria.

The bacteria used in making yogurt must be incubated between 41°C and 45°C. An eight-hour incubation seems to produce yogurt with the smoothest gel. Commercial yogurt makers often add gelatin or a vegetable gum to guarantee firmness, or body. The firmness of yogurt can also be increased by adding more milk solids. Fruits and other ingredients may be added for a variety of flavors. Yogurt also comes in a frozen dessert form similar to ice cream.

Cheese

Cheese is made by coagulating the casein protein in milk. Lactic-acid bacteria and enzymes promote coagulation. Once the curds form, the whey is drained away. Heating and cutting the curd helps this drainage. The curd is then treated. Finally, most cheeses are allowed to age, or ripen.

Ripening processes create physical and chemical changes in cheese. During ripening, the cheese is held in a temperature- and humidity-controlled environment for at least 60 days. Some ripening processes take many

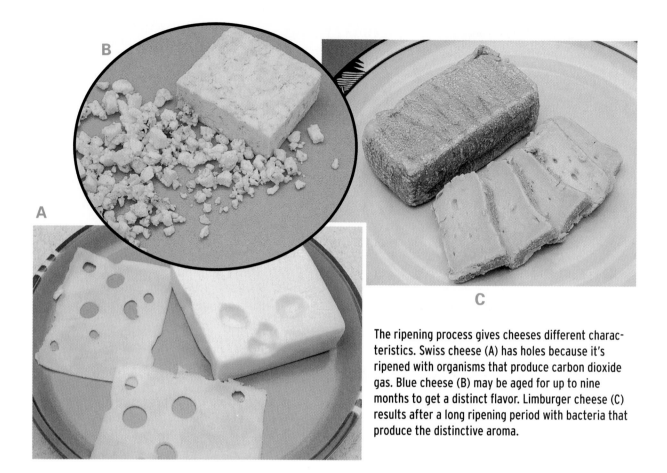

The ripening process gives cheeses different characteristics. Swiss cheese (A) has holes because it's ripened with organisms that produce carbon dioxide gas. Blue cheese (B) may be aged for up to nine months to get a distinct flavor. Limburger cheese (C) results after a long ripening period with bacteria that produce the distinctive aroma.

months. By controlling the ripening process with enzymes and other methods, producers can create the flavors, textures, and appearances that make so many cheeses distinctive. Ripened cheese is basically classified as strong and sharp or mild.

During processing, many variables affect the final cheese product. These include the following:

- Types and number of microorganisms present.
- Types and quantities of enzymes present.
- Degree of fermentation.
- Incubation temperature.
- Amount of liquid drained from the final product.
- Conditions of storing the cheese to ripen.

Depending on the processing, the moisture content of cheese will vary. Four categories of cheese are based on moisture content, as shown below. These range from least moisture (very hard) to most moisture (soft). A few examples are provided in each category.

- *Very hard cheeses*: Parmesan and Romano.
- *Hard cheeses*: Cheddar, Colby, and provolone.
- *Semisoft cheeses*: Muenster, Roquefort, and Stilton.
- *Soft cheeses*: Brie, Camembert, and mozzarella.

Although most cheese is ripened, some isn't. Cottage cheese, cream cheese, and mozzarella, are unripened cheeses. Unripened cheeses tend to have a high moisture content. Since moisture hastens spoilage, they are best enjoyed within two or three weeks of purchase.

How are milk products improved? Ask a ... Biochemist

If you like to solve difficult puzzles, then biochemistry may be for you," says Jim Nakai, a biochemist who works in the field of dairy science. "Biochemists have to delve deeply into the structure of molecules to figure out how organisms work and what affects them," Jim says. "Although biochemists work in practically every field that deals with living things, my rural background is what pointed me toward dairy science. Over the years, I've done research on bovine diseases and helped develop better techniques for processing cultured products. A difficult problem can take hours of study and experimentation to find an answer."

Education Needed:
Entry-level biochemists need at least a bachelor's degree, but many jobs in biochemistry take high-level research that requires an advanced degree. If you design or direct a research project, a Ph.D. is a must. Biochemists work in many fields, so students usually specialize by taking coursework in physics, molecular genetics, pharmaceutical sciences, or environmental science.

KEY SKILLS
- Laboratory skills
- Math skills
- Written communication
- Organization
- Detail-oriented
- Supervisory

Life on the Job: Biochemists work in industry, universities, hospitals, and government agencies. As you might expect, they do most of their work in a laboratory environment; however, those at the Ph.D. level are often policymakers for their laboratories. Biochemists plan and execute research as well as document experiments and report results.

Your Challenge

Imagine that you are working on your bachelor's degree in biochemistry. You're interested in finding an entry-level position with a university-funded research lab in the future.

1. What is networking and how would that be useful to you?
2. How could a mentor be helpful in this situation?

Storing Milk and Milk Products

Correctly storing milk and its products helps maintain its quality and safety. Careful handling at every step of processing is necessary. That chain includes the milk producer, the distributor, the food retailer, and you the consumer.

When buying fluid milk, you need to refrigerate it right away and return it to the refrigerator immediately after each use. Milk retains its quality for about one to three weeks when properly stored at 4° to 7°C. Leaving milk on a kitchen counter or on the table during meals causes rapid bacteria growth, which hastens spoilage.

Storing milk in a closed container helps prevent unpleasant flavors from developing.

Remember that riboflavin is sensitive to light, so milk should be kept in opaque (not transparent) cardboard or plastic containers.

Nonfat dry milk should be stored at room temperature in packaging that keeps moisture out. Whole dry milk doesn't store well. Unopened cans of evaporated milk can be stored at room temperature but must be refrigerated if opened.

Any refrigerated milk product you buy should also be refrigerated at home. An unrefrigerated product, such as a jar of processed cheese, can be stored at room temperature until opened. Then it should be refrigerated.

Cheese keeps best if cold, although it tastes better if served at room temperature. Careful wrapping prevents cheese from drying out.

Cooking With Milk and Milk Products

For many people, milk is best served cold in a glass or poured on cereal. Skilled cooks use milk and milk products—whipped, heated, or combined with other ingredients—in a variety of delicious and nutritious creations.

Making Foams

The whipped cream dollop on apple pie and ice cream sundaes is, scientifically speaking, a foam—gas bubbles trapped in a liquid. Several factors affect the ability of cream to form that foamy topping.

Milk should be returned to the refrigerator right after use. Why shouldn't you pour leftover milk back into the original container with other milk?

Would homogenized cream or unhomogenized make better foam? Homogenized products use considerable protein to surround the many tiny fat globules. Since a foam must have protein available for support, whipping cream isn't usually homogenized.

- **Fat content.** The higher the fat level in the cream, the better the foam is apt to be. Higher fat means greater viscosity, or thickness, which improves the quality of the whipped cream. Cream containing 30 percent fat will produce a less stiff foam than cream of 40 percent fat.
- **Temperature.** Cold temperatures increase the viscosity of the cream. Temperatures below 7°C produce the best foams. For best results, the beaters and bowl used for whipping cream should also be cold.
- **Amount of cream.** Whipping small amounts of cream usually gives better results than whipping large amounts.
- **Sugar content.** Adding sugar decreases both the volume and stiffness of the foam, while increasing the time it takes the foam to form. It's best to add sugar after the cream has reached the desired consistency.

Heating Milk

Like boiling pasta, heating milk demands your attention. If you heat milk for too long or at a temperature that's too high, you'll be cleaning up a scorched, foamy mess. Milk is sensitive and highly reactive to heat.

Heat denatures and coagulates the whey proteins of fresh milk, causing them to precipitate (prih-SIP-uh-tayt). To **precipitate** means *to cause a solid substance to separate from a solution,* as shown in **Figure 23-3**. Some of the calcium phosphate in milk also precipitates with the protein. The solids, which settle on the bottom of the container in which the milk is

Figure 23-3
To precipitate means to cause a solid substance to separate from a solution. If you were heating milk, would stirring the liquid constantly be a good idea? Why?

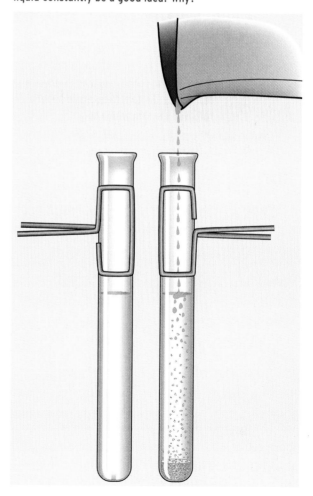

heated, scorch easily when hot. To prevent scorching, milk should be placed either over low heat or in a double boiler. With a double boiler, hot water in the lower pan gently heats the milk in the pan that sits on top.

When milk is heated in an open container, a skin forms on the surface. This is due to the concentration of casein as the water surrounding it evaporates. The skin tends to trap steam and build pressure that stretches the skin, allowing the milk to boil over.

If this skin is removed, milk solids are lost and a new film forms. A foam on the surface of hot milk minimizes skin formation. This is the reason hot chocolate is often whipped until foamy. A marshmallow or whipped cream served on hot chocolate is a tasty topping that also helps prevent skin from forming.

Finally, heating milk can cause it to curdle. Curdling generally occurs only at high temperatures, when other factors that affect milk are also present. High levels of salt in hot milk, a low pH, or the presence of certain enzymes can cause curdling. Fresh pineapple, for example, contains a curd-creating enzyme called bromelin.

It's best to use fresh milk, low temperatures, and nonacid foods when cooking with milk. This will prevent curdling in foods such as custards and cream soups. Adding sodium bicarbonate (baking soda) to the milk helps prevent curdling by raising the pH.

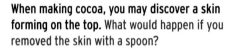

When making cocoa, you may discover a skin forming on the top. What would happen if you removed the skin with a spoon?

Separating Milk

Do you recall the childhood rhyme about Miss Muffet, who ate curds and whey? As you've learned, milk is both a solution and a colloidal dispersion. Normally, the protein molecules in milk, which are too large to dissolve, remain suspended as colloids. Miss Muffet's colloids had clumped into solid curds, which she ate along with the whey, or the remaining liquid. In this experiment you will see what may have produced Miss Muffet's meal.

Equipment and Materials

whole milk	vinegar	lemon juice	water
3, 250-mL beakers	100-mL graduated cylinder	stirring rod	

Procedure

1. Add 100 mL of milk to each of three, clean, dry, 250-mL beakers.
2. After carefully rinsing out your graduated cylinder, measure 30 mL of vinegar and add it to the milk in the first beaker. Stir.
3. Repeat Step 2, adding 30 mL of lemon juice to the second beaker, and then 30 mL of water to the third beaker.
4. Allow the milk to sit for 1 minute. Record your observations in your data table.
5. Record your observations after the milk sits for 5 minutes.

Analyzing Results

1. What effect did the vinegar, lemon juice, and water have on the milk?
2. Why do vinegar and lemon juice affect milk as they do?
3. What dairy products can you identify that have curds?

SAMPLE DATA TABLE

	Milk with vinegar	Milk with lemon juice	Milk with water
Appearance of milk initially			
Appearance of milk after 1 min.			
Appearance of milk after 5 min.			

Chapter Summary

● Milk contains substances in solution, in colloidal dispersions, and in emulsions.
● Milk contains all the major nutrients.
● Proper processing and storage increase the flavor, nutritive value, and shelf life of milk and milk products.
● Milk products may be fresh, canned, dried, or fermented.
● The ability of cream to form a foam depends on internal and external factors.
● Milk is sensitive to heat and shows dramatic reactions when heated.
● Fermented milk products are made from pasteurized milk inoculated with specific starter cultures.

Using Your Knowledge

1. How does casein react to acids?
2. How are casein and whey alike and different?
3. What makes milk a solution, a colloidal dispersion, and an emulsion?
4. Why is lactose a valuable nutrient in milk?
5. How do salts prevent milk from curdling?
6. How is milk processed? What is the purpose of each type of processing?
7. What does the phrase "whole milk" mean? How does whole milk compare to other classifications of milk?
8. What is the difference between sweetened condensed milk and evaporated milk?
9. How are cultured milk products made?
10. Explain what ripening is and why it's needed to make cheese.

Real-World Impact

The Value of Milk

Hippocrates, often called the father of medicine, stated that "milk is the most nearly perfect food." When compared to meat products, milk is also a very efficient way to obtain protein. This advantage is partly due to modern dairy breeding techniques, but also to the dairy cow's natural abilities. With their unique digestive system, cows eat plants that other animals can't digest. Cows often survive on land that is unsuitable for farming. The quality of plant protein is upgraded when converted into milk, contributing to its high protein-to-calorie ratio. Therefore, a growing global population may find itself relying more and more on products derived from cow's milk.

Thinking It Through
1. Why would dairy cows produce more protein for individuals on a set amount of land compared to beef cattle?
2. What is meant by the term "protein-to-calorie ratio"? Why is it significant?
3. What milk products can be stored without refrigeration? How might this impact global food needs?

Using Your Knowledge *(continued)*

11. To make a cream foam, would you get better results by using light or heavy whipping cream? Why?
12. What might happen if you leave an open pan of milk over high heat without stirring?

Skill-Building Activities

1. **Problem Solving.** Suppose you needed a type of cheese that could be stored for a long time. What would you select and why?
2. **Communication.** In a visual manner, such as a drawing, collage, or cartoon, illustrate how milk can be a solution as well as an emulsion.
3. **Management.** Suppose your family typically leaves a glass pitcher of milk on the table throughout the meal. After dinner, the remaining milk is always poured back into the milk carton. What would you suggest to your family and why?
4. **Problem Solving.** Explain what might have happened in these situations and suggest a solution for the next time: a) Mike pulled a bowl from the dishwasher and used it to whip cream, but the cream wouldn't foam. b) After making chocolate pudding, Emily discovered a thin, dark layer on the top.
5. **Critical Thinking.** Suppose a family member is making homemade cream of tomato soup. After the tomatoes are added to the milk, the soup begins to curdle. What happened?

Understanding Vocabulary

With a partner, take turns testing each other on the definitions of the vocabulary terms at the beginning of the chapter.

Thinking Lab

Cultured Milk Products

Do you like plain yogurt? If not, you can find yogurt flavored with strawberry, peach, lemon, or blueberry. Can't decide? Try a mixed fruit variety. Even sugar-free and low-fat versions are available. If yogurt choices seem wide-ranging, just think how many cheeses are available!

Analysis

Thanks to consumer demand, modern marketing, and food technology, you can choose from more varieties of cultured milk products than ever. As with many things, however, not all choices are equally good for you.

Organizing Information

1. Compare the nutrition facts on six to eight cultured milk products.
2. Which are lowest in calories, grams of fat, and cholesterol per serving?
3. Which have the highest daily value of calcium per serving?
4. Which would best meet your personal nutrition needs? Why?

The SHAKER TC
Seas® dressing b
spices to perfectio
just the right amou
poultry marinade, a
chilled vegetable sala
antipasto!

INGREDIENTS: SOYBEAN OIL, WATER, WHITE DISTILLED VINEGAR, SALT, SUGAR, DRIED GARLIC, HYDROLYZED VEGETABLE PROTEIN, SPICES, XANTHAN GUM (IMPROVES MIXING), CALCIUM DI-SODIUM EDTA (A PRESERVATIVE), OXY-STEARIN (PREVENTS CLOUDINESS UNDER REFRIGERATION), ARTIFICIAL COLORING.

Objectives

- Identify common food additives and their uses.
- Compare natural and synthetic additives.
- Explain how additives are regulated.
- Identify general and specific uses of preservatives.
- Compare methods for adding nutrients to foods.
- Describe how additives make foods more appealing.
- Describe how additives aid food processing.
- Evaluate the pros and cons of using food additives.

Terms to remember

enrichment
food additive
goiter
GRAS list
nutrification
restoration
stabilizer

"**M**ore nutritious than ever!" "New improved flavor!" "Stays fresher longer!" When you read such claims about foods, you might wonder how manufacturers make all these improvements. Food additives are often responsible.

What are food additives? Are they helpful or harmful? This chapter will help you answer these questions and more.

What Is a Food Additive?

If you read far enough on a food label ingredient list, you'll see such substances as BHT, sorbic acid, and guar gum mentioned. These are all examples of **food additives**, *any substance a food producer intentionally adds to a food for a specific purpose.* If it weren't for additives, food would be quite different than it is.

Producers use around 3,000 additives to preserve and improve foods. Acids, vitamins, and seaweed—any of these can be an additive, depending on the use. In fact, if you eat processed foods, and even many fresh ones, you're probably consuming some type of food additive. Many common food additives and their uses are shown in **Figure 24-1** on pages 380-381.

Natural and Synthetic

People have been using natural additives, such as salt and sugars, for thousands of years. These additives occur naturally in food or specific parts of plants. They are separated from the food or plant to make use of particular properties.

Increasingly, technology allows scientists to create new substances to add to foods. These artificial, or synthetic, additives are made in a laboratory; they aren't found naturally in food. Their chemical "ingredients" are the same as any that occur in nature, but the chemicals are joined or modified in the food science lab to produce new substances.

Figure 24-1

Common Food Additives

Additive	Formula	Uses
Acesulfame-K	$C_4H_4NO_4SK$	Artificial sweetener; used in chewing gum and dry beverage mixes
Acetic Acid	$C_2H_4O_2$	pH control; acid in vinegar; used in processing cheese
Ammonium alginate	Polysaccharide	Natural thickener for ice cream and yogurt; made from seaweed
Ascorbic acid (vitamin C)	$C_6H_8O_6$	Natural nutrient in citrus fruits; antioxidant; used in cured meats; prevents fruit from browning
Aspartame	$C_{14}H_{18}O_5N_2$	Artificial sweetener
Azodicarbonamide	$C_3H_4O_2N_3$	Bleaching agent in flour
Benzoyl peroxide	$C_{14}H_{10}O_4$	Bleaching agent in cheese and flour
Beta carotene (vitamin A precursor)	$C_{40}H_{56}$	Natural nutrient in vegetables; yellow coloring agent in margarine, butter, cheese
BHA (butylated hydroxyanisole)	$C_{11}H_{18}O_2$	Antioxidant for fats, oils, cereals; retards rancidity
BHT(butylated hydroxytoluene)	$C_{15}H_{24}O$	Antioxidant for fats, rice, gum; retards rancidity
Biotin	$C_{10}H_{18}N_2O_3S$	Nutrient; vitamin supplement
Calcium alginate	Polysaccharide	Stabilizer
Calcium propionate	$C_6H_{12}CaO_4$	Preservative; mold inhibitor in bread
Calcium silicate	$CaSiO_3$	Anticaking agent in salt, powdered sugar, baking powder
Carob bean gum	Polysaccharide	Stabilizer and thickener
Carrageenan	Polysaccharide	Emulsifier in evaporated milk, sour cream, cheese foods
Guar gum	Polysaccharide	Stabilizer for ice cream, soups
Mannitol	$C_6H_{14}O_8$	Anticaking; nutritive sweetener
MSG	$C_5H_8NNaO_4$	Flavor enhancer in soups, Chinese foods
Potassium iodide	KI	Nutrient added to salt
Potassium nitrite	KNO_2	Curing; pickling; coloring agent
Potassium sorbate	$C_6H_7KO_2$	Preservative; mold inhibitor
Propionic acid	$C_3H_6O_2$	Preservative; mold inhibitor
Propyl gallate	$C_{10}H_{12}O_5$	Antioxidant; preservative in gum
Saccharin	$C_7H_5NO_3S$	Artificial sweetener

Figure 24-1

Common Food Additives (continued)

Additive	Formula	Uses
Silicon dioxide	SiO_2	Anticaking agent; beer production
Sodium alginate	Polysaccharide	Stabilizer
Sodium citrate	$C_6H_5Na_3O_7$	pH controller; meat curer
Sodium nitrate	$NaNO_3$	Curing; pickling agent; meat curer
Sodium propionate	$C_3H_5NaO_2$	Fungicide used in baked products
Sorbic acid	$C_6H_8O_2$	Fungistatic agent in baked goods
Sorbitol	$C_6H_{14}O_6$	Nutritive sweetener
Sucralose	$C_{12}H_{19}O_8Cl_3$	Artificial sweetener; used in carbonated beverages and many foods
Tartaric acid	$C_4H_6O_6$	pH controller used in soft drinks
Tocopherols (vitamin E)	$C_{26}H_{44}O_2$	Natural nutrient; antioxidant; rancidity deterrent
Tragacanth gum	Polysaccharide	Stabilizer; thickener; texture additive
Vanillin	$C_8H_8O_3$	Flavoring agent (synthetic vanilla) in ice cream, baked goods, candies
Xylitol	$C_5H_{12}O_5$	Nutritive sweetener in dietary foods

Additives give these jellies their bright colors.

Regulating Additives

Various government groups regulate additive use. On a global level, the Food and Agriculture Organization (FAO) and the World Health Organization (WHO) provide oversight. These groups together have established strict standards about what additives may be used and how they are tested.

The Food and Drug Administration

In the United States, the Food and Drug Administration (FDA), a branch of the Department of Health and Human Services, is the agency most responsible for assuring that the food consumers buy is safe to eat, nutritious, and honestly represented. Monitoring additives involves all of these duties.

Gaining FDA acceptance for an additive is no simple task. The agency has exacting guidelines. A manufacturer must submit evidence from extensive tests showing that the substance is effective and doesn't cause short- or long-term harm, including cancer and birth defects. If satisfied, the FDA determines how the additive may be used and in what amount, using a 100-fold margin of safety. This means that a food can contain no more than 1/100 of the amount of an additive found to be safe in laboratory animals.

Another FDA standard is called the Delaney Clause, which is an article included in the 1958 Food Additive Amendment. This clause states that no substance shown to cause cancer in humans or animals may be added to food.

With so many additives possible, how do manufacturers avoid testing every spice or seasoning they want to use? In 1959 the FDA created a special class of additives that are "generally recognized as safe," or GRAS. The **GRAS list** contains *substances, such as spices, natural seasonings, and flavorings, that are considered safe for human consumption and not regulated as additives.* The GRAS list currently numbers about 670 items. It is updated periodically as new facts about additives come to light.

The FDA also limits the type and amount of unintentional, or indirect, additives a food may contain. These substances get into the food supply by accident. They have no purpose and some, such as molds and pesticides, can be harmful.

Extensive testing must be done before an additive is approved for use. Who makes the final decision about which additives are safe?

BUTTER, DEXTROSE, SOY LECITHIN – AN EMULSIFIER, AND VANILLIN – AN ARTIFICIAL FLAVOR), HIGH FRUCTOSE CORN SYRUP, SUGAR, DEXTROSE, LACTOSE, CORNSTARCH, WHEY, BAKING SODA, SALT, MOLASSES, EGGS, CARAMEL COLOR, NATURAL AND ARTIFICIAL FLAVOR, ANNATTO EXTRACT (VEGETABLE COLOR).

Some people are surprised to discover how many additives are used in the foods they buy. What benefits are there to knowing what's in the food you eat?

How Additives Are Used

Food additives have many uses; however, decreasing a food's nutritional value and deceiving the consumer are not among them. Additives are not meant to conceal damage, spoilage, or low quality.

In general, the true purposes of additives fall into four categories. Additives are used in foods to improve storage properties, increase healthfulness, make food more appealing, and improve processing and preparation.

Improving Storage Properties

Two hundred years ago, fresh foods weren't a luxury; they were a way of life. Some products could be salted or dried for longer storage, but most were eaten as they became available. Diets were largely limited to locally produced foods. A California orange sent to a market in Boston would be inedible long before it arrived.

Food suppliers need to get food items to consumers before mold and contamination ruin the product. What impact do you think preservatives have had on food manufacturing?

Ingredient Labeling

Another tool the FDA uses to hold manufacturers accountable for additives is the food label. Manufacturers are required to list all food ingredients on the label. By checking ingredients and additives on the label, you can decide whether to buy a food.

To help people make informed choices, the FDA has strict rules on how additives are identified on labels. Suppose you want products made with only natural ingredients. The word "flavored" on a label tells you that the food uses only natural flavorings. These might be a variety of flavors, but all occur naturally. A food labeled "artificially flavored" contains some, and possibly only, synthetic flavorings.

Today, that orange can be treated with preservatives. Preservatives are usually chemicals used to prevent mold and bacteria from spoiling food. Preservatives can slow down the process that makes many fatty foods turn rancid. They can prevent certain fresh fruits and vegetables from turning brown. Commonly used chemical preservatives include sodium nitrite, sorbic acid, sodium bisulfite, and sodium nitrate.

Preservatives are generally chosen because they are economical and don't affect a food's flavor, color, or texture. Occasionally a preservative has an added purpose. Sodium nitrite, for instance, is used to color and flavor such cured meats as bacon and corned beef. Although some natural substances can be used as preservatives, they aren't considered additives. These include salt, organic acids (such as acetic acid in vinegar), sugar, and spices.

Increasing Healthfulness

Increasingly, additives are included to boost a food's nutritional profile. Nutritional additives have the following main purposes:

- **Fortification.** Fortification is adding nutrients that are not normally found in a food. Milk is fortified with vitamin D. Vitamin A is added to margarine, and vitamin C is added to many soft drinks.
- **Restoration.** *Nutrients that are lost in processing are returned to the food with the process called* **restoration**. Restoration reestablishes the product's original nutritive value. Vitamin C, for example, is put back into canned citrus juice, such as orange juice, during restoration.
- **Enrichment.** Like restoration, **enrichment** involves *adding nutrients lost in processing*. Enriched foods, however, contain more nutrients than existed in the food before processing. Levels of thiamin, niacin, riboflavin, and iron are often increased. Cereal products are likely candidates for enrichment because many natural vitamins and minerals are lost during processing. Enriched flour is common in the United States, England, and Canada, but banned in France.

Researchers recently discovered that folic acid is needed to prevent birth defects of the nervous system. Since getting the recommended amount of folic acid from food is difficult, bread, pasta, and grain products are now fortified with this B vitamin.

Nutrition *Link*

Antioxidants: Fountain of Youth?

Vitamins are multiuse additives that improve food as well as health. The antioxidant vitamins are believed to have special value.

Improving food. Vitamin C (ascorbic acid) helps cured meats keep their red color. Beta carotene, the vitamin A precursor, is a common coloring agent in margarine. These two vitamins, as well as vitamin E (tocopherol), are also antioxidants. They slow or prevent oxidation. As foods react with oxygen, cells lose electrons. When vitamins and lipids break down, a food's nutritional value is lessened and spoilage can occur. Colors and flavors can change, making food less appetizing.

Antioxidants act as a substitute. The oxygen reacts with the antioxidant instead of with the food, saving the food's quality and appeal. That's how vitamin C helps keep fruit flavorful and how vitamin E helps keep oil from going rancid.

Health benefits. Now research suggests that antioxidants may do the same for people. As body cells work, they produce "free radicals"—damaged molecules that are unstable because they lack one electron. To stabilize themselves, they may "snatch" an electron from a neighboring atom. The "donating" atom then becomes unstable. If that atom is part of a liver cell or brain cell, then the liver or brain suffers a tiny bit of damage. Over time, the damage builds and the organ wears down.

In theory, antioxidants slow this process by supplying electrons to free radicals, thus sparing the organ's cells. The organ remains healthier for a longer time. It functions better and is better able to withstand some diseases. Studies indicate that antioxidants may help prevent cancer, cataracts, and atherosclerosis (a-thuh-RO-skluh-RO-sus).

More research is needed, however, before antioxidants can claim the title "fountain of youth." Even if these vitamins do slow the effects of illness and aging, no one is sure how much is needed or what combinations are best. Since megadoses of vitamins can be dangerous, scientists advise what parents have been saying for generations: eat your (antioxidant-rich) vegetables, especially the deep yellows and dark greens. Including a few citrus fruits each day is a good idea, too.

Applying Your Knowledge

1. Why are sweet potatoes and broccoli valuable additions to a diet? What other foods are similarly helpful?
2. What might be some long-term impacts on society if the theory about antioxidants is true?

- **Nutrification. Nutrification** is *a process that adds nutrients to a food with a low nutrient/calorie ratio so the food can replace a nutritionally balanced meal.* Nutrified bars and shakes, for example, are staples in some diet plans.

Making Food More Appealing

Food that looks and tastes good is more likely to be eaten. As manufacturers know, the food is also more likely to be purchased. That's why additives are used to add appeal to food products.

Color

As you can see from some children's cereals, manufacturers use a rainbow of food colors to capture interest. Almost all soft drinks, cheeses, ice cream, jams, and jellies owe at least part of their coloring to additives.

Some colors are made from food—the browning of caramelized sugars, for example. However, nearly half of the most common colorings are created in laboratories. Each of these synthetic food colors is identified with a number. A hearty brown barbecue sauce gets some help from Red No. 40, while the green of pistachio pudding is an artful blend of Blue No. 1 and Yellow No. 5 (known to chemists as tartrazine). Synthetic colors are also called certified colors, because the maker and the FDA test every batch for purity and safety. Numbers are sometimes preceded by "FD&C," for the Food, Drug, and Cosmetic Act of 1938, which gives the FDA much of its regulatory power.

Flavor

More additives are used to flavor foods than for any other purpose. About 2000 natural and synthetic flavors are available.

For practical reasons, most added flavors are artificial. The demand for natural flavorings far exceeds the supply. In the United States, five times as many products are grape-flavored as are flavored by the Concord grape crop. Sometimes using natural flavors would make a food too costly to produce. It takes over two tons of cocoa beans to produce one ounce of cocoa oil. If not for synthetic cocoa flavor, imagine what you would have to pay for a cup of hot chocolate. Synthetic flavors are also better suited for mass production. They are uniform in strength and made to stand up to the rigors of processing.

Flavor enhancers are substances that impart no flavor of their own but bring out the flavor already in food. They are often made from the amino acid glutamate. When not bound to other

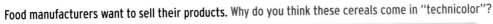

Food manufacturers want to sell their products. Why do you think these cereals come in "technicolor"?

Food scientists must develop flavors that people like. Why would products with natural flavors probably be more costly?

aminos, glutamate allows for more pronounced food flavor; mushrooms are a natural example. The enhancer MSG (monosodium glutamate) is a salt form of glutamate. Because some people react allergically to MSG, concerns have been raised about its use.

Sweeteners

Of all flavor enhancers, sweeteners are the most common. They improve the aroma and taste of food. Sweeteners are basically either *nutritive* or *nonnutritive*.

Nutritive sweeteners metabolize to produce calories. Table sugar (sucrose), brown sugar, maple syrup, molasses, and honey are natural, nutritive sweeteners. A nutritive sweetener that is very similar to sucrose is *sorbitol* (SORE-buh-tahl). This glucose-based, sugar alcohol has as many calories as sucrose but tastes only half as sweet. Dieters and people with diabetes use sorbitol because it doesn't metabolize as well as sucrose. Sorbitol absorbs more slowly from the intestinal tract than su-

crose does, so the blood sugar level may not raise as high. Sorbitol is used in "sugarless" gum because it doesn't cause dental cavities as sucrose and other sugars do.

Nonnutritive sugars are also called artificial sweeteners. They have no calories but still taste sweet. Food scientists have created these additives to produce low-calorie and calorie-free products. A combination of artificial sweeteners may be used. The following sweeteners are currently approved by the FDA:

- **Sucralose.** This sweetener is made from sugar but is 600 times sweeter. It is created by replacing three hydroxyl groups on each sugar molecule with three chlorine atoms. The chlorine stabilizes the molecule at high heats, making sucralose suitable for baking. It also leaves the molecule inert, so the sweetener produces no calories.
- **Saccharin.** Made from petroleum products, saccharin is 300 times as sweet as sucrose. It remains stable in many products under extreme processing conditions. If used in

With dieting so common in society, manufacturers continually search for better ways to sweeten products without adding calories. Eliminating any aftertaste and finding the right level of sweetness are two of the challenges.

any great amount, however, it leaves a bitter aftertaste.

- **Aspartame (AS-pur-taym).** Aspartame is one of the most thoroughly tested products in marketing history. It is about 200 times sweeter than sugar, supplies no calories, and leaves no aftertaste. Aspartame metabolizes in the body to produce methanol (methyl alcohol) and two amino acids: aspartic acid and phenylalanine. Aspartame is unstable when heated so it can't be used in baked or cooked products. It loses its sweetness in beverages, which is why many diet sodas have a use-by date.

- **Acesulfame-K (ay-see-SULL-fame-KAY).** This sweetener is a synthetic derivative of acetoacetic acid (a-SUH-toh-uh-SEE-tik). It has little disagreeable aftertaste and is often used in combination with other sweeteners. It is heat stable and used in cooked and baked products.

Improving Processing and Preparation

Additives are also used in processing and preparing food. Some give products a more appealing texture. A **stabilizer**, for instance, is *a substance that keeps a compound, mixture, or solution from changing its form or chemical nature.* Without stabilizers, the fat in peanut butter separates from the protein, creating an oily pool over a stiff paste.

Thickeners are stabilizers that contribute smoothness or body to a food. Ice cream is creamy, in part, because thickeners prevent crystals from forming as it freezes and stabilize the foam as the ice cream is whipped. Other thickeners make frostings less sticky.

Stabilizers and thickeners are usually natural additives. Many are starch-based. Some are made from proteins, including pectin, casein, sodium caseinate, and gelatin. Another common source is the gum extracted from some bushes, trees, and seaweed.

Other additives, called buffers, are used to achieve the desired pH in preparing and preserving foods. Buffers help maintain the balance of hydrogen to hydroxide ions. Citric acid, sodium citrate, and lactic acid are examples. Bases such as sodium bicarbonate can also act as buffers.

Other processing aids include leavenings and emulsifiers. Anticaking agents, such as calcium silicate and silicon dioxide, prevent clumping in powdered foods by absorbing moisture.

Concerns About Food Additives

"Contains no preservatives." "Made with only fresh ingredients." "Totally natural." Phrases like these, used to promote food products, are signs that not everyone is sold on using food additives. Opponents believe that some additives cause "more trouble than they're worth."

The effect of consuming large quantities of artificial sweeteners still isn't clear. That's why nutritionists recommend drinking diet soft drinks in moderation. What other beverages might be overlooked if soft drinks are chosen too often?

If you buy natural peanut butter, you'll see a layer of oil on top after the product sits. Why won't this layer form on other types of peanut butter?

Long-Term Effects

One concern is that not enough is known about the long-term effects of using additives. Many people worry about substances that seem safe now but might be proved otherwise with time and scientific advances.

Nitrites are a good example. For centuries, these compounds have been used to preserve and flavor cured meats. They also inhibit growth of the bacteria that cause botulism. Researchers now know that under intense heat, nitrites can react with amines found in meats. This reaction forms nitrosamines (ny-TROH-suh-meens), which are suspected of causing cancer. No case of cancer in humans can be traced directly to nitrosamines, however, and nitrites' ability to prevent botulism is a valuable one. Thus, small amounts of nitrites are allowed in cured meats. Nitrates are also used in curing meats, but their use, too, is carefully limited by the FDA.

Likewise, saccharin was removed from the GRAS list when extensive testing showed it could produce cancer in laboratory animals. The sweetener was very popular among people with diabetes, however, as well as many others

The flavor of processed meats appeals to many people. Why is it a good idea to eat such foods in moderation?

who had used it for years without ill effects. Therefore, Congress has repeatedly passed special legislation to suspend the ban on saccharin, which may be sold with a warning on the label.

Food Allergies and Sensitivities

If you're allergic to apricots, you can just avoid eating them. If you have an asthmatic reaction to the sulfites used to preserve dried apricots, the situation is not so simple. This problem affects only a small percentage of people, but their reaction can be severe. Other people are sensitive to MSG and a few react badly to the preservative BHT.

MSG is an additive that has traditionally been used to enhance the flavor of Chinese foods. Although some restaurants use the additive, some don't. A person who is allergic to MSG can request that a dish be prepared without it.

Food Inspector

What makes food safe for the public? Ask a ... Food Inspector

As a food inspector, Rob Weigel works for the federal government. He has a full-time assignment in a large meat processing plant. According to Rob, "My work could easily make me the 'bad guy' at the plant. That's why I 'wear two hats,' one as an authority and one as a public relations person. Federal regulations for food processing and sanitation are complicated. My job is to make sure that people in the plant where I work understand the rules and comply with them. That means I have to enforce the standards but build a spirit of cooperation with company personnel at the same time."

Education Needed: As a food inspector, experience is usually more important than a bachelor's degree. Many inspectors in meat processing have worked as a butcher or in a slaughterhouse or processing plant. Chefs and cooks who've been responsible for determining product quality could also move into this career. Their experience with proper food preparation, handling, and sanitation is useful.

KEY SKILLS

- Decision-making skills
- Physical fitness
- Verbal communication skills
- Personal relations

Life on the Job: Food inspectors work with plant managers to evaluate production processes and find innovative ways to meet standards while keeping costs low. They visually examine food products, making sure they are high quality and fit for consumption. Some inspectors travel to plants within a territory; others are stationed only in one plant. Food inspectors typically stand for long periods of time and do some heavy lifting.

Your Challenge

Suppose you inspect fruit and vegetable processing plants. One plant manager hints that he might help you out in some way if you overlook a problem you've discovered.

1. What would you say and do as part of an ethical response?
2. What follow-up might be needed?

Some foods that you eat and drink are fortified with vitamins and minerals. How can you judge whether you're consuming the right amounts of nutrients?

The FDA has responded by reducing the amount of sulfites allowed in foods. Those and other additives must be included on food labels. Critics point out that labels may not be on hand when you eat a food and may be unhelpful in any case. If you must avoid milk products, a label can't tell you whether you should eat a food stabilized with casein, a milk protein.

Poor Eating Habits

Opponents warn that nutritional additives may actually lead to less nutritious diets. If your breakfast cereal is fortified with a dozen vitamins and minerals, you might be tempted to skimp on other healthful foods that contain those nutrients. In doing so, you would miss out on needed fiber, protein, and other, possibly unknown, nutrition essentials. For example, some newly identified substances in plants, called phytochemicals, seem to play a role in preventing cancer. Their exact benefit is still under study.

On the other hand, the argument continues, you may be getting too much of a nutrient. Frequently, breakfast cereal, milk, and orange juice are vitamin-fortified. Adding nutrients where they don't occur naturally makes getting balanced nutrition more confusing.

Unneeded Additives

Have you noticed that three-day-old apples in the supermarket look much shinier than apples freshly picked from a tree? Apples, oranges, eggplant, and lemons are some of the produce treated with a light coat of oil-based waxes. These waxes are FDA-approved as preservatives; they help maintain freshness by sealing in moisture. Making the fruit look more appealing to consumers is a bonus for sellers.

Which tastes better, an apple with or without a shiny exterior? Which one will people more likely buy? The producer's motivation to use additives can be economic as well as practical.

Critics see such practices as a troubling trend. They warn that by conditioning buyers to set unnatural standards for foods, producers discourage people from eating what's really good for them. You may come to prefer the artificially colored, artificially flavored strawberry in a toaster pastry to the real, more nutritious strawberry itself. Moreover, using additives raises the price of the product.

The Value of Food Additives

Supporters of additives offer their own arguments. They see many problems and little good coming from a ban on these substances, especially given the many national and international safeguards against abuse.

Safety

Proponents stress that producers and consumers alike now rely on additives to help ensure a safe food supply. Preservatives extend the shelf life of some products. A loaf of bread stays edible because of the mold inhibitor calcium propionate and the BHT that keeps fats from going rancid. The additive ethylenediamine tetraacetic acid (ETH-uh-leen-DY-uh-meen TET-ruh-uh-SEE-tik), or EDTA, also helps prevent rancidity. EDTA chemically traps metal ions that might enter and spoil foods that are processed with metal machinery.

Improved Nutrition

Far from complicating good nutrition, supporters say, additives prevent diseases caused by malnutrition. In fact, the first case of fortification in the United States was prompted by high rates of **goiter** (GOY-tur), *an enlargement of the thyroid gland caused by a lack of iodine.* Iodine was first added to table salt in 1924. Iodized salt has almost eliminated goiter in this country.

Other "success stories" include the addition of vitamin D to milk in the 1930s to combat rickets, a bone-deforming childhood disease. Fortifying flour and cornmeal with iron and niacin greatly reduced occurrences of pellagra (puh-LA-gruh), a vitamin B deficiency disease, which had been the leading cause of death in the southern United States.

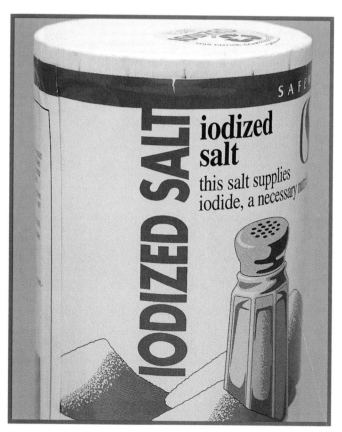

Salt is the primary source of iodine in food. Since it's an additive, you may not get the iodine you need in your diet unless you buy the container that says iodized on it.

Preservatives, too, contribute to improved nutrition. By helping to solve problems of long-distance travel and long-term storage, preservatives allow people to enjoy a more varied diet. Also, color and flavor enhancers make nutritious foods more appealing, which encourages people to eat them.

World Hunger

With the world's population topping six billion as the century turns, there seems to be a hungry mouth for every bit of food produced. With additives, foods can be shipped safely over long distances. They can be stored in less than ideal conditions without losing quality. More food is available to the people who need it.

Supporters suggest that some fears about using additives are a reaction against using "chemicals." All life, however, is made up of chemicals. Used responsibly, chemical additives are no more dangerous than the food itself.

The bottom line regarding additives, as with other decisions, is that consumers can and must educate themselves about what goes into the foods they eat. Every day research gives new information. Anyone who has a question or concern about additives can contact the FDA or local government agencies.

Concerns about the use of additives must be balanced with their advantages. Suppose this is the question: "Is it better to use preservatives in order to send food to a hungry population or to withhold food because you want the food to remain pure?" What would your answer be?

Testing for Food Additives

SAFETY FIRST
Review these safety guidelines before you begin this experiment.

As you learned in Chapter 19, one method of preventing browning in dried fruits is by sulfuring, a technique in which fruits are exposed to sulfur dioxide gas for varying amounts of time. Once the process has been completed, there is no way to tell visually if a fruit has been exposed to the sulfuring process. In this experiment, you will use a series of chemical reactions to determine whether or not sulfur dioxide was used to preserve the fruit.

Equipment and Materials

dried fruit (two kinds)
distilled or deionized water
hydrogen peroxide solution

400-mL beaker
stirring rod
dropper (or dropping bottle)

250-mL beaker
filter paper
barium chloride solution

100-mL graduated cylinder
funnel

Procedure

1. Place 2-4 pieces of one type of fruit in the 400-mL beaker and cover the fruit with distilled or deionized water.
2. Let the mixture stand for 15 minutes, stirring often with a stirring rod. As the fruit absorbs the water, you may need to add more water to the beaker.
3. Drain the liquid from the fruit through filter paper in a funnel into a clean, dry, 250-mL beaker. This liquid is called the filtrate.
4. Add 50 mL of 3-percent H_2O_2 (hydrogen peroxide) to the filtrate.
5. Using a dropper, add several drops of $BaCl_2$ to the filtrate. If a white solid forms, SO_2 was used to preserve the fruit. If no white solid forms, sulfur dioxide was not used.
6. Repeat the above steps, using another fruit and record all information in your data table. Check with other lab groups to determine the results for fruits you didn't test. Add to your table.

Analyzing Results

1. Which of the fruits had been exposed to the sulfuring method of preservation?
2. Why are some fruits exposed to SO_2?
3. What could happen to fruits during shipping or after sitting in a store for long periods of time if SO_2 were not used?
4. What other types of additives may be in the fruits you tested?
5. SO_2 is also used to change the color of foods. Raw sugar is brown, but when bleached with SO_2, it becomes white. Why do you think sugar is bleached?

SAMPLE DATA TABLE

Kind of Fruit	White Solid (yes/no)	Sulfur Dioxide Used (yes/no)

Chapter Summary

- Approximately 3,000 food additives serve many functions in the food industry.
- The United States Food and Drug Administration closely regulates the use of food additives.
- Additives may be created synthetically or derived from sources found in nature.
- Preservatives help extend the shelf life of food products and prevent spoilage.
- Nutritional additives restore or improve a food's nutrient value.
- Additives can be used to improve the color and flavor of food.
- Natural and synthetic sweeteners are common flavoring additives.
- Some additives are used during processing to improve a food's texture.
- While additives have a proven worth in foods, some people worry about the unanswered questions and unintended consequences of their use.

Using Your Knowledge

1. In general, what is a food additive?
2. How are natural and synthetic additives different?
3. How does an additive gain acceptance from the FDA?
4. What is the Delaney Clause?
5. Why is the GRAS list useful?
6. What are four basic uses for additives?
7. What is sodium nitrite and why is it used?
8. Compare four techniques producers use to increase the nutritional value of food.
9. Describe three artificial sweeteners.
10. Would marshmallow creme be possible without stabilizers? Why or why not?
11. Explain the arguments for and against using nutritional additives in food.
12. According to supporters of food additives, what would result from eliminating preservatives?
13. How would you advise someone who is worried about the chemicals in food?

Real-World Impact

Good Foods, Globally

Since 1962, the Codex Alimentarius Commission has helped nations agree on food safety and trade regulations, including the use of additives. Each member nation sends representatives, who shape policy affecting the international food supply. New proposals undergo a lengthy process of discussion and public comment before being accepted or rejected. Although no nation is forced to accept these standards, the World Trade Organization uses them to set its own policy and resolve disputes.

Thinking It Through

1. Should the standards of Codex be mandatory for all nations? Why or why not?
2. How might cultural and economic differences among countries affect Codex standards?
3. Why would international trade be easier for those who follow Codex guidelines?

Skill-Building Activities

1. **Math.** Suppose a food additive has been approved by the FDA. It has been proven safe in laboratory animals when used in amounts up to 300 mg per cup. How much of this additive would the FDA allow in a cup of the product?

2. **Consumer Skills.** Bring in labels from three dissimilar food items, such as cereal, a dairy product, and canned soup. Identify the additives listed on the labels and explain the purpose of each one.

3. **Communication.** Imagine that a food manufacturer wants to market a new soft drink fortified with eight vitamins and minerals. It would be available in regular and diet varieties. Assume the role of either a food additive opponent or advocate. Write the speech you would give to a consumers group, explaining why they should or should not endorse the product.

4. **Technology.** Visit the Food and Drug Administration on the Internet at www.fda.gov/ and try to learn more about food additive research and regulation. Report your findings to the class.

5. **Social Studies.** Investigate another nation's policy regarding the use of food additives. What agencies are responsible for regulation? How does the country's policy compare to that of the United States?

6. **Science.** Research the process of creating additives in laboratories. What professions have input? What roles do safety, economics, and ethics play?

Understanding Vocabulary

Write a question about each of the "Terms to Remember" in this chapter, using facts from the chapter. The answer to each question should be one of the terms. Take turns with other class members asking your questions.

Thinking Lab

Natural Versus Synthetic

Synthetic additives are born in the laboratory. Natural additives, while they are derived from sources found in nature, undergo considerable change at the hands of scientists. Amino acids in plants are separated to form hydrolyzed protein. Seaweed components are chemically altered in the thickener alginate (AL-ji-nayt). Sugar molecules in cornstarch are rearranged and "reborn" as the sweetener, high-fructose corn syrup. Such changes produce additives with useful properties in the food industry.

Regardless of their beginnings, however, all additives are just chemicals to the human body. Natural or synthetic, both are treated the same.

Analysis

More people today are concerned about the ingredients in the foods they choose for themselves and their families. Many claim a preference for "natural" foods. Responding to this trend, manufacturers often label and advertise foods as "all natural" and "nothing artificial added."

Thinking Critically

1. Why do you think some people are concerned about consuming synthetic additives but not natural ones?
2. Will knowledge of chemistry make people more or less comfortable with synthetic additives? Why?
3. Do you think the terms "natural" and "artificial" should be restricted in promoting foods? Explain your answer.

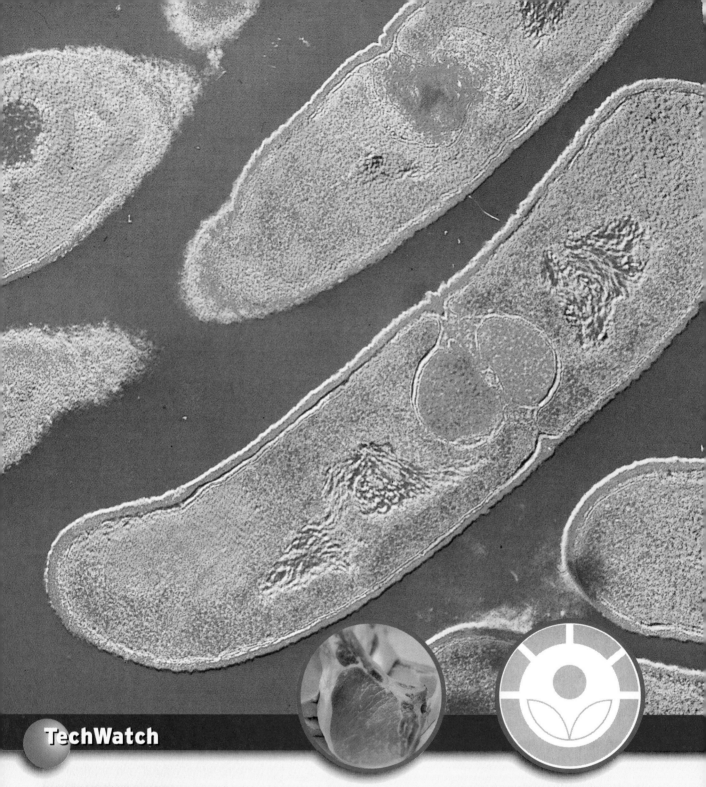

TechWatch

IRRADIATION OF MEAT

After USDA approval, the meat packing industry began irradiating meats in 2000. Irradiation is currently the only known method to eliminate deadly *E. coli* O157:H7 bacteria in raw meat. Irradiation also significantly reduces levels of *Listeria*, *Salmonella*, and *Campylobacter* on raw products. Because irradiation doesn't destroy some bacteria, especially spoilage organisms, consumers still need to handle and prepare irradiated meat and poultry properly. The USDA requires that all irradiated meat and meat products be labeled with the radura international symbol for irradiation as well as a statement that the product was treated by irradiation. For unpackaged meat products that don't have labels, this information and logo must be displayed where the meat is sold.

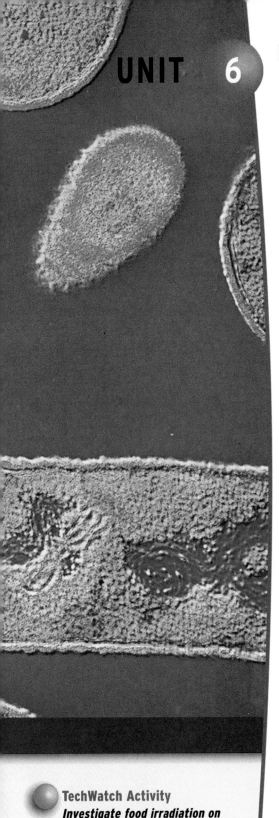

The Microbiology of Food Processing

TechWatch Activity
Investigate food irradiation on the Internet and in other resources. Report to your class on public reactions to the process of irradiation.

Keeping Food Safe

Objectives

- Name and describe the microorganisms that cause food spoilage.
- Differentiate between food intoxication and food infection.
- Identify sources and symptoms of foodborne illnesses.
- Explain the role of various government agencies that keep the food supply safe.
- Demonstrate steps to prevent the spread of foodborne illnesses.
- Assess the safety of food preparation methods.

Terms to remember

adulterated
botulism
cross-contamination
food infection
food intoxication
foodborne illness
parasites
pathogenic
pesticides
residue
spores

"Wash your hands before you eat!" This standard request has been around for years, and the words still offer good advice for avoiding illness. Clean hands, however, may not offer enough protection if something isn't right with the food you eat.

According to estimates, every year in the United States alone, at least 6 million people get sick from something they ate. Some estimates put the number as high as 80 million. Nine thousand people die. These numbers are alarming. They clearly point to the responsibility of everyone along the food supply line to make sure that food is safe. As you learn about the science of food contamination, you can share that responsibility by making sure that the foods you and others serve and eat are wholesome.

Foodborne Illness

Approximately 250 foodborne illnesses have been identified. A **foodborne illness**, sometimes called food poisoning, is *an illness caused by eating food that has been contaminated in some way.* Contamination can come from such things as microorganisms and their toxins, animal parasites and their eggs, viruses, and toxic chemicals.

These culprits often do their work in secret. When foodborne illness strikes, you might not notice any change in how a contaminated food looks, smells, or tastes. You will soon feel the effects of food poisoning, however, and its assorted symptoms. Severe cases require hospitalization, but even your last bout of "stomach flu" could have been a foodborne illness.

Microorganisms and Foodborne Illness

When food spoils and causes illness, microorganisms are often to blame. These microorganisms are typically mold, yeast, and bacteria. You're already familiar with the good work that some of these organisms do. Yeast leavens bread and ferments foods. Certain bacteria aid digestion. In contrast, however, some microorganisms can be the unwelcome cause of illness.

- **Molds.** These aerobic microorganisms, which give spoiled foods a furry look, are at home in pH ranges from 2–8.5. They prefer a moist environment and moderate temperatures ranging from 20°C, which is room temperature, to 35°C.

  Some molds can grow, though more slowly, when temperatures drop to 0°C. Thus, you may find tufts of mold fuzz on bread you keep too long in the refrigerator. Because you can see and discard moldy foods, these microorganisms generally produce more waste than danger. Certain molds on improperly stored nuts, grains, and legumes, however, produce mycotoxins that can be harmful.

- **Yeast.** Yeast is a one-celled plant. Like mold, most yeasts are aerobic, although those used in fermentation can grow without oxygen. Moisture and a pH range of 4–6.5 are essential for yeast growth.

  Flourishing in an acid environment, yeast commonly forms on top of pickles and sauerkraut. It reacts with the organic acids in these foods, making the environment less acid. Other organisms that prefer a higher pH can then continue the spoilage. Yeast thrives at temperatures between 20°–38°C. Growth is inhibited by low temperatures and sometimes by sugar levels of 65 percent or more. However, some yeasts prefer high concentrations of sugar, salt, or other solutes. Foods with a high sugar content— syrups, dried fruits, and concentrated fruit juices, to name a few—are prime breeding ground for these yeasts. Yeasts can "ride out" harsher conditions by forming **spores**, *microorganisms in a dormant, inactive, or resting state*. Yeast spores can survive

Mold on food tells you that the food should be discarded.

temperatures up to 100°C. Yeasts are probably the least threatening microorganism to food supplies. That's because they are routinely killed when food is processed and prepared and no risk occurs in preservation.

- **Bacteria.** By far the most threatening microorganisms in the food supply are bacteria. Although about 20 types of bacteria can cause foodborne illness, about eight are major causes. These single-cell organisms are found in air, soil, water, and food. They can grow both aerobically and anaerobically, depending on the type. Thus, some bacteria flourish even in a sealed container. Like yeast, bacteria grow within a fairly limited pH range of 4.5–7. They need moisture and do best at temperatures ranging from 20°–50°C. Some resistant varieties are completely destroyed only at very high temperatures, but most are killed in boiling water. When bacteria cause foodborne illness, two general results occur: food intoxications and food infections. These are summarized in **Figure 25-1** on page 404 and discussed below.

Food Intoxication

Food intoxication is *a bacterial foodborne illness that occurs when microorganisms grow in food and produce a toxin there. The toxin causes the illness when the food is eaten.* A few examples of toxin-producing bacteria are described here.

Clostridium Perfringens

Certain amounts of the bacteria *Clostridium perfringens* (klahs-TRID-ee-um pur-FRIN-jens) are in your intestinal tract all the time. Fortunately, they don't grow there. To make you sick, they must be ingested in large numbers. Once in the body, *C. perfringens* form spores that release toxins in the intestinal tract. The bacteria can produce toxin only in the body. Since the toxin isn't found in the food, these bacteria have qualities of both an intoxication and an infection.

A typical setting for *C. perfringens* poisoning is the holiday meal or family gathering, where foods are served, then left out as people return for extra helpings. Meanwhile, the temperature of these foods hovers around 50°C. Your favorite casserole or stew, as well as the turkey, meat gravy, and dressing, becomes a welcome habitat for the anaerobic *C. perfringens*.

Likewise, these bacteria can grow rapidly during low-temperature, long-time warming and cooking, such as in a slow cooker or a cafeteria heating table. They pose a problem for the food service industry, earning the nickname "the buffet germ" because they grow in conditions associated with large-scale feeding. Thus, outbreaks of this type of food poisoning usually involve large groups of people.

C. perfringens poisoning causes stomach cramps, diarrhea, and occasional nausea. Symptoms generally appear 4–22 hours after eating and last from one to five days. This illness is rarely fatal. Most cases are never reported or are misdiagnosed as the flu or another viral infection.

Sealed cans protect from most bacteria. What kind of bacteria could grow in these cans? Why?

Figure 25-1

Common Food Intoxications and Infections

Illness	Cause	Symptoms/Effects	Likely Sources	Prevention
Clostridium food poisoning	Toxin produced by *Clostridium perfringens*	Diarrhea, stomach cramps, chills, headache	Meat, poultry, hot foods cooled to room temperature	Keep hot foods hot; refrigerate uneaten foods promptly
Staph poisoning	Toxin produced by *Staphylococcus aureus*	Severe vomiting, diarrhea, stomach cramps	Moist, cooked meat dishes or starchy foods cooled to room temperature	Refrigerate uneaten foods promptly
Botulism	Toxin produced by *Clostridium botulinum*	Nausea, vomiting, diarrhea, fatigue, double vision, muscle paralysis, respiratory failure	Low-acid, canned foods, honey (for infants)	Boil home-canned foods; do not give infants honey
Salmonellosis	Infection by *Salmonella*	Nausea, diarrhea, abdominal pain, headache, fever, weakness	Meat, poultry, egg, and milk products; cross-contamination	Cook foods thoroughly; follow sanitation rules carefully
E. coli poisoning	Infection by *Escherichia coli O157:H7*	Diarrhea, stomach cramps, hemolytic uremic syndrome, kidney failure, brain damage	Undercooked ground beef; unpasteurized milk; cross-contamination	Cook beef thoroughly; follow sanitation rules carefully
Listeriosis	Infection by *Listeria monocytogenes*	Fever, chills, nausea, loss of balance, miscarriage	Soft cheeses; ready-to-eat deli meats	Heat ready-to-eat meats to steaming; at-risk groups should avoid likely foods

Since *C. perfringens* occur almost everywhere, they are hard to keep out of the food supply. The problem is compounded by the bacteria's heat–resistant spores. High and low temperatures can halt bacterial growth before it reaches dangerous levels.

Staphylococcus Aureus

Staphylococcus aureus (STAF-uh-lo-kahk-us OR-ee-us) bacteria, better known as staph organisms, are found in the nasal passages and on the skin. These bacteria can spread through the air in drops of moisture when you talk, sneeze, cough, or even breathe. Staph organisms usually enter food from a human or animal source.

Staph bacteria grow best in moist meat dishes and starchy foods. Sliced roast beef or ham, improperly stored, can become a staph incubator, allowing the bacteria to multiply and produce toxin. Even prolonged heating doesn't destroy this toxin.

Like *C. perfringens*, staph bacteria do not cause illness—their toxin does. Symptoms of food poisoning caused by staph toxin are generally felt sooner and more acutely than other

A picnic is the perfect place for *C. perfringens* bacteria to lurk. Why do these bacteria grow in such circumstances?

types. Diarrhea, vomiting, and stomach cramps may appear within 1–7 hours after eating and may last for 24–48 hours.

Clostridium Botulinum

Botulism, *a very serious—and sometimes fatal—form of food poisoning*, is caused by toxin of the *Clostridium botulinum* bacteria (klah-STRID-ee-um BAH-chuh-LY-num). These bacteria occasionally grow in contaminated fish, but improperly home-canned foods are most often to blame. An adult may become ill by eating spoiled food containing the botulism toxin rather than the cells themselves. Some infants, however, get botulism from eating the *C. botulinum* spores. Other bacteria that inhibit the growth of these spores aren't well established in their systems. Honey is a common source of the spores.

Early symptoms of botulism include general weakness, blurred vision, dry mouth, difficulty swallowing or speaking, and shortness of breath. The illness may progress to complete paralysis, respiratory failure, and even death. Although

Staphylococcus aureus bacteria are found on the skin and in the nasal passages. How could you prevent the spread of these bacteria?

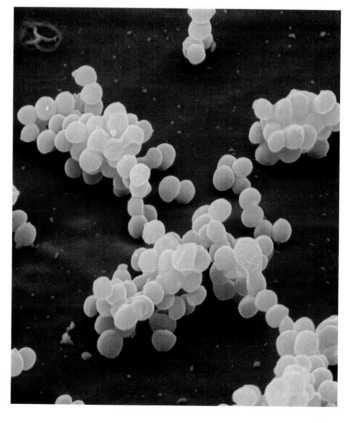

very few cases of botulism occur each year, its life-threatening nature makes prevention of utmost importance.

Since the botulism toxin is destroyed by boiling food for 10–15 minutes, people who eat home-canned food can take this precaution before eating. To prevent infant botulism, children under twelve months old should not be fed honey.

Food Infections

Food infections are another *bacterial foodborne illness that occurs when pathogenic microorganisms enter the body along with the food eaten.* Organisms that are **pathogenic** *(path-un-JEN-ik) are disease-carrying.* If these organisms are present in food that is eaten, they can multiply in the body and cause illness. Some examples of common food infections are described here.

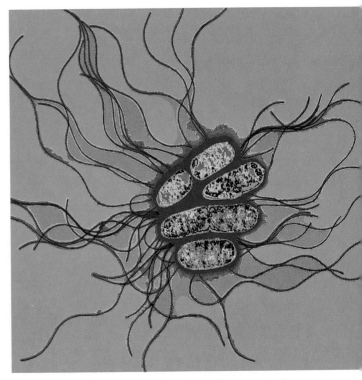

According to the Centers for Disease Control, as many as 2 million clinically significant Salmonella infections occur in the United States each year. Two thousand may be fatal. About 2,000 different strains of Salmonella are involved.

These bacteria cause botulism, a very serious food poisoning. What are some sources of botulism?

Salmonella Enteritidis

Eggs are a nutritious food. If eaten raw or served sunny side up or soft-boiled, however, they could harbor a dose of pathogenic bacteria along with their protein and minerals.

Raw and undercooked foods—including poultry, meat products, fish, shellfish, and eggs—are favorite settings for *Salmonella enteritidis* (sal-muh-NEL-uh en-teh-RIT-uh-dis). These bacteria cause salmonellosis (sal-muh-nel-OH-sis), the most common food infection.

Milk can also be a source of *Salmonella*. In 1985 salmonellosis struck 16,000 people in the Chicago area after they drank contaminated low-fat milk.

Since the bacteria grow at a phenomenal rate at room temperature, *Salmonella* infection most often occurs after eating perishable food that has been left unrefrigerated. Problems can also arise from **cross-contamination**, *the transfer of bacteria from one* *food to another food or location*. If you touch a raw chicken leg carrying *Salmonella* and then handle raw vegetables while preparing a salad, you could transfer the bacteria to the vegetables. One salmonellosis outbreak was linked to ice cream ingredients that were transported in a tank that had last carried unpasteurized eggs.

Generally, it takes large numbers of *Salmonella* to cause illness. Symptoms may appear as soon as five hours or as long as three days after the contaminated food is eaten. The time depends on how much food was eaten, the level of contamination, and the rate of digestion. Symptoms include nausea, vomiting, abdominal pain, and diarrhea. Some people also experience headaches, fever, drowsiness, muscle weakness, and chills.

Like other bacteria, *S. enteritidis* are killed by high temperatures.

Escherichia Coli 0157:H7

Ironically, one variety of *Escherichia coli* (eh-shur-EECH-ee-uh COH-ly) lives in human intestines, where it helps digest food and metabolize vitamins. In dramatic contrast, *E. coli* 0157:H7 is a hardier strain that is difficult to detect. It can also be deadly. This organism has been a major cause of illness, death, and food recalls in the last decade.

E. coli 0157:H7 attacks the cells lining the intestinal wall, causing cramps and diarrhea. It can lead to the kidney disease called hemolytic uremic syndrome (HE-muh-LI-tik yu-REE-muk) (HUS), which damages blood vessels, destroys platelets, and disrupts blood flow to vital organs. *E. coli* poisoning can also leave its victims with acute kidney failure or brain damage. *E. coli* can kill.

As these bacteria normally live in the intestines of cattle and other animals, poisoning usually comes from eating undercooked, contaminated beef. However, contaminated, fresh-pressed apple cider and unpasteurized milk have also caused outbreaks. In addition, *E coli* can be passed among people or foods that carry the bacteria.

Listeria Monocytogenes

Listeria monocytogenes (lis-TIR-ee-uh mah-noh-SY-toh-juh-neez) are often found in soil and water. Ingesting these bacteria can cause a

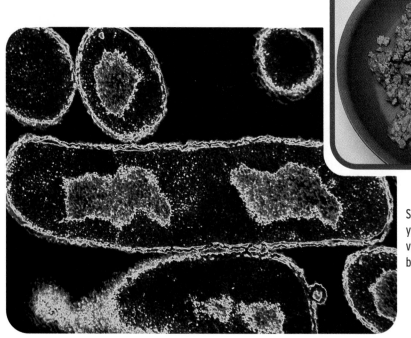

Several food poisoning outbreaks in recent years have been linked to *E. coli*. To prevent the illness, hamburger should never be served undercooked.

potentially serious illness called listeriosis. Healthy people may feel some discomfort from this disease, but those with weakened immune systems are more susceptible to its worse effects. At special risk are newborns, older adults, and pregnant women, who may experience miscarriage or stillbirth.

L. monocytogenes can lurk in the body for up to eight weeks before illness sets in. Listeriosis begins with flu-like symptoms, such as fever, chills, and sometimes nausea. If the infection spreads to the nervous system, a person may suffer from headaches, stiff neck, confusion, loss of balance, or convulsions. Listeriosis can be successfully treated with antibiotics.

L. monocytogenes can live harmlessly in the intestines of animals. From there, they spread to meat and dairy products. Soft cheeses, such as Brie and Mexican-style varieties, are an especially attractive environment. Since these bacteria are destroyed during pasteurization, look for "pasteurized" on the soft cheese labels.

The bacteria can also contaminate ready-to-eat foods after processing, either within the processing plant or on route to your plate. Ready-to-eat foods suspected in outbreaks of listeriosis include hot dogs, cold cuts, fermented or dry sausage, and other deli-style meat and poultry. These bacteria can survive at greater temperature extremes than most other organisms. Ready-to-eat foods must be heated to steaming hot to destroy the bacteria.

Animal Parasites

Bacteria are not the only organisms that cause foodborne illnesses. **Parasites**, *organisms that grow and feed on other organisms,* are another cause. Although such illnesses are rare in the United States, they can occur.

Raw pork may contain a parasite called *Trichinella spiralis* (trih-kuh-NEL-uh spuh-RAL-is), a tiny worm that lives in pork muscle. If infected pork is eaten without thorough cooking to destroy the larvae (to 77°C, 170°F), the parasite grows in the intestine and causes an illness called trichinosis (tri-kuh-NO-sus). Mild cases are marked by diarrhea, fever, fatigue, and muscle pain. Severe infections can fatally damage the heart and brain. Wild game is also subject to this parasite.

Viruses

Viruses, too, can cause foodborne illnesses. The Hepatitis A virus, for example, is excreted in feces of the infected person, who then contaminates food or water that is consumed by others. Because the infected person isn't aware of the illness at first, the disease can spread before precautions are taken. Symptoms include fever, nausea, and abdominal discomfort, followed in several days by jaundice. Approximately 23,000 cases of Hepatitis A are reported in the United States each year.

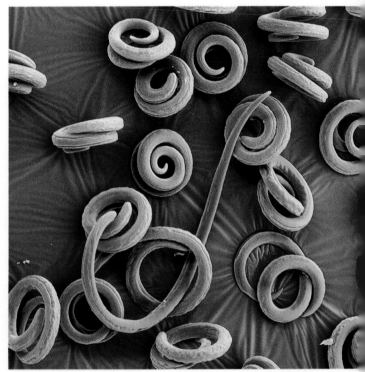

Trichinella spiralis, sometimes found in raw pork, is killed if the meat is cooked to an internal temperature of 77ºC (170ºF). How can you be sure that temperature is reached?

Restaurant food handlers must be very careful about sanitation because of this illness. If they contract Hepatitis A, they must take time from work to get well.

Other Causes of Foodborne Illness

Foodborne illnesses have still other causes. The following are two of these:

- **Chemicals.** Most toxic chemicals are unintentionally added to foods. Foods may pick up metals from processing equipment. Animals, especially fish, may ingest toxic chemicals that pollute their environment. Chemical contamination can occur in the kitchen as well. Bleaches, drain cleaners, and chemical traps for pests, if used improperly, could actually sicken the people they are meant to protect. Perhaps the biggest worry to many is **pesticides,** *chemicals used to kill insects, animal pests, and weeds.* If these products are misused or not completely removed from foods during processing, health risks could occur.

- **Natural Toxins.** Other toxins are chemicals produced naturally by a food itself. The toxins of certain types of mushrooms, for example, can cause death within hours. Potatoes that are exposed to light may produce a chemical called solanine (SOH-luh-neen), which is toxic to mammals. This toxin is in the green portion under the skin and can be removed with peeling.

Government and Food Safety

For the grocery shopper in 1900, the best resources for choosing safe foods were experience, common sense, and a trustworthy butcher or produce seller. Few laws protected consumers from food that was **adulterated** (uh-DULL-tuh-ray-tud), or *made impure by the addition of improper ingredients.*

Over the last century, government at local, state, and national levels has played a role in ensuring the safety of the food supply. The United States government is a leader in the effort. The Wiley Food and Drug Act of 1906 was passed to enforce purity in food processing. It

Some foods produce their own toxins. Eating them can be deadly. If you can't be sure that a mushroom is edible, it's best to eat only those you buy in a supermarket.

was replaced in 1938 by the Federal Food, Drug, and Cosmetic Act. The responsibility for safeguarding the commercial food supply falls largely to the Food and Drug Administration (FDA) and the United States Department of Agriculture (USDA).

Food and Drug Administration

You've read how the FDA monitors food safety by regulating the use of additives. To the FDA, safe means free from any hazards, and one of the main hazards is foodborne disease. To reduce the chances of a contaminated food getting into your home, the FDA does the following:

- **Oversees food processing.** It establishes sanitation and safety regulations and guidelines for commercial processors. FDA inspectors visit and take product samples from processing sites.

- **Tests for environmental contaminants.** These can be toxic chemicals or metals, or industrial pollutants. In one highly publicized decision, the agency ordered the re-

The FDA is responsible for seeing that commercially processed food is safe to eat.

moval of swordfish from the market nationwide after unacceptable amounts of mercury were found in swordfish samples. Once the mercury levels dropped, the FDA again allowed the fish to be sold. Because swordfish are prone to mercury contamination, the FDA continues to monitor the supply.

• **Sets levels of pesticides that can be used on crops.** The FDA works with the USDA and the Environmental Protection Agency (EPA) to determine levels of pesticides that are effective for farmers yet safe for consumers.

• **Checks levels of pesticide residue in harvested foods. Residue,** or residual, *is the matter remaining after a chemical or physical process.* Many pesticide residues are toxic to humans. The FDA does spot checks to make sure allowable limits of residues are not exceeded.

• **Regulates food labeling.** Since the Food, Drug, and Cosmetic Act defined what a label

Governmental agencies regulate the use of pesticides in agriculture to help keep the food supply safe.

is—"a display of written, printed, or graphic matter upon the immediate container"—the FDA has pushed to make it more accurate and informative. Sodium content on nutrition labels was mandated in 1986 to accommodate the number of people on sodium-restricted diets. In 1992, with the FDA's strong urging, Congress passed legislation responsible for the detailed "Nutrition Facts" panel on almost all food labels. You often find nutrition information posted over fresh produce and cuts of meat in the supermarket.

United States Department of Agriculture

The term "Department of Agriculture" may call to mind cornstalks and tractors. Actually, the USDA is one of the most far-reaching agencies of the federal government, overseeing food production at every stage from farm to table.

To the food producer, the USDA is a source of help and information for farming and ranching issues, from identifying pests in crops to answering questions about agricultural biotechnology. Consumers who need to know the safest way to prepare turkey and stuffing can find that information through a USDA pamphlet, the agency's web site, or their hot line. Research on bioengineered food is currently underway in USDA laboratories.

The food safety squad of the USDA is the Food Safety and Inspection Service (FSIS). The FSIS oversees the nation's meat, poultry, and egg production.

Hazard Analysis Critical Control Point System
One of the FSIS's tools for preventing foodborne hazards is the Hazard Analysis Critical Control Point system (HACCP, pronounced "hassip" for short). This process-control system was developed for NASA in preparing for space flights and has since been adopted by many industries. The FDA requires HACCP for low-acid canned food and has proposed it for the seafood industry. The USDA has similarly adopted HACCP for the meat and poultry industry.

While food safety inspection relies on spot checks of manufacturing conditions and random sampling of final products, HACCP focuses on problem prevention. Companies review their food production processes to determine the "critical control points." These are points in the process where hazards can be prevented, controlled, or eliminated.

A HACCP program typically involves several steps. Initially, HACCP workers identify potential hazards and the critical control points. They can then decide how to prevent each hazard and monitor the control point. Workers also plan corrective actions if monitoring should show that a hazard still exists. Effective record keeping, in order to document the company's efforts and provide a way to verify its success, is included in the HACCP system.

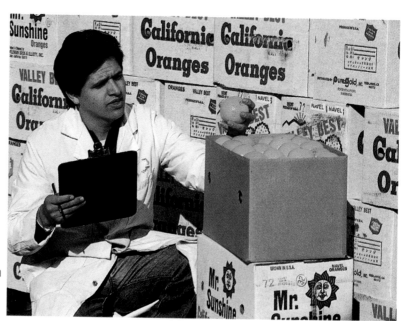

The USDA inspects foods. How can spot checks assure the safety of huge quantities of food?

Careers

Health Inspector

Why is restaurant food safe to eat? Ask a . . . Health Inspector

Brian Terlesky is a local health inspector. "There are so many sanitation issues that most people never think about, even in their own homes," he says. "Have you ever put uncooked chicken on the refrigerator shelf above fruit or vegetables that will be eaten raw? Most people do this without thinking, but that's how foodborne illnesses like salmonellosis usually occur. This is the kind of thing I'm on the lookout for when I conduct inspections."

Education Needed: A bachelor's degree is not required to be a local health inspector, but a strong science background is recommended. Coursework in food science and microbiology is useful. Most local health inspectors must complete a certification program.

Life on the Job: Health inspectors promote food safety throughout the country. They may work for city and county health departments, state public health agencies, or for the federal government. Many laws and guidelines at different levels of government dictate their

Key Skills
- Analytical ability
- Eye for detail
- Communication skills
- Time-management skills
- Problem-solving ability

duties. Local health inspectors like Brian regularly visit food service establishments to identify and document conditions that could lead to foodborne illnesses. Using knowledge and technical equipment, they evaluate refrigeration and work areas for possible food safety hazards. Inspectors recommend safer food handling practices to restaurant managers and make written reports for the local health agency.

Your Challenge

Imagine that you are a local health inspector evaluating the sanitation of a fast food restaurant.

1. What are five food safety hazards you would look for?
2. If you found these problems, what recommendations would you make to the manager?

President's Council on Food Safety

The President's Council on Food Safety was established in August, 1998. Its purpose is to coordinate the resources and focus the efforts of numerous public and private groups to improve the safety of the food supply. Its goals include education and research as well as regulation and enforcement.

The council's most prominent members are the Secretaries of Agriculture and of Health and Human Services. However, it seeks a partnership with other voices in the community who can help make and carry out food safety policy. Consumer advocates, food scientists, and leaders of Native American tribes are among the many people called upon to make the venture a success.

International Agencies

The United Nations has established two organizations to address issues related to the supply of wholesome food throughout the world. The Food and Agricultural Organization (FAO) and the World Health Organization (WHO) identify problems and recommend improvements in how countries feed their citizens.

The FAO was formed in 1944 with the twin objectives of eliminating hunger and improving nutrition worldwide. WHO was founded four years later. WHO's overall goal is fostering international cooperation in all health-related fields. To that end,

the agency spends much of its budget on solving problems of nutrition. WHO and the FAO work closely together to promote nutrition research worldwide. They produce joint reports on the nutritional state of the planet.

As these and other international agencies have learned, hunger is an obstacle to food safety. When quantities of food are barely enough to sustain life in underdeveloped countries, people are less concerned with the quality of whatever food is available. Governments may feel pressured to allow unsafe foods and processing methods as long as they get people fed. Some government and company officials may even try to profit by supplying adulterated foods to desperate people.

In a country that has economic problems, feeding the population adequately can be difficult. What might the impact on food safety be?

Preventing Foodborne Illness

In a sense, the number one cause of foodborne illness is something more than the causes you've read about so far. It's negligence. Someone, somewhere in the food supply line, neglects some basic rules of sanitation and food handling.

Later you'll read about how proper food processing and preservation can help prevent the growth of spoilage microorganisms in food. Right now you can learn ways to protect yourself from foodborne illness.

Practicing Cleanliness

A first line of defense against the spread of pathogenic bacteria is to keep them off the food handler. Hand washing—a good, 20-second scrub with soap and hot water—is one of the simplest, most effective, and possibly most ignored means to do this. Wash your hands before and after you work with food, especially when handling raw meat, poultry, or eggs. Wash them after using the toilet, coughing, sneezing, or touching any object not involved in food handling.

Avoid giving bacteria that are on you a "free ride" to the food. Wear clean clothes, tie up long hair, and take off jewelry.

Strive for a spotless work area also. Bacteria lurk on surfaces that contact food and in food by-products. When you wipe counters with a disinfectant, clean up spills, and take out the trash, you wipe out microbe colonies.

Storing Food Safely

A food-friendly setting gives edibles a proper home. To create such a setting, perishable items should be refrigerated at 4°C or below. Closed, opaque containers protect against possible damage by light, oxygen, and microbes.

Frozen foods keep well in a freezer set at –18°C. Fresh and cooked foods intended for the freezer may need additional, airtight wrapping to prevent loss of quality from intense cold.

Nonperishable items should be kept cool and dry at room temperature. Place them in resealable containers to maintain freshness and quality. Avoid storing food near water pipes, an area prone to pests. Foods kept near an oven, radiator, or other heat source may be easier targets for mold and bacteria.

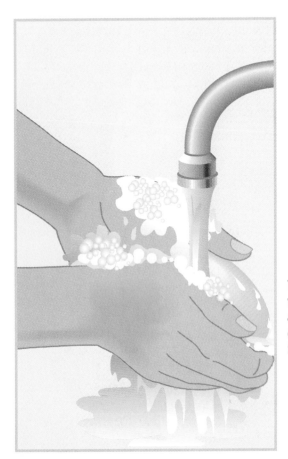

To wash your hands properly, use soap and warm water. Rub all surfaces vigorously before rinsing and drying well. In public places, turn the water off with a paper towel instead of your clean hand.

Declining Nutritional Value

Spoiled or deteriorating food can be unsafe to eat. Even when safety isn't an issue, nutritional value may be. All food deteriorates from the time it's harvested, slaughtered, or manufactured. Along with deterioration comes loss of nutrients. The rate of deterioration and nutrient loss depends on many factors: light, oxygen, radiation, temperature, humidity, industrial contaminants, and the actions of enzymes and organisms. Processing, storage, and preparation also have an effect.

- **Processing.** When eaten fresh and fully ripe, fruits and vegetables are at their nutritional prime. What happens, however, when fresh produce is shipped, handled, and sits in the supermarket for several days? Frozen foods that are processed immediately after harvesting may actually retain more nutrients than such produce. Typically, canned foods are slightly less nutritious than fresh or frozen due to processing.
- **Storage.** To retain nutrients, foods must be stored correctly. Specific procedures apply to different foods. For example, juices can oxidize and lose vitamin C if stored in open containers.
- **Preparation.** Long cooking times, high heat, and pre-preparation can break down tissue, and thus nutrients, in food. Broccoli loses about ten percent of its vitamin C when prepared by microwaving, but 33 percent after boiling for 10 minutes. Cutting green beans before boiling increases vitamin C loss from 45 to 70 percent. Beef cooked at 150°C loses about 40 percent of its thiamin. At 230°C, over 50 percent of thiamin is destroyed.

Applying Your Knowledge

1. Suppose you noticed that eggs often sat outside the supermarket for a long time before being placed in coolers. What would you do?
2. Which group of vitamins do you think are more likely to be lost during cooking: water- or fat-soluable? Why?

Finally, store disinfectants, pesticides, and other hazardous products in a separate area, away from all food. Keep these supplies in their original containers, never in used food containers. You may never mistake white cleansing powder for Parmesan cheese, but it has happened.

Handling Food with Care

Foods are constantly exposed to microbes. To limit a food's encounters with harmful microorganisms, use only clean containers, utensils, and serving dishes.

Hazardous products should be stored away from food and in properly marked containers. What could happen if this isn't done?

To prevent bacteria on raw meat and poultry from cross-contaminating other foods, use a separate cutting board for these foods. Research suggests that plastic cutting boards may be better than wooden for cutting meats and poultry. Wash any raw meat drippings from dishes with hot, soapy water before reusing them.

Because cutting boards can shelter bacteria in nooks and crannies, they all demand special care. Cutting boards should be washed thoroughly with hot, soapy water, rinsed, and dried after each use. They should also be sanitized regularly with a solution of one part chlorine bleach to eight parts hot, soapy water, then air-dried. Replace old cutting boards.

Bacteria are fond of cooked foods as well. Hot foods should be kept hot—above 60°C—until served. Otherwise, they should be chilled rapidly, in the refrigerator.

Why the hurry? Slowly cooling foods fall into that zone where bacteria grow—4° to 60°C. Refrigeration doesn't kill bacteria; it just keeps them from multiplying. Bacteria that grow on a food sitting at room temperature will be there when the food is refrigerated—and when it's eaten. In fact, cooked foods cause foodborne illness more frequently than raw foods do.

Preparing Foods Properly

Not every safe method of preparing food preserves its quality. Green beans boiled until they resemble seaweed are probably safe, but not very appetizing or nutritious.

On the other hand, most methods that preserve quality can be safe. Those same beans would be far more healthful and no less safe if appropriately steamed, stir-fried, or cooked in a microwave oven.

Since bacteria are sensitive to high temperatures, cooking a food carefully kills them. **Figure 25-2** shows the internal temperatures required for safe cooking of different meats.

Cleanliness as a way to ensure safety extends to foods also. Wiping off canned foods before opening them removes dirt. Scrubbing unpeeled produce and peeling off outer layers of leafy greens generally reduces bacteria and residue.

The bacteria that cause most food poisoning are literally everywhere. You cannot see, smell, or taste them or the toxins they produce, even as you're eating the contaminated food. Only later, when symptoms appear, do you suspect their work. The best strategy is to give these organisms only a short stay around food.

Figure 25-2

Internal Temperatures for Meat and Poultry		
Food	Internal Temperature	
	°F	°C
Beef		
Medium-rare (some bacterial risk)	145	63
Medium	160	71
Well-done	170	77
Pork, Lamb, Veal (Roasts, Steaks, Chops)		
Medium	160	71
Well-done	170	77
Poultry		
Whole chicken, turkey	180	82
Turkey breasts or roasts	170	77
Stuffing (cooked beside bird)	165	74
Ham		
Fresh (raw) or shoulder	160	71
Precooked (to reheat)	140	60
Ground Meat and Poultry		
Turkey, chicken	165	74
Beef, veal, lamb, pork	140	60
Leftovers		
Meat in soups and stews	165	74
Casseroles	160	71

Growing Cultures

Although single microorganisms are not visible, colonies of microorganisms, called cultures, can be seen. Cultures are grown in the laboratory on a gelatinous substance called nutrient agar (AH-gar), which promotes bacterial growth. In this experiment, you'll discover just how widespread microorganisms really are.

Equipment and Materials

sterile petri dish filled with agar   cellophane tape   marking pen   masking tape

Procedure

1. On the bottom of a petri dish containing nutrient agar, use a felt-tip pen to draw two intersecting lines that divide the dish into quarters. Number the quarters 1–4.
2. Using the procedure outlined in Steps 3–7, test the three surfaces in the group assigned to you by your teacher.
   Group 1: your shoulder; sink bottom; cutting board
   Group 2: your hair; clean dish; countertop
   Group 3: refrigerator shelf; washed fingertip; unwashed fingertip
   Group 4: tabletop; floor; doorknob
3. Obtain a 10-cm strip of cellophane tape. Fold over about 2 cm of the tape to make a nonstick end to hold.

4. Holding the tape at the folded end, put the sticky side on one of the three surfaces on your list. Pull the tape from the surface, and immediately place it on the agar surface on the quarter of the dish marked number 1. Remove and discard the tape.
5. In your data table, record the source from which you took the sample and the number of the area in the dish where you put it.
6. Repeat Steps 3, 4, and 5 for each of the other two surfaces on your list, putting each sample in a new section of the dish.
7. As a control, obtain a fourth piece of tape and touch it to the agar without letting the end that touches the agar touch any other surface, including your fingers. Repeat Step 5.

8. Incubate the petri dish at room temperature for three or four days in the place designated by your teacher.

9. Observe the petri dish daily. Describe in your data table and on the chalkboard any growths that have appeared on the agar.

10. In your data table, copy the information on the growths reported by the other groups.

SAMPLE DATA TABLE

Surface Tested	Area Number	Agar Day 1	Agar Day 2	Agar Day 3

Chapter Summary

- Certain microorganisms can cause changes in food that damage its quality and safety.
- Foodborne illness can be caused by microorganisms and their toxins, animal parasites and their eggs, viruses, and toxic chemicals.
- Bacteria that cause food intoxication include *Clostridium perfringens*, *Staphylococcus aureus*, and *Clostridium botulinum*.
- The major cause of food infection is the bacteria *Salmonella*. *E. coli* and *Listeria monocytogenes* are other causes.
- Symptoms of food poisoning range from mild to fatal.
- The safety and quality of the food supply is monitored at the local, state, national, and international levels.
- Foodborne illness can be prevented by proper sanitation and food handling.

Using Your Knowledge

1. Describe three microorganisms that cause food spoilage.
2. How can chemical toxins be introduced into the food supply?
3. Why might *Clostridium perfringens* be a special hazard for cafeteria diners?
4. What might you suspect if someone had trouble breathing or speaking after eating homemade green beans?
5. What is the difference between food intoxication and food infection?
6. How does the FDA regulate pesticide use?
7. How does HACCP work to safeguard the food supply?
8. Why would improving the world food supply help improve its safety?
9. When should you wash your hands while handling food?
10. Suggest ways to avoid cross-contamination when using cutting boards.

Real-World Impact

Increasing Threats from Bacteria

With increased food trade around the world, bacteria have become a greater threat. One food product can rapidly become available to thousands of people. Tracing a problem to its source can be a huge task for health inspectors. As improved disinfectants, pesticides, and animal vaccines aim to stop the spread of bacteria, newer and tougher strains of bacteria appear. To stop the cycle, many authorities suggest having a single federal agency monitor food safety; using only essential pesticides and animal vaccines; and avoiding antibacterial soaps that may kill bacteria unnecessarily.

Thinking It Through
1. What might be the impact on foodborne illness if two countries sign a trade agreement?
2. If new, stronger bacteria appear faster than researchers can develop products to combat them, what might the impacts be?
3. Why might a single federal agency be better than several at monitoring food safety?

Using Your Knowledge *(continued)*

11. If the cheese dip was refrigerated after sitting on the counter all afternoon, is it safe to eat? Why or why not?
12. Can you safely cook hamburger until it looks done? Why or why not?

Skill-Building Activities

1. **Research.** Locate information and report to your class on one of these topics: a) the effects of a botulism case that was in the news; b) the history and effects of *E.coli*; c) the USDA safe handling instructions for raw or partially cooked meat and poultry products.
2. **Problem Solving.** Suppose you are in charge of a school banquet that will serve as many as 200 people. Make a list of ten basic food safety rules that you would set up for people to follow.
3. **Social Studies.** Obtain a copy of the WHO publication *World Health*, which highlights current global health issues. Read one or two articles in the magazine and summarize them for the class.
4. **Workplace Skills.** Suppose you have a part-time job in a fast-food restaurant. You've noticed that safe food-handling procedures aren't being followed. What would you do? Why?
5. **Demonstration.** With a partner, plan and present to the class a demonstration of three ways cross-contamination could occur.

Understanding Vocabulary

With a partner, write each of the "Terms to Remember" on small pieces of paper. Place them upside down on a desk. Take turns drawing terms and defining them for each other.

Thinking Lab

Fighting Fire with Fire

Microorganisms were the earth's original inhabitants. Adaptability is one reason they've survived. When an antibacterial agent is developed, bacteria frequently develop a resistance. Thus, foodborne illness remains a health threat.

Analysis

Two promising technologies pit one bacterium against another. Preempt, which contains 29 harmless bacteria, is sprayed on baby chickens to fight *Salmonella*. As the chicks clean themselves, they ingest the Preempt, which grows in their intestines and prevents *Salmonella* from attaching there. The chickens could still pick up the bacteria during processing, so consumers must continue to handle poultry carefully.

Another technology uses poisonous proteins produced by bacteria themselves. These proteins, called *bacteriocins* (bak-TIR-EE-uh-suns), kill only bacteria that resemble the ones that produced them. Thus, they can be used selectively, only against specific microbes. This reduces the chance for unintended consequences, such as causing illness, but also limits usefulness. The bacteriocin that kills *Listeria*, for example, is harmless against *E. coli*.

Bacteriocins are very stable and survive boiling and freezing. They can be introduced into food products as attachments to a dead, harmless, "parent" bacteria. Also in the experimental stage is a plastic film coated with bacteriocin to protect foods from the outside.

Thinking Critically

1. What does "adaptability" mean as applied to microorganisms?
2. How do these technologies compare to traditional methods of combating foodborne illness?

CHAPTER 26

The Dehydration of Food

Objectives

- List benefits of dehydrated foods.
- Describe the role of air temperature and circulation in dehydration.
- Explain how pretreating foods improves dehydration.
- Compare different pretreatment methods.
- Describe different methods of dehydration.
- Demonstrate how to store dried foods.
- Compare different methods of rehydrating foods.

Terms to remember

caseharden
dehydrator
dehydrofreezing
rehydration
steam blanching
sulfiting
sulfuring
syrup blanching

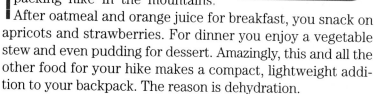

Imagine taking a weekend backpacking hike in the mountains. After oatmeal and orange juice for breakfast, you snack on apricots and strawberries. For dinner you enjoy a vegetable stew and even pudding for dessert. Amazingly, this and all the other food for your hike makes a compact, lightweight addition to your backpack. The reason is dehydration.

Dehydration, which is any process that drives water out of food, makes such convenience commonplace—while backpacking or at home. This chapter introduces you to the science and technology behind the process of drying foods. While meat and fish can be dried, the main focus in this chapter is on fruits and vegetables.

Benefits of Dehydration

As you can see, food dehydration is useful—to backpackers as well as many others. Relief agencies that provide food during a crisis, for example, can use dehydrated foods to get them safely to those in need.

The benefits of dehydration can be summarized as follows:

- **Longer storage.** Drying preserves food for longer storage without spoilage. Dried apple slices will keep long after fresh apples have turned soft and shriveled.
- **Smaller size and weight.** Drying decreases the weight and bulk of food, making it easier to transport and store. A dozen dried apricots cost less to ship than a dozen fresh ones, and they store in much less space.

Dehydrated foods offer convenience to the consumer. Because dried foods on average have about one-fifteenth the bulk of the original product, they are also easy to transport and store.

- **Convenience.** Dried foods are time-savers. Instant coffee and potato flakes can be prepared in a fraction of the time needed to brew regular coffee or cook and mash fresh potatoes.

Dehydration in History

Dehydration has a long history as a means of food preservation. Dried foods sustained ancient Greeks and Egyptians on sea voyages. Even the dried wheat placed in tombs 3,000 years ago could still be eaten today.

Sun drying is the oldest method of dehydration and still the most practical one for many people throughout the world. The French first used a mechanical method to dry food in 1795, but many more methods have been developed since then.

Like many technologies, dehydration advanced through warfare. Because it is lightweight, compact, and long lasting, dried food is excellent for military rations. Large-scale dehydrating began in order to preserve food for the armies that fought in World War I.

Today the food industry uses many methods to dehydrate food. Their equipment includes different driers designed for specific uses. Many consumers dry food themselves at home.

Typical home methods use the sun, the oven, or a special appliance for dehydrating food.

Principles of Dehydration

Wherever food is dried, the principle is the same: add heat to remove water—but not too much water. As you can see in **Figure 26-1**, some foods, especially fresh fruits and vegetables, have a much higher water content than other foods. After drying, food should contain 15-20 percent of its original moisture. If food is any drier, its color, flavor, texture, and nutrient value all suffer. Dried fruit should be chewy and almost leathery; vegetables should be brittle or crisp.

Controlling Temperature and Circulation

Dehydration is a natural process, but it's only safe if carefully handled. Air temperature and circulation must be monitored to avoid bacteria growth and spoilage.

Figure 26-1

Water Content of Foods	
Food	Percentage of Water
Cucumber	95
Tomato	94
Zucchini	94
Strawberry	90
Mushroom	90
Papaya	89
Broccoli	89
Peach	89
Carrot	88
Pineapple	86
Apple	84
Eggs	75
Banana	65
Chicken breast, cooked	65
Whole wheat bread	38
Raisins	15
Popcorn	4
Peanuts	2

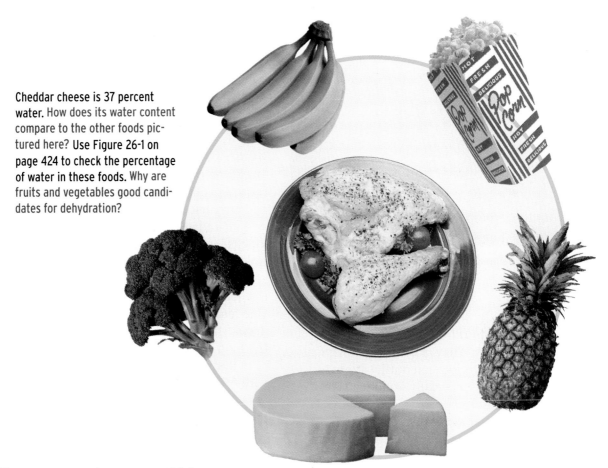

Cheddar cheese is 37 percent water. How does its water content compare to the other foods pictured here? Use Figure 26-1 on page 424 to check the percentage of water in these foods. Why are fruits and vegetables good candidates for dehydration?

Temperatures that are too high may **case-harden** the food, *or cook the outside of it. The food forms a hard outer layer, trapping moisture inside.* This moisture can then support bacteria growth. Controlling air temperature prevents this condition.

Temperatures used during drying aren't high enough to kill bacteria, but they are high enough to remove moisture from the food. Removing just the right amount of moisture makes the food an unsuitable environment for bacteria growth. Bacteria don't usually grow at moisture levels under 16 percent.

If food is dried too slowly, however, bacteria have time to multiply and cause spoilage. Good air circulation helps solve this problem. The more rapid the air movement, the faster moisture is swept away. The air must come from an outside source, so the moisture removed from the food is not recirculated. If this happens, the air eventually becomes saturated with moisture, and the food doesn't dry. In commercial dehydration, air humidity is closely monitored to ensure quality.

Preparation for Dehydration

Both commercial processing and home drying begin the same way—by choosing unblemished fruits and vegetables that are at their peak of ripeness, and not past it. Otherwise, one piece of spoiled food can ruin an entire batch. Foods chosen should be as clean as possible before drying.

Most foods are sliced into thin pieces to speed the drying process. Slices have more surface area per unit of mass than larger pieces do. Moisture in the food escapes more rapidly because it has a shorter distance to travel to get to the surface. Fruits that are dried whole, such as grapes and plums, are poked with holes that allow moisture to escape.

Pretreatment

In preparing to dry foods, pretreatment is another step that improves the finished product.

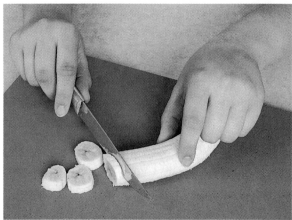

Fruit is sliced before dehydrating. What impact does thickness have on the process?

Pretreated foods retain more fat-soluble vitamins and flavor. They suffer less enzymatic browning and deterioration during storage.

Pretreating is especially beneficial for vegetables. They deteriorate more rapidly than fruit during storage because enzymes in them continue to function. In fruits, a higher sugar content and acidity lessen enzyme action. In fact, some fruits, such as grapes, figs, and plums, don't need pretreating.

Introducing sulfur during pretreatment helps prevent the unsightly discoloration of enzymatic browning. Sulfer acts as an antioxidant, preventing oxygen in the air from reacting with compounds in the fruit. The amount of sulfur needed is harmless to most people. Both sulfiting and sulfuring are processes that use sulfur. Blanching is a pretreatment method that can be used by those who don't wish to use sulfur.

Sulfiting

When **sulfiting** is used for pretreatment, *the food is soaked in a solution of water and sodium metabisulfite or sodium bisulfite.* Mixing either of these compounds with water releases sulfite ions. These ions slow oxidation and enzymatic browning.

Since food can become mushy if soaked too long, sliced food is soaked for only 10 minutes. Halves of fruit or large pieces of vegetables soak for 30 minutes maximum. After soaking, the food is drained thoroughly before drying. One disadvantage of sulfiting is that it extends drying times 15-20 percent due to the water the food absorbs while soaking.

Sulfiting is sometimes used commercially with fruit and light-colored vegetables. Because of possible allergic reactions, commercial use of this treatment is limited and must be indicated on packages.

Sulfuring

The method used by most commercial fruit dryers to prevent browning is **sulfuring**. In this

Enzymatic browning can be prevented when drying fruits at home by soaking the fruit in orange juice. What might happen if you soak the fruit too long?

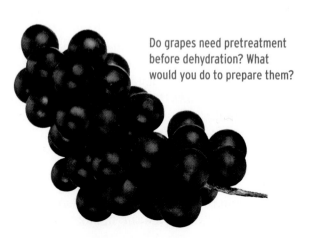

Do grapes need pretreatment before dehydration? What would you do to prepare them?

Some people like to dry garden vegetables for later use. Steam blanching is an effective method to stop enzymatic action that would change the texture and flavor of the vegetables.

process, *food is placed on large trays that are stacked together and covered. Burning sulfur creates fumes that circulate around the food. The fumes contain sulfur dioxide gas.*

Sulfuring usually takes from one to four hours. The sulfur dioxide used in the process is irritating and harmful to the lungs, so the fumes should not be inhaled. The food is dried outdoors, either by sunlight or in a dryer, to help the fumes disperse.

The amount of sulfur used depends on the weight and type of food to be processed. Pears, peaches, and apples, for example, need twice the sulfur per kilogram as apricots.

Sulfuring is considered the best commercial pretreatment method. The food maintains its original shape and color and remains pliable. Vitamins A and C are not affected. In addition, sulfuring shortens drying time, repels insects, and inhibits mold growth.

Blanching

Blanching is another pretreatment that stops enzyme activity. It also shortens the time needed for drying by relaxing the tissue walls so moisture can escape more quickly. This method is used in both home and commercial dehydration.

The most common form of blanching is water blanching. Prepared food is placed in boiling water to denature enzymes. This checks the ripening process and prevents undesirable flavor changes.

Better results can be obtained with **steam blanching**, or steaming. Here, *food is placed in a perforated basket over boiling water and blanched by the steam. A set of two pans that stack for steaming may also be used.* Because less water is added to the food, drying time is shorter, and smaller amounts of water-soluble vitamins and minerals are lost.

Blanching is commonly used for vegetables. Besides preserving them, blanching shortens their cooking time because they are already partially cooked. A similar process used for fruits is **syrup blanching**, in which *cut fruit is soaked in a hot solution of sugar, corn syrup, and water, then drained and dried.* The result is a sweetened, softened fruit. This may or may not be an advantage, depending on how you feel about the fruit's natural flavor, sweetness, and texture. A definite disadvantage is the destruction of vitamin C by the high heat in a fruit like pineapple.

Methods of Dehydration

As with fermentation, the first dehydrated food may have come about by accident. Over the millennia, the process has been refined and diversified.

Sun Drying

Sun drying, the oldest method of drying food, is still popular today, both for home and commercial drying. You may have noticed the term "sun dried" on commercially prepared packages of such dried fruit as peaches, apricots, raisins, and figs. The process is relatively low in cost and technology: all that's needed are quality food, drying trays, and a break with the

NutritionLink

Dried Food: A Nutritional Powerhouse

Flavor and convenience aren't the only reasons that "trail mix" is a favored snack of hikers, bicyclists, and other all-day athletes. Rid of most water, those raisins, dried dates, bananas, and coconut are a condensed, concentrated source of vitamins and minerals.

Most forms of processing reduce nutrient value in some way, however, and this is true of drying. Actually, more nutrients are lost during pretreatment than in drying, especially quantities of vitamins A and C. Blanching is most destructive; sulfuring the least. As **Figure 26-2** shows, however, dried foods consistently beat the fresh and canned competition.

Applying Your Knowledge

1. How do you think equal masses of fresh, canned, and dried fruits compare in calories?
2. Why do you think foods generally lose more nutritional value during cooking and canning than from dehydration?
3. How do you think dried foods should fit into an overall eating plan?

Sun drying, the oldest method of drying food, is used today in commercial fruit drying operations.
Photo provided by Valley Sun Products of California

Figure 26-2

Nutrient Comparison of Foods			
Peaches	Fresh	Canned	Dried
Amount	2	250 mL	250 mL
Mass	200 g	244 g	160 g
Calcium	18 mg	10 mg	71 mg
Vitamin A	266 RE*	110 RE	624 RE
Riboflavin	0.1 mg	0.07 mg	0.3 mg
Niacin	2 mg	1.5 mg	8.5 mg
Vitamin C	14 mg	7 mg	29 mg
Apricots	Fresh	Canned	Dried
Amount	3	250 mL	250 mL
Mass	107 g	258 g	130 g
Calcium	18 mg	28 mg	87 mg
Vitamin A	289 RE*	449 RE	1417 RE
Riboflavin	0.04 mg	0.05 mg	0.21 mg
Niacin	0.6 mg	1 mg	4.3 mg
Vitamin C	11 mg	10 mg	16 mg

*RE=Retinol Equivalent

weather. The sun's ultraviolet rays have a sterilizing effect, which inhibits the growth of microorganisms.

Sun drying requires a stretch of pleasant days with temperatures of 29°C or higher and relatively low humidity. As with other drying methods, good air circulation and ventilation are essential. Of course, foods must be protected from whatever environmental hazards the area is prone to. They should not be left out overnight.

This ancient drying method has been enhanced by a modern innovation: solar dryers. These devices concentrate the sun's warmth and protect the food from insects and pollution.

Room Drying

Room drying has been used since pioneers trekked westward across the United States,

with varying degrees of success. In this process, as you might infer, food is set out and allowed to dry at room temperature. The spot selected must be warm with good air circulation, yet relatively free of airborne contaminants.

Room drying is the least effective of home dehydration methods. The food produced is usually of low quality, because maintaining the right conditions for the entire drying process is nearly impossible.

Oven Drying

Oven drying is a popular method with those who dehydrate small amounts of food at home. Temperatures between 38°–60°C are best for oven drying. Temperatures above 63°C destroy more of the vitamins and may caseharden the food.

Oven drying is faster than sun drying and more controllable than room drying, but it has disadvantages of its own. Ovens lack needed ventilation, even with the door open, so food must be rotated by hand. Sulfured fruit is never oven-dried due to the irritating fumes produced, which may also discolor the oven surface. Such limitations make it hard to get high-quality results.

Dehydrators

Dehydrators offer by far the safest and most efficient means of creating dried foods. A **dehydrator** is *an appliance that provides a steady supply of circulated, heated air to dry foods.*

By creating a sanitary and consistent environment, dehydrators generally produce the best quality home-dried foods. Home dehydrators are equipped with thermostats, so they can be set for, and maintain, the exact temperature needed. They have a fan that circulates air continually. Since they require very little monitoring, they can operate 24 hours a day. Many models are constructed so the number of drying trays can be changed.

Commercial dehydrators are basically larger versions of those used in homes. They function in the same way, although several types are

Food dehydrators such as these are used commercially.

used, depending on the food. Drum and vacuum driers are commonly used to dry purees and liquids. Food cut into pieces is almost always dried in kiln or tunnel driers. Milk is dehydrated in a spray drying system. As the milk is sprayed into a fine mist, it is brought into contact with hot air, which causes quick removal of moisture. This process is shown in **Figure 26-3**.

Dehydrofreezing

Dehydrofreezing is a method used by the food processing industry. As the name implies, it *preserves food through a combination of partial drying and freezing.*

Food treated by dehydrofreezing has the same moisture content as dehydrated food but does not mold. It occupies less than half the freezer space of food that is only frozen. Fruits and vegetables processed by dehydrofreezing have good flavor and color. They reconstitute in about half the time required for fully dried food.

Storing Dried Food

Once food is dried to the correct moisture content, it must be sealed in glass jars, heavy plastic bags, or metal cans. This prevents moisture in the air from reentering the food.

After purchasing dried food, you'll generally find that it lasts longest if kept dry and stored at room temperature. Refrigeration isn't recommended, due to the moisture present. Dried food needs to be checked periodically for mold. If any is present, of course, the food must be discarded.

Figure 26-3

In commercial spray dryers, a liquid food is sprayed into the dryer where the hot air evaporates the liquid, leaving the dry food product. This process is used in the dairy industry.

Careers

Vice President of Manufacturing

How do food processing companies stay competitive?   Ask a . . .

Vice President of Manufacturing

saiah Patterson is vice president of manufacturing for a company that produces dehydrated foods. Isaiah says, "Since I came up through the ranks, being a vice president means more responsibility than I've ever had. I help make high-level decisions that could make or break our company. Right now we're looking at the possibilities of microwave drying, which could save us time and energy. When foods are dried with microwaves, they heat internally. Vapor pressure from within pushes the moisture out and speeds the drying process. Changing our manufacturing methods isn't an action I take lightly. On the other hand, I have to make sure we stay current with technology in order to be competitive."

KEY SKILLS
- Leadership
- Communication
- Organizational skills
- Problem-solving skills
- Self-motivated
- Supervisory skills

Life on the Job: A vice president of manufacturing must guide development of manufacturing operations that produce the goods a company sells. This can be a high-pressure position. Since large companies often have plants in different parts of the country and the world, this career can involve extensive travel. Many manufacturing VPs learn a foreign language.

Education Needed: A vice president of manufacturing must hold at least a bachelor's degree, usually in a field directly related to the company's business. A general business background in planning, analysis, and sales is valuable, since most vice presidents are part of the top management team that determines a company's goals and strategies.

Your Challenge

Suppose you're a VP in a large company. Changes to less labor-intensive processes may lead to a need to reduce staff.

1. If a more efficient drying method calls for fewer employees, how would you handle the situation?
2. What steps could you take to find work for all employees, while still using more efficient methods?

Rehydration

If dehydration refers to removing water from food, you can conclude that **rehydration** is *replacing water that was previously removed.* Rehydration is also called reconstitution (ree-kahn-stih-TOO-shun).

As any raisin or date lover knows, dried fruit is quite tasty as it is; however, many recipes call for dried fruit to be rehydrated. This is usually done by one of the following two methods:

- Bring water to a boil; then remove from the heat. Soak the fruit in the water for five to ten minutes; drain and pat dry.
- Place the fruit in a steamer over boiling water for three to five minutes. Pat the fruit dry with a paper towel.

Once rehydrated, fruit should be refrigerated and used fairly quickly, before it spoils.

Vegetables take longer to rehydrate than fruit because they lose more water in drying. The process is a bit faster if the vegetables

---

EXPERIMENT 26

Dehydrating Beef

SAFETY FIRST
Review these safety guidelines before you begin this experiment.

Moisture can be removed from fruits and vegetables to minimize the growth of bacteria and molds that cause them to spoil. When this technique is used with meat, the resulting product is called jerky. In this experiment you'll compare dry-cured, brine-cured, and unsalted air-dried beef. Do not eat the samples prepared.

Equipment and Materials

flank steak	coarse salt	paper towel	heavy bottle or mallet	two utility clamps
balance	250-mL beaker	skewer	two ring stands	

Procedure: Day 1

1. Obtain a beef strip and lay it on a clean cutting board. Pound it with a heavy bottle or mallet until it is very thin.
2. Mass the strip and record the result in your data table.
3. Place the strip in a 250-mL beaker and add the brine solution provided by your teacher until the meat is completely covered. Label the beaker with your name and set aside until day 2.

Procedure: Day 2

4. Obtain two more strips of beef and lay them on a clean cutting board. Pound them with a heavy bottle or a mallet until they are very thin.

5. Mass each of the strips and record results in your data table.
6. Place one of the strips on a clean cutting board. Sprinkle salt heavily on one side of the strip and pound it in. Turn the strip over and pound salt into the other side.
7. Remove the strip that has been soaking in the brine and pat dry with a paper towel. Hang all three strips on a shish kebab skewer. Place the skewer between two utility rings attached to ring stands. Leave until day 3.

Procedure: Day 3

8. Carefully remove the strips from the skewer, determine their masses, and record in your data table. Return them to the skewer for another day.

were blanched, since cell walls weakened by blanching admit water more readily. A long soaking, preferably overnight, makes vegetables more tender.

Vegetables vary in how they should be rehydrated. When using a recipe, follow directions for rehydration carefully. Remember, though, that soaking vegetables for more than 2 hours at room temperature invites bacteria growth. It's safer to soak vegetables in boiling water, then refrigerate them in the water to soak for several hours more. Soaking times vary with the type of vegetable and size of pieces.

Many dried foods must be rehydrated for use. What are some foods that would need to be rehydrated? What are some that wouldn't?

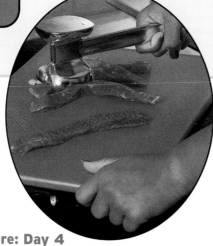

11. Try breaking the jerky. Then describe its appearance and texture in your data table.

Analyzing Results

1. Which method gave the toughest jerky? Which gave the most crisp?
2. What effect does salt have on the drying process?
3. Which strip lost the largest percentage of its mass during the drying period?
4. Do you think any of the strips need further drying? Why?
5. How might you determine if any of the strips still contain water?

Procedure: Day 4

9. Again, remove the strips from the skewer, determine their masses, and record the results in your data table.
10. Calculate and record the percentage of mass lost, using the formula:

$$\text{percentage of mass lost} = \frac{\text{fresh mass - dried mass} \times 100}{\text{fresh mass}}$$

SAMPLE DATA TABLE

Beef Strips	Fresh Mass	Mass Day 3	Mass Day 4	Percent Mass Lost	Appearance and Feel
Brine-cured					
Dry-cured					
Unsalted					

Chapter Summary

- Dehydration is a preservation method that removes water from food.
- Dehydrating food lengthens its shelf life and adds convenience.
- Before drying, food is sorted for quality, sliced, and pretreated.
- Sulfiting, sulfuring, and blanching reduce flavor loss, enzymatic browning, and deterioration during storage.
- Proper air temperatures and circulation are essential to safe, successful dehydration.
- Dehydration at home can be done by the sun, in an oven or room, or in a dehydrator.
- A dehydrator produces the best results by providing a sanitary and consistent environment.
- Dried food should be stored in airtight containers to prevent reabsorption of moisture.
- Reconstitution is another word for rehydration.

Using Your Knowledge

1. What are the benefits of food dehydration?
2. Since drying temperatures aren't high enough to kill bacteria, why is dehydrating an effective way to preserve food?
3. A friend likes dried fruit slices that are crunchy on the outside and moist inside. What would you tell this friend?
4. How would you cut apples for drying? Why?
5. Which food is likely to supply more nutrients: a quarter-cup of plums or a quarter-cup of prunes? Explain.
6. Why is sulfur sometimes used to pretreat foods?
7. Why might a food that has been pretreated by sulfiting take extra time to dehydrate?
8. A friend gives you some home-dried apricots, which you find too soft and sweet. What is probably responsible?
9. Would room drying be a good choice for someone living in a small, crowded apartment in a big city? Why or why not?

Real-World Impact

Solar Drying

People who prefer sun power for drying foods can order solar dryer kits from catalogs or even make their own with just a few parts. Racks or screens hold the food inside a box-like chamber. A dark-colored heat absorber, often metal, sits above the food. Over the absorbing surface is a piece of clear glass. The sun's heat is absorbed while protecting the food from direct rays. In places around the world where produce is abundant but power plants are not, sun drying foods may be a family's source of income. Dried foods can keep people alive during droughts when crops won't grow and during war when armies cut food supplies.

Thinking It Through
1. What are the advantages of using a solar dryer over open-air drying?
2. What safety hazards might be associated with solar drying foods at home?

Using Your Knowledge *(continued)*

10. How is a dehydrator an improvement over older methods of drying foods?
11. Should dried banana slices be stored in the refrigerator? Why or why not?
12. Should fruits and vegetables be rehydrated by the same method? Explain.

Skill-Building Activities

1. **Teamwork.** With a team of four, design a food dehydrator. Create an owner's manual that explains its use and care. Include examples of drying times for different fruits and vegetables.
2. **Management Skills.** Experiment to see how well various containers store dried food. You might include glass jars, plastic containers, and plastic bags. What types of containers are most effective? Why?
3. **Math Skills.** One small apple has a mass of 95 grams fresh, 17 grams dried. You need to ship 1000 apples, at a rate of 75 cents per kilogram. How much money would you save by shipping dehydrated apples instead of fresh ones?
4. **Consumer Skills.** As a class, conduct a blind taste test of rehydrated foods and their fresh or freshly made varieties. Compile the results. What comments did classmates make regarding the appearance, taste, texture, and aroma of the foods sampled? Overall, which did they prefer? What types of food are most successfully rehydrated? What factors seem to affect the appeal of rehydrated foods?
5. **Research.** Use the Internet and other resources to gather information about commercial dehydration equipment. Prepare a presentation for the class.

Thinking Lab

Food for Crises

For consumers, food dehydration means raisins in their cereal and hot cocoa on a moment's notice. To food manufacturers, dehydration may represent new food products as well as new markets in places where fresh foods can't be shipped.

Analysis

Most people appreciate dehydrated food for its convenience and economy. For groups in crisis situations, dried food is nothing less than a lifesaver. Its long shelf life, nutrient-density, and portability make it valuable survival food.

Thinking Critically

1. For what populations in crises would dehydrated foods be valuable?
2. What added precautions might be needed to ensure the safety and usefulness of dehydrated food in crisis situations?
3. Imagine you're putting together a food package for aid and rescue crews to deliver in an emergency. What dehydrated items would you include? Why? What other items would be needed?

Understanding Vocabulary

Look up "hydr-" and "hydro-" in the dictionary. What do these word prefixes mean? How are they used in "Terms to Remember" for this chapter? What other words can you identify that use these prefixes with similar meaning?

The Canning of Food

In an age when food science and technology offer genetically modified tomatoes and dry ice cream, the old-fashioned canned food remains a staple in kitchens the world over. For many people, canned foods are an economical choice for meeting nutritional needs.

As a quick scan of any supermarket shows, canning is a major commercial enterprise. It's also a popular and practical method for preserving foods at home. Whether the purpose is to stock a supermarket shelf or your own pantry, successful canning depends on creating a sealed, airtight environment that prevents pathogenic microorganisms from growing in food. This chapter highlights the main steps in that process.

The Discovery of Canning

It was 1795, and the Emperor Napoleon had a problem: he needed to feed his massive French army on their long campaigns. Soldiers on the march couldn't always depend on ready meats and produce as their civilian peers did. Napoleon ordered that a 12,000-franc prize be awarded to anyone who could preserve food from one harvest to the next.

In 1809, after ten years of work, Parisian candy maker Nicolas Appert unveiled his technique for preserving foods. He heated cooked food in glass bottles that he'd sealed with cork. Appert walked away 12,000 francs richer. The government even helped fund his research to refine the technology.

With Appert's achievement, the era of food canning began—a step that would eventually lead to modern food preservation. Within a year after Appert's work, Englishman Peter Durand patented the tin canister, or can, which quickly became the most popular container for the job. Due to widespread use of the can, *the process of preserving food by heating and sealing it in containers for storage* came to be known as **canning**. To this day, even food preserved in glass jars is canned, not "jarred."

Even products in jars are called canned foods.

Two other names are remembered for their contributions to canning. John Mason invented the thread-top jar so widely used today. Alexander Kerr devised a lid with a rubber gasket that makes vacuum sealing possible.

It took another forty years before anyone could explain why canning worked. In the 1850s, Louis Pasteur's work showed that the high temperatures used in canning killed the bacteria responsible for food spoilage. That same principle, better understood, underlies the process today.

Home Canning

For many people, canning food is the final, satisfying step in growing and enjoying produce from their own garden. Others, especially those with special dietary needs, appreciate the control over quality and additives that home canning gives them. A variety of foods are canned at home, with fruits and vegetables the most popular choices.

Successful canning rests on three fundamentals: using the necessary equipment; choosing the best method for packing the food; and using the correct processing method. This is as true for the home canner as for the giant corporation.

Home-Canning Containers

In a sense, the materials you use for home canning are like those used to build a house. All must create a safe and durable environment for the inhabitants. Special three-part containers are used to hold home-canned foods. They have the following parts:

- **Canning jars.** Jars for canning are made of tempered glass that can withstand high temperatures and rough treatment. A mouth surface that is smooth and free from chips and scratches allows the jar to seal thoroughly. Widemouth jars are the most popular because they are easy to fill.
- **Lids.** The lid that seals the jar is a flat metal disc with a thin rim of sealing compound on the underside.
- **Bands.** The lid is held in place by a threaded band, which screws on the top of the canning jar.

Packing Food

To begin the canning process, thoroughly cleaned food is prepared and placed in clean and sanitary containers. One technique used is the **raw-pack** method. As its name suggests, *uncooked food is placed in a container that is then filled with boiling water or juice and*

Jars for home canning typically come in sizes ranging from 1/2 pint to 1/2 gallon. What type of food might be canned in a decorative jar?

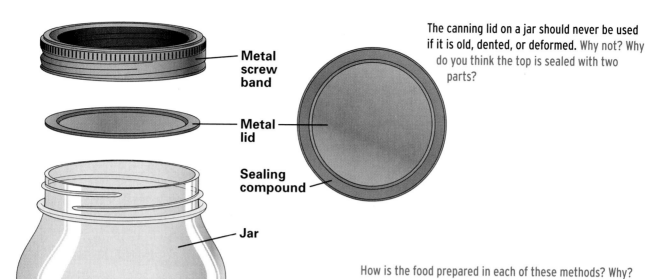

Metal screw band

Metal lid

Sealing compound

Jar

The canning lid on a jar should never be used if it is old, dented, or deformed. Why not? Why do you think the top is sealed with two parts?

closed with a lid and a ring band. This is also called the **cold-pack** method. Food tends to hold its shape better and stay firmer when processed using raw-pack. This method is usually recommended for foods that may fall apart when cooked, because of their relatively low density.

Foods may also be packed with the **hot-pack** method. *Food is heated in syrup, water, or juice to at least 77°C. While still hot, the food is packed into a container that is closed with a lid and ring band.* Hot-pack is used for food that can be precooked slightly to allow a closer fit. Precooked spinach, for example, takes less space than raw. Unsweetened fruit is usually canned with this method.

Processing Methods

Once food has been packed, it's ready to be processed. Processing meets the double duty of making canned foods safe by killing harmful microorganisms and also by sealing the containers. Two main methods are used for processing: water bath and pressure processing. Each one takes a different style of pan, or canner. The method you choose depends on the particular food to be processed.

How is the food prepared in each of these methods? Why?

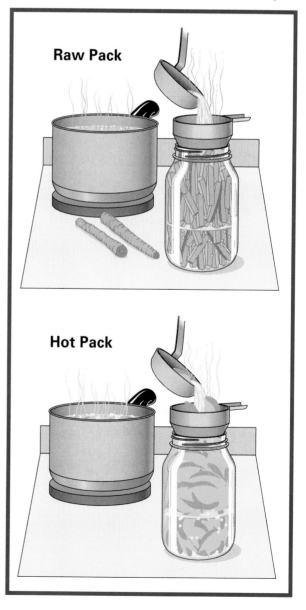

Raw Pack

Hot Pack

Water-Bath Processing

Before using water-bath processing, most foods are packed into jars with the hot-pack method. This packing method is preferred with water-bath in order to preserve food safely.

In **water-bath processing**, *the containers of food are heated in boiling water in a canning kettle*. The jars are placed on a rack in the kettle and covered with water. The kettle must be deep enough to allow about 3–5 cm of water above the containers and 3–5 cm of space above the water so it can boil freely. As the water heats, it circulates under and around the jars. In order to kill any harmful organisms, the food is processed until its coolest spot registers at least 88°C. Following recommended timing ensures the correct temperature.

After the food has been processed, the jars are removed from the water and cooled. During cooling, the sealing compound on the lid forms an airtight seal that prevents microorganisms from entering the jar. The food cannot be contaminated and spoil unless the seal is bad or the processing faulty. Thus, properly canned food has a long shelf life.

A water-bath is safe only with high-acid foods, those with a pH under 4.5. Examples are cucumbers, tomatoes, berries, fruit, pickled red cabbage, and sauerkraut. Because high acidity prevents the growth of many of the most troublesome microorganisms, these foods can be processed at lower temperatures than other foods, although a boiling temperature is necessary.

Pressure Processing

With pressure processing, either the raw-pack or hot-pack method is used to fill containers. In **pressure processing**, *the containers of food are heated under pressure in a pressure canner.* This canner also has a rack for the jars. Water is added, but only to a height of about 5–8 cm.

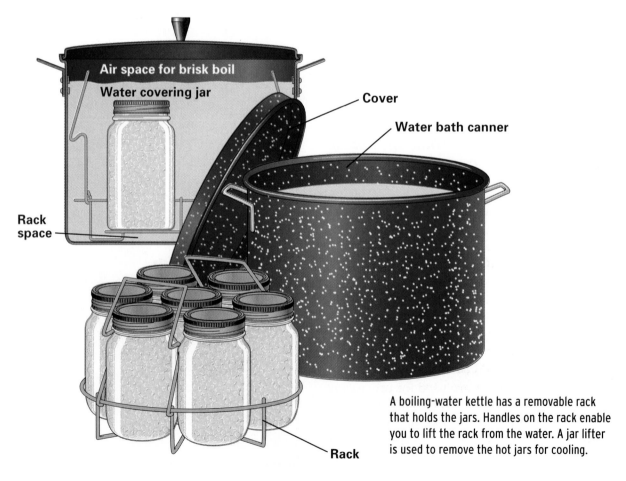

Air space for brisk boil

Water covering jar

Cover

Water bath canner

Rack space

Rack

A boiling-water kettle has a removable rack that holds the jars. Handles on the rack enable you to lift the rack from the water. A jar lifter is used to remove the hot jars for cooling.

Careers
Toxicologist

Are food supplies safe from chemicals?
Ask a . . . Toxicologist

Dominic Villegas is a toxicologist for a fresh produce company. Dominic says, "Concerns about chemicals in food are common. That's why companies employ people like me. It's difficult and costly to raise produce today without chemicals that control pests and weeds, so we do use them. But we're also careful about testing them. I help decide what chemicals can be used. I look at how safe they are and in what amounts. We're not only concerned about the health of consumers but also about the farm workers who raise our produce."

Education Needed: Toxicologists hold a bachelor's degree at a minimum, but many have an advanced degree in toxicology or a biological science that provides laboratory experience. Many supplement their science major with courses in technical writing and law, which relate to many jobs in toxicology. Agriculture, pharmaceutical science, and environmental health are popular areas for specialization. People with two- and four-year degrees can work in toxicology as lab assistants and research technicians.

KEY SKILLS
- Attention to detail
- Quantitative and statistical skills
- Technical writing
- Communication
- Knowledge of laws

Life on the Job: Toxicologists work in industries, academic institutions, and the government. Some do consulting work. Toxicologists evaluate substances, products, and work practices for possible health hazards. They interpret government regulations for their company and file reports.

Your Challenge

Suppose you were thinking about a career in food science, perhaps as a toxicologist. You learn that a research lab is located in your area.

1. How might that research lab be a resource for you?
2. What specific actions could help you take advantage of the research lab?

Jars of food to be processed can be placed in a rack within a pressure canner. A gauge on the lid shows how high the pressure is in the pan. Why would following the manufacturer's instructions for use be important?

In the air above the heated water, steam forms. A rubber ring inside the lid seals the canner so no air or steam can escape during processing. Because the steam can't escape, pressure increases. The high pressure causes water to boil at a higher temperature than it does at normal atmospheric pressure. A steam pressure gauge on the lid tells you how high the pressure has become. A safety valve on the lid is used to slowly release small amounts of steam so the pressure doesn't become too high. Jars are removed once the pressure drops. Here, too, the lids of the jars seal as the food cools after processing.

Low-acid foods, with a pH above 4.5, require pressure processing. These foods include peas, corn, carrots, beets, lima beans, asparagus, spinach, pumpkin, meat, fish, and poultry. Such foods must be heated to temperatures above 140°C. The high temperature kills the bacteria in the food. If you are unsure about a food's acid level, choose pressure processing. This ensures the destruction of all harmful bacteria and their spores.

Processing Times

With both processing methods, the amount of time needed depends on the food and the containers used. Longer heating times are required for larger pieces of food. Large containers need more time than small ones, and glass containers must be processed longer than metal ones. **Figure 27-1** gives you a general idea of how long some home-canned foods must be processed before they are safe to eat.

Because so many factors affect canning success, a reliable canning guide is required reading for every home canner. Canning guides tell you the best packing and processing method for a particular food, how long the food should be processed, and other helpful details. The health of anyone who eats home-canned food is too valuable to be left to chance.

Different foods take different processing methods. How would you process a high-acid food such as tomatoes and a low-acid food such as corn?

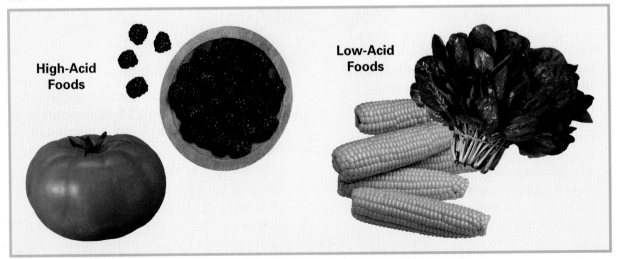

High-Acid Foods

Low-Acid Foods

Figure 27-1

Examples of Food Processing Times

Food	Pack	Jar Size	Processing Time
Pressure Canner Method			
Lima beans	hot	0.47 L	40 minutes
		0.94 L	50 minutes
Corn, kernels	hot	0.47 L	55 minutes
		0.94 L	85 minutes
Peas, green	hot	0.47 L	40 minutes
		0.94 L	40 minutes
Water-Bath Method			
Cherries	hot	0.47 L	10 minutes
		0.94 L	15 minutes
Peaches, halves	hot	0.47 L	20 minutes
		0.94 L	25 minutes
	cold	0.47 L	25 minutes
		0.94 L	30 minutes
Tomatoes, whole	hot	0.47 L	35 minutes
		0.94 L	45 minutes

Obviously a task this immense requires that processing be done on a large scale. **Retort canning** is simply *pressure processing carried out by commercial food canners.* The name comes from the **retorts**, *the huge pressure canners used commercially.* These canners, in turn, are named for a closed vessel used in laboratories to distill or decompose substances by heat.

Food processed by retort canning is considered commercially sterile. This doesn't mean all microorganisms have been destroyed. It does mean that all toxin-forming organisms that could cause illness and those that could cause spoilage under usual storage conditions have been eliminated. While most commercially canned food has a long shelf life, it loses quality if kept for over two years.

Heat Transfer

For canning processes to work, commercially as well as at home, the heat that kills pathogenic microorganisms must reach all the food being sterilized. How does heat energy transfer from its source and into food? Here's a quick review of the two processes that allow this to happen: conduction and convection.

Commercial Canning

The commercial canning industry began in Philadelphia in 1874 with the development of commercial canners. Today, commercial food canning is a major industry that provides safe food for consumers.

Retort Canning

The next time you're in a supermarket, try counting the cans of corn or tomato sauce. Then imagine processing enough of one item to fill the shelf space in supermarkets across your region of the country. That gives you some idea of the goal of the commercial canning plant.

Commercial canning is done in huge pressure canners called retorts.

In conduction, energy passes from particle to particle through molecular collision. When two substances are in contact, the molecules of the warmer substance transmit energy to the molecules of the cooler one until all molecules are the same temperature. Therefore, a food will heat as it sits on the surface of a hot container. If the food is immersed in a liquid, heat energy will be conducted from the heat source, to the liquid, and then to the food.

Figure 27-2 shows that convection is an assistant to conduction when heating food. With convection, currents form as a fluid or air heats. Suppose you are processing green beans in a container of water. As the water at the bottom of the container heats, the water molecules begin to move faster and push apart. The heated layer of water becomes less dense and lighter, making it rise above the cooler water. The coolest layer of water at the top then sinks to the bottom, where it begins to heat and rise. The water continues to rise and fall in currents until all the water is heated to the same temperature. As you can see, these water currents will force increasingly hot water next to the beans. Therefore, the hot water can keep transferring heat to the beans by conduction.

Agitation Retorts

Peas, corn, and other loosely packed foods are heated by both conduction and convection. Heat transfers from the surrounding juice and liquid. Because convection is not equally effective in every food, some commercial establishments use **agitation retorts**. These are *canners that mechanically increase the movement of the food to shorten processing time.*

To find out whether a food has been heated sufficiently, canners check the **cold point**. This is the *point in the food that is the last to reach the temperature considered safe for killing microorganisms in that food.*

In a can heated only by conduction, the cold point is at the center of the can. This is because colliding molecules transfer heat evenly from the outside of the can to its center. When both conduction and convection are used, the cold point is lower, as you can see in **Figure 27-3**. In this case, movement of the fluid from the bottom of the can to the top leaves a spot slightly below the center that is last to heat. The time it takes the cold point to reach the required temperature determines the overall processing temperature and time.

Pouch Retorts

Some foods—spaghetti precooked in a sauce, for example—are "canned" in a pouch. **Pouch canning** is *similar to retort canning except that flexible tin packages replace cans and jars.* Heat penetrates these pouches more rapidly than it does traditional containers. This shortens processing times, which produces a higher quality food while using less energy.

Retort pouches have been used in Europe for many years. They are common in the United States military services and have also gained popularity in the general food market. Pouches are usually lighter and take up less space than rigid containers, which is especially appealing to manufacturers and grocers. Consumers like the convenience of the easily opened packages. Unlike cans, however, these pouches are not recyclable. The federal government has established durability standards to ensure the safety of any pouches sold in the United States.

Water currents Figure 27-2
Liquid heats by convection as different temperatures cause the liquid to rise and fall in currents. The food in liquid heats by conduction as the hot liquid comes in contact with the food.

Where Is the Cold Point?

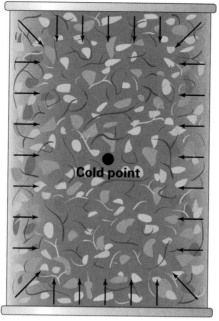 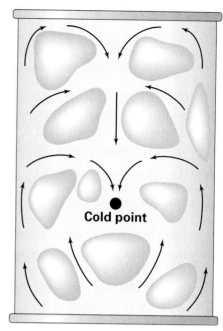

Figure 27-3
The cold point in the can is the last place to heat. Why is the cold point lower in one can?

Cold point

Heating by conduction only

Cold point

Heating by convection and conduction

Aseptic Processing

Do you wonder how those little packages of liquid coffee creamer on restaurant tables stay fresh? How does a paperboard foil container keep juice safe without refrigeration? The answer is **aseptic processing** (ay-SEP-tik), in which *food is first sterilized by heat and then placed in sterilized containers.*

Aseptically processed food is very quickly heated until sterile in a plate-shaped or tubular heat exchanger. This sometimes takes only a few seconds. The food is cooled quickly to protect its quality, then placed in a container. All of this takes place in a sterile filling zone.

While the technology necessary for aseptic processing has existed since World War II, the procedure became widespread only during the 1980s. As mechanized processing grows more streamlined, aseptic processing is becoming more popular.

Advantages and Disadvantages

Aseptic processing has a number of advantages. First, products can be stored without preservatives at room temperature for long periods of time. Even delicate foods, such as milk and orange juice, can be held for months without refrigeration or artificial preservatives.

Pouch canning produces a high-quality product because the processing time is short. What other advantages are there? What are the disadvantages?

Finding Nutrition in Canned Food

Canned foods have a reputation for being less nutritious than their fresh or frozen varieties. As the following shows, this reputation is earned:

- High temperatures used in canning eliminate heat-sensitive nutrients. You won't usually find much vitamin C or thiamin in canned foods, for instance, especially those processed in a pressure canner. The one exception is tomatoes. A high-acid food, tomatoes can be processed at low temperatures. As a result, canned tomatoes retain more vitamin C than other canned vegetables.
- When canned foods are processed in liquid, water-soluble vitamins and minerals dissolve. The concentration of these nutrients is the biggest difference between canned and fresh food.
- Packing foods in sweetened syrup or oils adds empty kilocalories.
- Nutrients are lost in long storage. The higher the storage temperature, the more nutrients are lost. Light too can destroy some nutrients.

Working Around the Shortcomings. To get the most nutritional value for your canned food dollar, you can try these suggestions:

- Serve the canning liquid with the food or use it to prepare other foods, so vitamins and minerals aren't wasted.
- Choose fruits packed in juice or tuna in water. These have fewer kilocalories and less fat.
- Store food that is in glass containers in the dark to help preserve light-sensitive nutrients.
- Read labels to learn what nutrients, as well as additives and ingredients, each food contains. Such nutrients as carbohydrates and fats are not affected by the canning process.

Applying Your Knowledge

1. Why do you think fats withstand the canning process better than some vitamins?
2. Because they are usually less expensive, canned items are often the first choice of people with limited income. What might be some consequences of this situation?

The Institute of Food Technologists considers aseptic processing to be the most significant food science innovation of the last fifty years because it substantially reduces the time and temperature needed for sterilization. That, in turn, increases nutrient retention and flavor.

This stands in sharp contrast to traditional canning. For example, milk canned by traditional means is heated so long that the sugars in the milk react, causing a change in color and flavor. Aseptically processed milk has a flavor and nutritional value more similar to that of fresh milk.

Aseptic processing was developed in the 1940s but used more after the 1980s when improved methods for sterilizing the packaging were devised. Lower packaging cost is an advantage of this processing method.

Another advantage of this processing is cost. Aseptic processing is less expensive than other canning methods.

One disadvantage of aseptic processing is that total control of the sanitary conditions must be maintained during processing. This requires special equipment.

Botulism Poisoning

It may sound dramatic, but proper canning really is a matter of life and death. Neglecting details or taking shortcuts when home canning can result in poorly sealed containers that admit pathogenic bacteria. Commercial canners must also make sure their procedures prevent the entry of bacteria into foods.

You've read about the microbe *Clostridium botulinum*, which causes botulism, the most serious foodborne illness known. Scientists say that 250 mL—one cup—of pure botulinum toxin would be enough to kill every human being on earth.

Properties of *C. Botulinum*

The rod-shaped bacteria called *C. botulinum* live in soil, water, and feces. Like many similar microorganisms, these bacteria can form spores under adverse conditions. The spores resist chemical treatment, heat, and other environmental changes that would destroy active cells. Unless deliberate action is taken to destroy it, *C. botulinum* can exist indefinitely. Because commercial processors are very careful, *C. botulinum* seldom survives retort canning.

Most cases of botulism result from improperly home-canned foods. *C. botulinum* bacteria grow anaerobically, or without air. When food is heated during canning, the air is driven out and the container is sealed. Any spores that survive the heat can still grow in the airless interior of the container. The spores germinate in the food to produce new bacteria, which grow and produce a deadly toxin in the canned food.

When home-canned food is heated and sealed, air is driven out of the container. Will this prevent *C. botulinum* from growing inside? What could happen?

When you find a can that is damaged or has a bulging top or sides, what should you do?

Air is also removed when food is heated during cooking. Therefore, the interior of the food can become an anaerobic environment. Without sufficient heat during cooking, *C. botulinum* spores may survive in the airless interior. If leftover food is stored at room temperature, the spores can then grow and produce toxin.

EXPERIMENT 27

Evaluating Canned Peas

SAFETY FIRST
Review these safety guidelines before you begin this experiment.

Y ou will recall that young vegetables contain mostly sugar, which turns into starch as the vegetables age, or mature. Young peas are sweet and tender, while mature peas have a tough starchiness. Since young peas are more desirable, the most important factor in evaluating or grading peas is maturity.

There are several ways to evaluate canned peas. One method involves placing peas in a salt solution. Young tender peas float, while the mature peas sink from the density of their starch. Another way to check the maturity of peas is to observe the liquid, or brine, in which the peas are packed. The starch in mature peas dissolves into the brine, creating a cloudy effect. Finally, the skin of mature peas is often split and broken, while young peas are apt to have smooth, unbroken skins. This experiment lets you use all three methods to evaluate various brands of canned peas.

Equipment and Materials
canned peas	paper towel	100-mL beaker
200-mL salt solution	250-mL beaker	

Procedure
1. Count 50 peas from the source indicated by your teacher.
2. Pour 200-mL salt solution provided by your teacher into a 250-mL beaker.
3. Add the peas to the salt solution, and observe for 30–60 seconds. Count how many peas sink to the bottom. If nearly all the peas sink, count how many remain floating and subtract from 50. Record this information in your data table.

4. Remove the peas from the salt solution, and place them on a paper towel. Count how many peas in your sample have broken skins. Record this information in your data table.
5. Place 50 mL of pea brine in a 100-mL beaker, label it with the brand name, and place in the area designated by your teacher. Note the appearance of the brine, compare it to the other brands, and record the information in your data table.

Avoiding Botulism

Like many pathogenic microorganisms, *C. botulinum* may lurk undetected in foods. This is why no food should be eaten that has been stored or prepared unsafely. In some foods, a foul, rancid odor or an unusual appearance is the telltale sign of some type of spoilage. Such foods should also be thrown out.

Have you ever heard that you shouldn't eat foods from bulging cans? "Puffy" cans, cracked jars, and loose lids often indicate that pressure is building in the container. Such pressure buildup is a sign of microorganism activity—in both home canned and commercially canned foods. The properties of *C. botulinum* make it a likely suspect.

Because the toxin is so potent, such cans should be discarded, not opened. If purchased, you may want to report the product to a local health official. Although instances of botulism are relatively rare, a cautious approach to processing canned foods as well as preparing and storing foods is wise.

6. Obtain unit cost information for your brand of peas from your teacher, and write it in your data table.
7. Write your information on the chalkboard. In your data table, copy the information about the other brands of peas.

Analyzing Results

1. Which brand of peas had the largest number of peas that sank? The smallest number?
2. Which brand of peas had the largest number of peas with split skins? The smallest number?
3. Was there any relationship between the number of peas that sank and the number that had broken skins?
4. Was there any relationship between the number of peas that sank and the cloudiness of the brine from which they came?
5. Which brand of peas seemed to be of the highest quality?
6. Were the peas with the highest unit cost the ones of the highest quality?

SAMPLE DATA TABLE

Brand of Peas	Number That Sank	Number with Broken Skins	Appearance of Brine	Unit Cost

CHAPTER 27 Review and Activities

Chapter Summary

- Equipment and containers used in canning are designed to kill microorganisms and seal foods for long-term storage.
- Food is packed for canning with either the hot-pack or cold-pack method.
- Foods are processed in either a water-bath or a pressure canner.
- Conduction and convection are methods of heat transfer.
- Commercial, or retort, canning is used industrially to process foods in cans and pouches safely.
- With aseptic processing, foods can be stored without refrigeration.
- Properties of *C. botulinum* make it a particular threat in home-canned foods.
- Learning the signs of *C. botulinum* contamination helps prevent botulism.
- Care must be taken to choose and store canned foods for maximum nutritional value.

Using Your Knowledge

1. Why don't canning jars crack during processing?
2. How do jars, lids, and bands help keep canned foods safe?
3. How would you decide whether to hot-pack or cold-pack a food for canning?
4. Compare water-bath processing with pressure processing.
5. If you're not sure what the pH of rutabagas is, which processing method would you choose and why?
6. Which takes longer to process, a 227 g can of peas or a 227 g can of sweet potatoes? Why?
7. How does convection assist conduction when heating food during the canning process?
8. When might a commercial canner use an agitation retort?
9. Why is the cold spot in a different location in some canned foods?
10. Which food is likely to retain more nutrients, vegetable stew in a can or in a retort pouch? Why?

Real-World Impact

Feeding an Army

The old military "C-rations" were convenient but they lacked appeal. With retort pouches came the meal ready-to-eat, or MRE. This pouch has a thin aluminum layer between two layers of the plastic polyolefin. MREs are light yet sturdy and resist pests and bacteria. They last from six months to three years and come with a flameless heater. MREs can even hold baked goods. MREs have been modified for humanitarian needs for the Red Cross and other international aid groups. The rugged containers can be air-dropped when land travel is too dangerous or impossible.

Thinking It Through

1. MREs are intentionally high calorie for the military. Why do you think this is so?
2. Why do you think military planners are concerned that MREs be appealing as well as convenient and nutritious?

Using Your Knowledge *(continued)*

11. How is aseptic canning different from retort canning?
12. Why are canned foods the most common source of botulism?
13. Why are cracked jars with loose lids associated with botulism?
14. If buying a can of fruit cocktail, what tips for choosing and storing the product will help you get the most nutrition from your purchase?

Skill-Building Activities

1. **Management.** Obtain a catalog for home canners. Working with a partner, choose items that you would use for safe home canning. Explain your choices in class.
2. **Communication.** Interview an older relative or neighbor who has experience canning foods. Why does this person home-can foods? What changes in equipment and methods has he or she seen? How are these changes related to newer information on canning and food safety? What aspects of canning have remained the same? Write your interview as an article on changes in home canning.
3. **Economics.** Research the economic impact of the canning industry. Which are the largest canners in the industry? How many people do they employ? How does the company affect other segments of the economy?
4. **Creative Thinking.** On separate slips of paper, write the names of five common canned foods. Form groups of five. Draw one slip of paper from each group member, to give each member a new set of foods. Try to plan a day of nutritiously balanced, appealing meals, using as many of the canned foods on your slips of paper as possible.

Thinking Lab

Pressure Canner Investigation

A pressure canner is essential equipment for anyone who cans low-acid foods at home. It's also useful for preparing foods quickly and nutritiously for a meal. Both canning and cooking employ the same principle: increasing the temperature by creating steam in an airtight environment.

Analysis

At a glance, a pressure canner may look like nothing more than a modified saucepan. On closer inspection, you'll see how the canner is carefully designed for its specialized role.

Organizing Information

1. Obtain the owner's manual for a pressure canner. Use the manual to identify the canner's parts.
2. In a three-column chart, list each major part, its use, and how it contributes to the pressure canner's function in home canning.
3. What might happen if the steam vent on a pressure canner became clogged?

Understanding Vocabulary

Write each term in the "Terms to Remember" list for this chapter in a sentence that shows you understand its meaning.

Food Preservation Technology

Objectives

- Describe the effect of freezing on foods.
- Identify and describe commercial freezing methods.
- Demonstrate how to choose and package foods for freezing at home.
- Explain the role of sublimation in freeze-drying.
- Explain how irradiation preserves foods.
- Assess arguments for and against irradiation.
- Evaluate the suitability of containers for commercial food packaging.
- Compare modified-atmosphere packaging with aseptic packaging.

Terms to remember

electromagnetic spectrum
electromagnetic waves
flash frozen
gray
immersion freezing
indirect-contact freezing
irradiation
kilogray (kGy)
lyophilization
modified-atmosphere packaging (MAP)
radiation

A soldier eating canned food in Napoleon's army was on the "cutting edge" of food preservation technology. Today, canned foods are used routinely. The first astronaut to sip reconstituted orange juice made from a powder enjoyed space-age technology. Now, everything from lemonade to cheese sauce comes in a powder. Once again, food processors have turned state-of-the-art technology to commercial use.

Improving food preservation methods is not just a matter of convenience and profit. An ever-growing world population can't afford to waste nourishing foods to spoilage. When disaster strikes, foil pouches of a nutritious drink shipped halfway around the world can help save lives. This chapter looks at how foods are preserved, including some of the newer techniques that have a high impact on people everywhere.

Freezing Food

The frozen food industry was born in the 1920s when Clarence Birdseye, produced the first frozen fish. The industry has continued to develop with the growth of such technology as fast-freezing equipment, home freezers, and the microwave oven.

The convenience of frozen foods has literally changed society. In the 1950s, for example, the television and the TV dinner grew up together to become mainstays of United States culture.

The widespread use of frozen foods reflects how convenient they are for consumers.

The Science of Freezing

Freezing preserves foods by slowing the action of enzymes rather than destroying them. Vegetables and some other foods are blanched before freezing to denature enzymes, which helps retain a fresh flavor.

Foods that are free of microorganisms when frozen will remain so, since most microorganisms can't grow at such low temperatures. However, any microbes that are already present simply become inactive. As the food thaws and warms to room temperature, these microbes can multiply just as rapidly as if the food were fresh. Consequently, careful handling is as important for thawed foods as for fresh.

When food is frozen, water in the food goes through a phase change, turning into ice. As the molecules slow down, they organize into crystals. The ice that forms continues to cool as the temperature is lowered to the necessary storage temperature.

Have you ever tried to freeze lettuce? What happens? Because of its high water content and fragile cell structure, the ice crystals in frozen lettuce damage its cell walls. Cell fluids leak, leaving the lettuce limp and unappetizing when it thaws. Since many fruits and vegetables have a high water content, they don't freeze well.

Freezer Burn

If not packaged properly, even foods that do freeze well can experience a damaging effect known as freezer burn. Foods with freezer burn often have a brown or grayish color and look very dry. They may be flecked with fuzzy gray or white spots that look like icy mold. The foods have an off-flavor and a tough texture.

Freezer burn is caused by moisture loss from the food when the food is exposed to air. Loose packaging may be the culprit. Freezer burn can also occur if temperature fluctuations in the freezer cause the food to repeatedly thaw and refreeze. A temperature change as small as 3° above the ideal –18°C mark can create the problem. With each refreezing, ice crystals grow larger. Cells that were damaged during the food's initial freezing are further weakened.

Commercial Freezing

Unlike many canned foods, those that are commercially frozen undergo only limited changes in color, texture, size, and flavor. With the older

Food that isn't tightly packaged can develop freezer burn. This is commonly seen in meats and poultry but may also occur in other foods.

slow-freezing methods, larger ice crystals formed, causing cells in food to expand and distort. With the quick-freezing methods used today, ice crystals are smaller, causing less damage to cells. Low temperatures help foods retain more natural quality. Quick freezing also helps foods resist freezer burn.

The rapid freezing processes used today include the following:

- **Freezing in air.** This is the oldest and still the most common method of freezing food, and also the one used in home freezing.
- **Indirect-contact freezing.** With **indirect-contact freezing,** *food is placed on belts or trays. A refrigerant circulates through a wall beside the food. As the food passes close to the cold wall, it quickly chills and freezes.* This is the method used by most major frozen vegetable processors.
- **Immersion freezing.** With **immersion freezing,** *food is submerged directly in a nontoxic refrigerant to cause quick freezing.* The refrigerant does not change the food's color, taste, or aroma. This technique is used for seafood and many fruits.

The packaging used for commercially frozen food is designed to prevent water vapor from leaking in or out. Since food expands up to 10 percent when it freezes, materials must be tough yet flexible. Most packages are opaque because many foods lose flavor when exposed to light.

Frozen products represent a great investment for the food industry. The specialized equipment needed to freeze, transport, store, and display frozen food requires a great deal of energy. A rise in energy costs could increase the use of other food preservation methods and spur development of new ones.

Home Freezing

Understanding how foods react to intense cold can help when you stock your own freezer. Choosing foods to freeze is easier when you remember that water expands during freezing. Foods with high water content, including grapes, pears, and salad greens, lose shape and texture when frozen.

Some seasonings also don't fare well in the freezer. Fresh chives, black pepper, and even sweet basil turn bitter. Raw garlic, on the other hand, grows stronger. Artificial vanilla often develops an unpleasant flavor. Hard-cooked eggs get leathery, while egg custards can separate and leak.

Storage containers and wraps used for frozen foods at home must be moisture-proof and airtight to maintain quality. Special wrapping papers and bags can be purchased for this purpose.

Even properly selected, wrapped, and frozen, most food does have a maximum storage time. Typically, frozen foods last 6–12

The frozen food industry continually improves product quality. Newer research aims at preventing large ice crystals from forming during frozen food storage. Applying certain proteins to the product may make this possible.

Properly preparing food for home freezing helps maintain the food's quality and prevent freezer burn.

months in the freezer. If not used within that time frame, quality starts to deteriorate. By labeling and dating packages destined for the freezer, you won't have to guess how long they've been there.

Freeze-Drying

Here's a food science riddle: what food items are frozen but not stored in the freezer? The answer is freeze-dried foods. Freeze-drying is a commercial process that combines freezing and drying to preserve foods. *First, a food is frozen; then it's treated to remove the solvent from dispersed or dissolved solids. In most food, this means removing water.* The technical name for this freeze-drying process is **lyophilization** (lye-ahf-uh-luh-ZAY-shun).

For freeze-drying, food may be raw or cooked, but it must be high quality. The procedure is too expensive to use on anything else. Freeze-drying is carried out with complex equipment housed in highly specialized facilities. Computers monitor the many variables that affect success.

Sublimation

During freeze-drying, water in the form of ice is removed from frozen food in a special way, through sublimation. The ice changes directly from the solid to the gaseous state without first becoming a liquid. The result is a dried food. Something similar happens if you hang clothes to dry outside on a winter day. Water in the fabric freezes and then sublimes, leaving dry clothes.

Ice can sublime at normal atmospheric pressure, but rather slowly. You can speed the sublimation rate by lowering the air pressure over the ice.

For freeze-drying, food enters a special, low-pressure chamber. There it is **flash frozen**, or *frozen very quickly*, then held at below-freezing temperatures.

Freeze-dried food can only be produced commercially.

Nutrition *Link*

Fine Dining in Space

As humans ventured into space in the 1960s, keeping astronauts well nourished was a major concern. No one was sure how a lack of gravity would affect the physical act of eating. Would weightless food be hard to swallow? Would it collect in the throat?

Experiments showed that eating in space was no problem. Food crumbs, however, posed a threat to the sensitive equipment. Thus, food taken on the first space flights was sucked through straws. Unable to see or smell the food, astronauts had a limited desire to eat.

The space program has come far since the early launches, and nutrition hasn't been forgotten. With improved food preservation techniques, "space food" isn't boring. A menu in space might read: irradiated ham or pouched turkey and gravy as an entrée; irradiated bread, canned applesauce, and freeze-dried strawberries as choices for side dishes; and a pouch of chocolate pudding for dessert. Nutritious snacks might include peanut butter and granola bars, with cookies and candy for a space traveler's sweet tooth.

Considering that all food taken into space must meet certain requirements, the variety of meals is impressive. Food must have a shelf life of six months at a temperature of 37.7°C. Outer space is vast, but space inside is limited, so food products and packaging must be compact. Also, on the theory that "every little bit counts," foods must be lightweight to make liftoff easier. As a result, most foods and drinks are dehydrated. They must also taste good to ensure that they'll be eaten.

Applying Your Knowledge

1. Suppose you were creating menus for space travelers. What specially preserved, light-weight foods would you create?
2. Space travel may someday become routine for ordinary citizens. How might this affect the foods served in space?

The low temperature keeps the water frozen, and the low pressure speeds the rate at which ice crystals in the food escape as water vapor. The vapor is drawn off by low-temperature condenser plates.

In freeze-dried food, sublimation occurs first on the surface and continues inward. By the time the ice at the very center has sublimed, up to 99 percent of the food's moisture has been removed. Since the food was frozen rigid before it was dried, the end product looks something like a sponge, with pores and tunnels where ice crystals once were.

The conditions under which freeze-drying occurs must be monitored very carefully. Manufacturers want to maximize the rate of sublimation by not allowing the temperature to rise and cause the ice to melt.

Using Freeze-Dried Foods

If you like chicken and mushrooms in your instant soup, or strawberries in your muffin mix, you can appreciate freeze-drying technology. Lyophilization works well with these foods. Coffee, fruit and fruit juice, and even shrimp may all come to you freeze-dried.

Freeze-dried food can be stored for months or even years. Since moisture in the air can cause the food to quickly reconstitute, however, it must be protected in airtight packaging.

Like other dried food, freeze-dried products are very lightweight. This means that supermarkets pay less in shipping and storage. Freeze-dried food slips easily into a camper's backpack.

Freeze-dried food is reconstituted before it's eaten. Stored properly, reconstituted food rates higher in taste, appearance, texture, and nutritional value than foods that are just dehydrated.

Convenience foods aren't the only items created through freeze-drying. Flavors, cultured microorganisms, enzymes, and specialty chemicals are all preserved by the same process. These are sold to food processors, as well as to chemical, biotechnical, and pharmaceutical companies.

Irradiation

Imagine having to choose between eating a moldy orange or none at all. Some people face such dilemmas every day. Despite modern preservation methods, between one-quarter and one-third of the world's total food supply is still degraded or destroyed by pests, bacteria, and the resulting spoilage. Contaminated food is a waste of the world's resources and a threat to human health. Worse yet, it is sometimes the only food available to victims of famine, war, poverty, and other disasters.

Irradiation, a newer method of food preservation, offers a way to avert such hardships. In **irradiation**, *food is exposed to a controlled amount of radiation for a specific time to destroy organisms that would cause spoilage.*

Experiments with food irradiation began in 1943, with the goal of bringing fresh food to troops during World War II. Eventually the technique was used to sterilize some of the food taken into space, as well as meals for cancer and AIDS patients with weakened immunity. Today, food irradiation is gaining acceptance worldwide, although use in the U.S. has been building slowly.

How Irradiation Works

The word irradiation produces a negative image for some people. They may think of the sickness felt after radiation treatments or of dangerous levels of radioactivity from nuclear warfare or leaks from nuclear reactors. Irradiation, however, is simply energy from **radiation**, that is, *the transfer of energy in the form of waves.*

Freeze-dried foods are lightweight, which makes transporting them easy. After rehydrating the foods, they are ready to use.

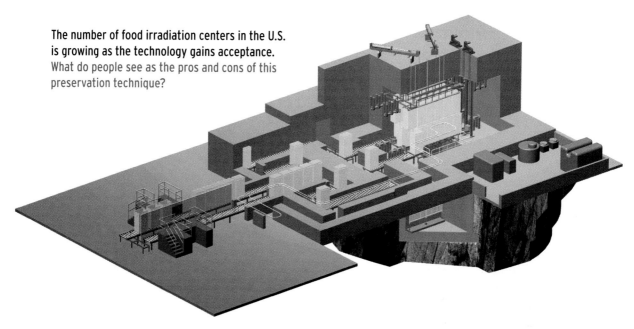

The number of food irradiation centers in the U.S. is growing as the technology gains acceptance. What do people see as the pros and cons of this preservation technique?

Gamma Rays

You're probably familiar with the microwaves that cook foods in a microwave oven. You may also know that ultraviolet rays are sometimes used in food processing to destroy bacteria and preserve food. Microwaves and ultraviolet rays are examples of **electromagnetic waves**, *waves that transfer energy through space.*

All electromagnetic waves travel at the same speed, but some transmit more energy than others do. The amount of energy depends on the length of the waves: the shorter the wavelength, the greater the energy. The **electromagnetic spectrum** includes *the entire range of electromagnetic waves ranked in order of wavelength.*

As you can see in **Figure 28-1**, long radio waves are at the low-energy end of the spectrum, followed by microwaves. Ultraviolet rays are further along the spectrum. At the high-energy extreme are gamma rays. These are the ones used in food irradiation.

Figure 28-1

The electromagnetic spectrum shows the range of radiation ordered by wavelength. Which waves have the most energy?

The Electromagnetic Spectrum

← Decreasing wavelength →

| Radio waves | Micro-waves | Infrared waves | | X rays | Gamma rays |

Visible spectrum   Ultraviolet waves

Basic Research Chemist

Are applied and basic research different?
Ask a . . . Basic Research Chemist

Erika Sweeney is a basic research chemist for a company that produces food preservatives. According to Erika, "Salt and spices have been used for thousands of years to preserve foods, and they're still used today. It's really only been in recent years that we've started to fully understand the chemical processes that explain why they're effective."

"Those of us who do basic research make fundamental discoveries about how preservation occurs—at the molecular level. Then the applied research lab uses our discoveries to develop new methods of preservation. That's the main difference between basic and applied research."

Education Needed: Most basic research chemists hold an advanced degree in chemistry. Depending on their area of interest, they may take additional coursework in microbiology, food science, or environmental science. Laboratory experience and a familiarity with statistical methods are a must for working in a basic research laboratory.

KEY SKILLS
- **Organizational skills**
- **Curiosity**
- **Ability to work alone**
- **Written and oral communication**
- **Attention to detail**
- **Mathematical ability**

Life on the Job: Basic research chemists investigate the composition and structure of matter as well as the molecular processes that determine the effect that one substance has on another. Someone with curiosity about the nature of the world and who enjoys working alone may find basic research highly rewarding.

Your Challenge

Suppose you're doing basic research in the lab when your project is cut from the department budget for next year. You feel you're close to making some significant breakthroughs.

1. How would you react to this situation?
2. What personal and work skills would help you manage this situation? How would they help?

Irradiation is also called cold food preservation because it raises food temperature only slightly. Thus, exposure to gamma rays doesn't cause food to undergo the chemical changes that occur during cooking. Instead, gamma rays inactivate enzymes and damage bacteria genetically. After exposure to gamma rays, microorganisms cannot multiply or produce toxin.

The Irradiation Process

During irradiation, food in metal carriers travels by conveyor belt into a metal-plated radiation chamber. This process is shown in **Figure 28-2**. Radioactive rods, mechanically lifted from a deep-water pool, expose the food to gamma rays and high-speed electrons called beta rays. Exposure time varies with the type of food.

The amount of radiation a food is exposed to during processing is commonly measured in grays and kilograys. These measures indicate the amount of radiation energy absorbed by living tissue. One **gray** *equals 1 joule of absorbed energy per kilogram.* A **kilogray** (kGy*), equals 1,000 grays.* A low dose of irradiation is less than 1 kilogray. A medium dose is 1–10 kilograys. Doses higher than 10 kilograys are restricted in size and used only for spices and dried vegetable seasonings.

The more radiation a food receives, the greater the preservative effect. One kilogray keeps fresh fruit from maturing too quickly. Three kilograys are needed to kill such microorganisms as *Salmonella* in poultry. Despite the use of irradiation, foods should still be treated as perishable.

Figure 28-2

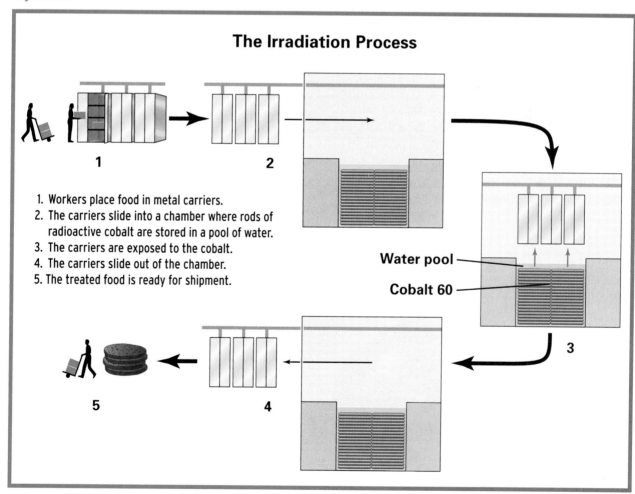

The Irradiation Process

1. Workers place food in metal carriers.
2. The carriers slide into a chamber where rods of radioactive cobalt are stored in a pool of water.
3. The carriers are exposed to the cobalt.
4. The carriers slide out of the chamber.
5. The treated food is ready for shipment.

Water pool

Cobalt 60

Uses of Irradiation

If you like turkey, mashed potatoes, and pumpkin pie on Thanksgiving, your feast may come courtesy of irradiation. Since the 1960s potatoes and onions have been treated to inhibit sprouting. Irradiation was approved for spices and fresh fruits and vegetables in the 1980s. Poultry and beef became eligible in the 1990s, partly due to increased cases of *Salmonella* and *E. coli* poisoning during that decade. In many countries, the process is used during pasteurization of milk.

Irradiation can also be used in a food preservation method called radiation sterilization. Food is first blanched to inactivate enzymes, then subjected to very high levels of radiation. This wipes out all forms of microorganisms. Properly sealed, food sterilized in this way has a shelf life of up to several years. Meat, for example, keeps for up to seven years.

Questions About Irradiation

Despite its growing use, some people still have concerns about the safety of irradiation. One worry is that the food itself may give off radiation. Careful testing has shown this to be untrue. The energy applied is not enough to split atoms, which is how radioactivity is produced.

Also, typical levels of radiation used have less effect on food quality or nutrient value than most processing and cooking methods. In

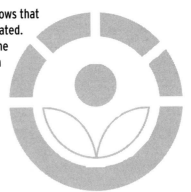

The Radura symbol shows that a food has been irradiated. It appears either on the food label or on a sign nearby.

fact, heat-sensitive nutrients that are destroyed by high-temperature preservation methods are retained during irradiation. High doses do reduce amounts of niacin and thiamin, some amino acids, and vitamins A, C, E, B_6, and B_{12}.

Another concern involves the increasing number of radiation facilities. Some people question the safety of workers in those sites and of people living nearby, as well as of those living near routes used to transport radioactive materials. Supporters of irradiation point to numerous safeguards and regulations in place, and to the industry's record of safety. They note that people are exposed to more radiation from the average hospital X-ray laboratory than from an irradiation center.

Regulating Irradiation

Emotional debates over irradiation's safety pose problems for lawmakers, food manufacturers, and consumers. The United States Food and Drug Administration (FDA) regulates irradiation as an additive. The agency has taken more reports on irradiation testing than any other food preservation method and has decided that the process meets the standard of the Delaney Clause. That clause states that no substance shown to cause cancer in humans or animals may be added to food in any amount. The FDA approves food for irradiation in the United States on a case-by-case basis.

Internationally, irradiation faces a situation like that of additives. Trade of irradiated products is often stalled by political and cultural conflicts. A worldwide standard for irradiated food has been adopted by Codex, the joint body

Irradiation provides consumers with products of better quality. Irradiated mushrooms, for example, have less deterioration.

The flavor of a food could change considerably from contact with packaging materials. That's one reason why packaging materials must be chosen carefully.

of the United Nations and the World Health Organization, with more than 130 member nations. Their statement, it is hoped, will result in a freer flow of food across national boundaries.

Packaging Food

The most advanced preservation techniques may not save food from spoiling if living organisms or moisture slip in through faulty packaging. That's why packaging materials and methods are almost as important as how a food is preserved.

Suitable Containers

A check of your kitchen might reveal foods packaged in a variety of containers—perhaps cans, glass jars, cardboard cartons, and plastic bags. Some packages use a combination of materials. Food packaging has changed greatly over the years. Some concerns that have influenced change include the following:

- **Safety.** Packaging needs to be nontoxic and provide a sanitary, protected environment for the contents. It must be sturdy enough to withstand shipping and handling so that seals are not damaged.
- **Food quality.** Contact with packaging materials can affect food flavors. Some containers are opaque to protect food from light. Others must provide protection from odors.

- **Convenience.** Consumers want packages they can open easily. Packages that are resealable and good for pouring are also valued.
- **Expense.** Lower packaging cost means lower production costs for manufacturers and lower food costs for consumers.
- **Environmentally safe.** More consumers want recyclable containers. As landfills quickly fill with waste products, concerns are growing about the quantity of material thrown away in society.
- **Marketing appeal.** Producers need packaging that is attractive to consumers, with room for information on the outside.

Modified-Atmosphere Packaging

One packaging technique the food industry uses is **modified-atmosphere packaging**, sometimes called MAP. *This process creates a specific gaseous environment for a food product to lengthen shelf life.* Generally, the air inside a package is replaced with a mixture of gases that often includes oxygen, carbon dioxide, and nitrogen. Oxygen helps preserve appearance, while carbon dioxide suppresses the growth of microorganisms. Nitrogen helps maintain the balance of the other two gases.

MAP is used with other techniques that control such factors as temperature, pH, and water activity. Combining these techniques can be ex-

Some containers go right from the freezer to the microwave oven for quick preparation. Why do so many consumers buy these products?

An advantage of MAP is that food isn't subjected to the extremes of heat and cold needed in other preservation methods. Damage to quality, texture, taste, and nutrition is minimal.

tremely effective in preventing microbial growth. This approach appeals strongly to consumers who want fewer additives and preservatives in their foods.

To use MAP successfully, food processors must meet these four technical requirements:

- Since foods require different gas mixtures to maintain quality, the gas mixture chosen must suit the food being packaged. For example, a carbon dioxide injection process is used to maintain the freshness, consistency, and taste of dairy products.
- Machines must be capable of mixing gases precisely.
- The MAP machine must create a tight seal to keep the atmosphere inside the container constant.
- The packaging room, machines, and workers must be clean and free of bacteria.

MAP technology is now widely used in marketing ready-to-eat products with extended shelf lives. Using this technology backed by a strong quality-assurance program, one company can offer preservative-free potato products with a 30-day shelf life. You may have noticed bags of lettuce or spinach and baby carrots in stores. These use MAP technology.

The meat industry may benefit most from MAP. Popular "family packs"—large packages of pork chops, chicken thighs, or other products—retain their quality much longer in modified-atmosphere packages.

Aseptic Packaging

Aseptic packaging is a storage system for beverages and liquid foods that have been aseptically processed, as described in Chapter 27. This packaging was widely used in Europe and Asia for several decades before introduction in the United States during the early 1980s.

Although these packages are not recyclable, they have many advantages. They require far less energy to manufacture, fill, ship, and store than any other comparable package on the market.

Since empty packages are stored flat or on rolls, one standard semitrailer truck can transport 1.5 million empty drink boxes as opposed to only 150,000 glass bottles. Since the boxes are compact and lightweight, more product can be shipped in fewer trucks. Finally, drink boxes preserve their contents without refrigeration, so they don't require refrigerated trucks, warehouses, or retail freezers or coolers. This saves both electricity and gasoline.

Aseptic packages use less packaging material than most comparable containers. A typical single-serve aseptic container is 96 percent product and 4 percent packaging. A glass bottle is 63 percent beverage and 37 percent packaging.

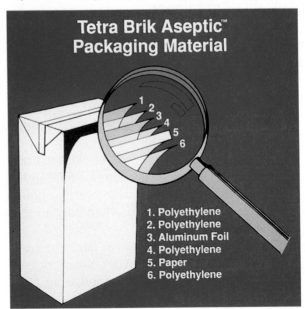

Tetra Brik Aseptic™ Packaging Material

1. Polyethylene
2. Polyethylene
3. Aluminum Foil
4. Polyethylene
5. Paper
6. Polyethylene

Comparison of Orange Juices

SAFETY FIRST

Review these safety guidelines before you begin this experiment.

The method used to preserve a food can affect the properties and quality of the food. In this experiment, you will use your knowledge of sensory evaluation to compare samples of fresh, frozen, freeze-dried, aseptically processed, and pouch-canned orange juices.

Equipment and Materials

5, 28-mL paper cups per student
masking tape
marking pen

samples of orange juice: fresh squeezed;
  reconstituted frozen; freeze-dried;
  pouch canned; and aseptically canned

Procedure

1. Compare the samples provided by your teacher on color, mouthfeel, and taste. Enter the information in your data table.
2. Your teacher will provide information on unit cost and the preservation method of each sample. Enter the information in your data table.

Analyzing Results

1. Which sample looked most like fresh orange juice?
2. Which sample "felt" most like fresh orange juice?
3. Which sample had the best taste?
4. Which juice would you rank the highest overall?
5. How does the juice that you ranked highest compare to the other samples in unit cost?
6. Which juice would you buy to drink regularly? Why?
7. Which juice would you take camping? Why?

SAMPLE DATA TABLE

Sample	Color	Mouthfeel	Taste	Unit Cost	Preservation Method

CHAPTER 28 Review and Activities

Chapter Summary
- Careful selection and packaging helps ensure that foods frozen at home retain quality.
- Commercial processors use advanced techniques to minimize the effect of freezing on a food's appeal and nutritive value.
- In freeze-drying, water is removed from a frozen food through sublimation.
- Freeze-dried food is lightweight, has a long shelf life, and is reconstituted before being eaten.
- Irradiation is a method of preserving food by exposing it to high-energy gamma rays.
- Irradiation has been approved for use for several decades but only now is beginning to come into commercial use in the U.S.
- Containers for packaging food help protect the safety of food.
- Modified-atmosphere packaging appeals to consumers who prefer fewer additives and preservatives in food.
- Aseptic packaging is energy efficient.

Using Your Knowledge
1. What happens to food when it is frozen?
2. What is freezer burn and why does it occur?
3. Why do manufacturers use quick-freezing methods today?
4. Describe the most common method of freezing vegetables used commercially.
5. Why would thawed tomatoes be better for cooking in a sauce than for slicing in a fresh salad?
6. Can frozen food be kept indefinitely? Why or why not?
7. How does sublimation take place in the freeze-drying process?
8. What is the relationship between radiation and food irradiation?
9. How does the irradiation process work?
10. Explain six influences on the types of packaging used to contain foods?
11. How is MAP used to preserve foods?
12. Why is aseptic packaging an energy saver?

Real-World Impact
Better Packaging for Today

New ideas in food packaging are continually under development in order to satisfy public priorities. People in small households may want to open a package, use part of the contents, and then reseal the container to save for later. For those with hearty appetites, packages that hold larger servings may be valued. Packages that can be heated in the microwave but have insulation that prevents burning the hand are another innovative idea. Increasingly, people are looking for conveniences that improve daily living.

Thinking It Through
1. Some food packaging is larger than needed to hold the food inside. Why would a manufacturer do this? What do you think of this practice?
2. Many food producers have switched to canning certain foods in jars. Why would this be appealing to consumers?
3. With your class make a list of how different types of foods and snacks are packaged? Which ones do you think show a wasteful approach to marketing the foods?

Skill-Building Activities

1. **Problem Solving.** With a partner, devise a more efficient way to package a food product. Present your plan, explaining how your idea is an improvement over the food's current packaging.
2. **Communication.** Research food irradiation and the controversy surrounding it. Examine the arguments on each side. Then write a persuasive paper for or against the use of irradiation.
3. **Leadership.** Plan an educational campaign for your school or community on how to reduce waste from packaging. With help from your teacher and classmates, put the plan into action.
4. **Science.** Investigate other uses of radiation in medicine and food processing. What electromagnetic waves are used? For what purpose?
5. **Management.** Imagine that you and a friend are going on a two-day, 30-mile backpacking trip. Plan a menu for the two of you. Remember that you'll need nutritious, safe food that will keep you going without weighing you down. Describe the forms and packaging of foods you'll take.

Understanding Vocabulary

With other students in a small group, create a quiz that tests the "Terms to Remember" for this chapter. Exchange quizzes with another group and complete them. Return the quizzes for evaluation.

Thinking Lab

The Irradiation Controversy

Should people have to pay for something they don't want and believe they don't need? Some critics say that's what food irradiation means to many consumers. Others point out that once consumers understand the benefits, they appreciate the process.

Analysis

According to the critic's argument, irradiation isn't needed because steps are already taken to ensure wholesomeness. If food producers, processors, inspectors, and consumers do their part, quality is almost certain. On the other hand, irradiation can't make up for human carelessness that causes contamination. It may even encourage negligence by providing a false "safety net" to catch mistakes.

Supporters of irradiation counter that the process is a valuable link in the chain of food safety. They aren't suggesting that irradiation take the place of careful food sanitation at any point. It's just one more weapon in the fight against waste, hunger, and foodborne illness.

Thinking Critically

1. The U.S. Centers for Disease Control and Prevention estimated that foodborne diseases cause about 76 million illnesses annually. What impact does this statistic have on the irradiation debate?
2. Packages of irradiated foods are labeled as such for consumers. How will this help settle the controversy?

Glossary

A

absolute zero. The temperature at which all molecular motion ceases and matter possesses no heat energy. (11)

accuracy. How close a single measurement comes to the actual or true value of the quantity measured. (4)

acid. A substance that ionizes in water to produce hydrogen ions. When acids dissolve in water, their molecules break apart and release hydrogen ions into the solution. (10)

activation energy. The energy that supplies the force needed for a reaction. (19)

active site. The area of an enzyme that fits the shape of a substrate. (19)

adenosine triphosphate (uh-DEH-nuh-seen try-FOS-fayt). Molecules used to carry and transfer energy in living cells. (14)

adipose tissue (ADD-uh-pohs). Pockets of fat-storing cells. (16)

adulterated. Made impure by the addition of improper ingredients. (25)

aerobic. Describes a chemical reaction that must take place in the presence of air or oxygen. (22)

aggregate (AG-ruh-git). A group or dense cluster of ions or molecules. (20)

agitation retorts. Canners that mechanically increase the movement of the food to shorten the processing time. (27)

albumen (al-BYOO-muhn). The inner white part of an egg. (17)

alimentary canal. A tubular passage functioning in the digestion and absorption of food and the elimination of food residue, commonly called the digestive tract. (13)

amine group (uh-MEEN). A group composed of two atoms of hydrogen and one atom of nitrogen and is written NH_2; contained in amino acids. (17)

amino acids. A type of organic acid that chains together to make protein. (17)

amylopectin (A-muh-loh-PEK-tun). A starch molecule with multiple branches, like the veins of a leaf. (15)

amylose (A-muh-lohs). A starch molecule with a linear structure, long and narrow like a line. (15)

anabolism (uh-NA-buh-li-zum). The combining of molecules during chemical processes in order to build the materials of living tissue. (14)

anaerobic. Describes a chemical reaction that can occur in the absence of air or oxygen. (22)

anaerobic respiration. Respiration that occurs without oxygen. (22)

anorexia nervosa (an-uh-REX-ee-uh ner-VO-suh). An eating disorder characterized by self-starvation, resulting in the intake of fewer kilocalories than needed for good health. (11)

antibodies. Very large proteins that weaken or destroy foreign substances in the body. (17)

aseptic processing (ay-SEP-tik). A food production method in which food is first sterilized by heat and then placed in sterilized containers. (27)

atherosclerosis (a-thuh-ro-skluh-RO-sus). A buildup of plaque along the inner walls of the arteries. (16)

atom. The smallest particle of an element that keeps the element's chemical properties. (7)

atomic mass. The average mass of a sample of atoms from an element found in nature. (8)

atomic number. The number of protons in the nucleus of an atom. (8)

B

baking powder. A leavening compound that contains baking soda, dry acids, and starch or some other filler; used to produce carbon dioxide for leavening in some baked products. (21)

baking soda. The chemical compound sodium bicarbonate used to produce carbon dioxide for leavening in some baked products. (21)

balance. A scientific instrument that determines the mass of materials. (3)

basal metabolic rate (BMR). A measure of heat given off per time unit; expresses the amount of energy used by the body at rest to maintain automatic, life-supporting processes. (14)

basal metabolism. Energy used by the body at rest to maintain automatic, life-supporting processes. (14)

base. A substance that ionizes in water to produce hydroxide ions. (10)

beaker. A glass container that has a wide mouth and holds solids and liquids. (3)

beriberi (ber-ee-BER-ee). A disease of the nervous system that causes partial paralysis, weakness, mental confusion, and death; caused by thiamine deficiency. (18)

beta carotene (BAY-tuh KAIR-uh-teen). An orange plant pigment that can be changed into vitamin A in the intestines and liver. (18)

bile. A greenish liquid that assists digestion by helping fat mix with water in the intestine. (13)

biodiversity. Cultivating a variety of plants and animals. (2)

biotechnology. Scientists' use of the tools of modern genetics in the age-old process of improving plants, animals, and microorganisms for food production. (1)

blanch. Briefly immersing food in boiling water in order to deactivate enzymes before freezing the food. (19)

boiling point. The temperature at which a liquid's vapor pressure equals the air pressure above the liquid. (9)

botulism. A very serious—and sometimes fatal—form of food poisoning, caused by toxin of the *Clostridium botulinum* bacteria; often caused by improperly home-canned foods. (25)

bound water. Water that cannot be easily separated in food; chemical groups in the food's molecules hold the water molecules tightly. (9)

brine. A water-and-salt mixture containing large amounts of salt; used in pickle making. (22)

brine pickling. A method in which brine is used for pickle making; the vegetable remains in the brine for several weeks, during which fermentation by lactic-acid bacteria takes place until an acceptably low pH is reached. (22)

buffer. A substance that helps maintain the balance of hydrogen and hydroxide ions in a solution; a system of buffers holds blood pH steady at a slightly basic 7.4. (10)

bulimia (byoo-LIM-ee-uh). An eating disorder marked by cycles of gorging on large amounts of food followed by self-induced vomiting and the use of laxatives. (11)

buret (byur-ET). A long, thin cylinder marked to 0.1 of a milliliter. (3)

C

calibrate (KAL-uh-brayt). To check, adjust, or standardize the marks on a measuring instrument. (3)

calorie. The amount of energy needed to raise the temperature of 1.0 g of water 1.0°C. (11)

canning. The process of preserving food by heating and sealing it in containers for storage. (27)

caramelization (kahr-muh-ly-ZAY-shun). The chemical browning reaction that can occur when a sugar is heated. (15)

carbohydrate. An organic compound that is the body's main source of energy. (15)

carboxyl group (kar-BAHK-sul). A group consisting of carbon bonded to oxygen by a double covalent bond, and to a hydroxyl group with a single bond; contained in fatty acids and written as –COOH. (16)

cardiac sphincter. The ring-like muscular valve that ends the esophagus and keeps food from rising from the stomach and back into the esophagus. (13)

carrageenin (kair-uh-GEE-nun). A vegetable gum added to evaporated milk before processing to stabilize the casein proteins. (23)

caseharden. The formation of a hard outer layer that forms on food and traps moisture inside; caused by cooking food at temperatures that are too high. (26)

casein. A protein that makes up about 80 percent of milk. (23)

catabolism (kuh-TA-buh-li-zum). The process of breaking down complex molecules into simpler ones during chemical reactions. (14)

catalyst (KAT-ul-ist). A substance that speeds up the rate of a reaction without being permanently changed or used up itself. (19)

cell respiration. A multistage process that releases energy from glucose when the glucose molecule in the cell is taken apart. (22)

chalaza (kuh-LAY-zuh). The twisted, rope-like structure in an egg that keeps the egg yolk centered. (17)

chemical bonds. The forces that hold atoms together. (8)

chemical equation. A written description of a chemical reaction, using symbols and formulas. (8)

chemical property. Describes the ability of a substance to react with other substances. (7)

chemical reactions. The change of substances into other substances. (8)

cholesterol. A lipid, specifically a sterol, found in humans and animals and made from glucose or saturated fatty acids; excessive amounts believed to contribute to plaque formation, which clogs arteries and causes heart problems. (16)

coagulation (ko-ag-yuh-LAY-shun). To change a liquid into a soft semisolid or solid mass. (17)

coenzymes. Heat-stable, organic molecules that must be loosely associated with an enzyme for the enzyme to function. (19)

cold-pack. A canning method in which uncooked food is placed in a container that is then filled with boiling water or juice and closed with a lid and a ring band; also referred to as raw-pack. (27)

cold point. The point in food that is the last to reach the temperature considered safe for killing microorganisms in that food during processing. (27)

colloidal dispersion (kuh-LOYD-ul dis-PUR-zhun). A homogeneous mixture that is not a true solution; contain relatively large solute particles called colloids. (9)

complete protein. A protein that contains all the essential amino acids. (17)

compound. A substance made of two or more different elements chemically joined together. (7)

concentration. The measure of the amount of a substance in a given unit of volume. (10)

conduction (kun-DUK-shun). Heat energy passed by the collision of molecules. (11)

continuous phase. Describes the substance that extends throughout a colloidal system and surrounds the dispersed phase. (20)

convection (kun-VEK-shun). Heat transferred by circulatory movement in a liquid or gas. (11)

covalent bond (ko-VAY-lunt). A chemical bond formed when atoms share electrons with each other. (8)

covalent compound. Compounds formed by the sharing of electrons in atoms. (8)

cream. Milk that is extra rich in emulsified fat droplets. (23)

creaming. A process in which some of the fat droplets come together in larger clusters that rise and float to the top of the milk. (23)

cross-contamination. The transfer of bacteria from one food to another food or location. (25)

culture. A controlled bacteria population that is added to milk in order to start the fermentation process. (23)

curds. The casein clumps that separate from the liquid when milk coagulates. (23)

cytoplasm (SY-tuh-pla-zum). A colloidal substance consisting of organic and inorganic substances, including proteins and water found in a living cell. Cytoplasm is the main component of both animal and plant cells. (14)

D

Daily Values (DVs). Information listed on the Nutrient Facts panel that helps determine how the nutrients in a food serving fit with what you can or should have for the day. (12)

data. The information gathered during an experiment. (5)

decimal system. A numbering system in which numbers are expressed in units of ten. (4)

deductive reasoning. A thinking process that reaches a conclusion about a specific case based on known facts and general principles. (5)

deficiency disease. A disease caused by the lack of a specific nutrient. (18)

dehydrator. An appliance that provides a steady supply of circulated, heated air in order to dry foods. (26)

dehydrofreezing. A method used to preserve food through a combination of partial drying and freezing. (26)

denaturation (dee-nay-chuh-RAY-shun). A process that changes the shape of a protein molecule without breaking its peptide bonds. (17)

density. A substance's mass per unit of volume. (9)

dependent variable. A factor that changes as a result of what happens to the independent variable. (5)

Dietary Reference Intakes (DRIs). Updated recommendations on nutrient intake as determined by several agencies in the United States and Canada. (12)

digestion. The chemical and mechanical process of breaking down food to release nutrients in a form your body can absorb for use. (13)

dispersed phase. Describes the substance that is distributed throughout another (the continuous phase) in a colloidal dispersion. (20)

double-acting baking powder. A leavening agent that contains two acids, one that reacts with cold liquid and one that reacts with heat, which produce carbon dioxide in two stages. (21)

double bond. A covalent bond in which each atom donates two electrons to form the bond. (16)

E

electromagnetic spectrum. The entire range of electromagnetic waves ranked in order of wavelength. (28)

electromagnetic waves. Waves that transfer energy through space. (28)

electron. A negatively charged particle that moves around the nucleus of an atom. (8)

element. The simplest form of matter that can exist under normal laboratory conditions. (7)

emulsifier (ih-MUL-suh-fy-ur). A substance composed of large molecules that are polar at one end and nonpolar at the other, allowing them to hold together molecules of different substances. (9)

emulsion (ih-MUL-shun). A mixture of two liquids containing droplets that don't normally blend with each other. (9)

energy. The ability to do work. (11)

enrichment. Using additives to increase the nutritive value of processed foods beyond the level contained in the food before processing. (24)

entry-level jobs. Jobs that don't require experience or training. (2)

enzymatic browning. The discoloration of fruits and vegetables caused by enzymes, as when peeled apples turn brown. (19)

enzymes. Special proteins that help chemical reactions take place in living organisms. (12)

epiglottis (e-puh-GLAH-tus). The thin, elastic flap at the root of the tongue that shields the windpipe when you swallow and prevents food and water from entering. (13)

equivalence point. The point at which neutralization occurs when an acid and a base neutralize each other. (10)

Erlenmeyer flask (UR-lun-my-er). A cone-shaped container with a narrow neck and a broad, flat base. (3)

esophagus. The tube-like passage connecting the mouth to the stomach. (13)

essential amino acids. Nine amino acids (out of 20) that the body needs but cannot make itself. (17)

essential nutrients. Nutrients that the body cannot make itself but that are needed to build and maintain body tissue. (12)

experiment. A way to test a hypothesis in order to verify or disprove it. (5)

F

fat-soluble vitamins. Vitamins that dissolve in lipids, rather than water. (18)

fatty acids. The organic acids in triglycerides; fatty acids are compounds that consist of a carbon chain with attached hydrogen atoms and a carboxyl group. (16)

fermentation (fur-men-TAY-shun). A biological reaction that slowly splits complex organic compounds into simpler substances; used to preserve and prepare some foods. (21)

flash frozen. A method used in the food industry to freeze foods very quickly. (28)

flavor. The distinctive quality that comes from a food's unique blend of appearance, taste, odor, feel, and sound. (6)

foam. Air bubbles incorporated and trapped in a protein film by whipping, as in meringue. (17)

food additive. Any substance a food producer intentionally adds to a food for a specific purpose. (24)

food infection. A bacterial foodborne illness that occurs when pathogenic microorganisms enter the body along with the food eaten. (25)

food intoxication. A bacterial foodborne illness that occurs when microorganisms grow in food and produce a toxin in the food. The toxin causes the illness when the food is eaten. (25)

food science. The study of producing, processing, preparing, evaluating, and using food. (1)

foodborne illness. An illness caused by eating food that has been contaminated in some way. (25)

fortification. The addition of a nutrient to a food during processing, as in adding vitamin D to milk. (23)

free water. Water that readily separates from foods that are sliced, diced, or dried. (9)

fresh-pack pickling. A method in which the food being pickled is not fermented as in brine pickling. Instead, the food is placed in brine only for a few hours or overnight, if at all. (22)

G

garnish. A decorative arrangement added to food or drink, such as a sprig of parsley or curled carrot slice. (6)

gelatinization (juh-LA-tun-uh-ZAY-shun). An irreversible thickening process; hydrogen bonds form between starch and water molecules, causing starch granules to absorb water and swell. (15)

glucose. The basic sugar molecule from which all other carbohydrates are built. (15)

gluten (GLOO-tun). An elastic substance formed by mixing water with the proteins found in wheat; produced when dough is kneaded. (17)

glycogen (GLY-kuh-jun). The form of carbohydrates stored in the muscles. (14)

goiter. An enlargement of the thyroid gland caused by a lack of iodine. (24)

graduated cylinder. A tall, cylindrical container used for measuring the volume of liquids. (3)

GRAS list (Generally Recognized as Safe). A list of additives that the FDA considers safe for human consumption; includes spices, natural seasonings, and flavorings, which are not regulated as additives. (24)

gray. Equals one joule of absorbed energy per kilogram; commonly used to measure food irradiation. (28)

H

hard water. Water that contains calcium or magnesium ions. (9)

heat. Energy transferred from an object of higher temperature to one of lower temperature. (11)

heat of fusion. The amount of heat needed to change 1.0 g of a substance from solid to liquid phase. When the situation is reversed, and the liquid phase becomes a solid, this heat energy is released. (9)

heat of vaporization. The amount of heat needed to change 1.0 g of a substance from the liquid phase to the gas phase. (9)

high-quality protein. A protein that contains all the essential amino acids. (17)

homeostasis (hoe-mee-oh-STAY-sis). A healthy and relatively constant internal environment within the body. (14)

homogenization (ho-majh-un-ih-ZAY-shun). A process that reduces the fat globules in milk to a smaller and approximately equal size, and distributes them evenly. (20)

hormone. A chemical messenger that affects a specific organ or tissue and brings forth a specific response, as with the hormone insulin's function in the body. (15)

hot-pack. A canning method in which food is heated in syrup, water, or juice to at least 77°C. While still hot, the food is packed into a container that is closed with a lid and ring band. (27)

hydrogen bond. An attractive force between any molecules in which hydrogen is covalently bound to a highly electronegative element. (9)

hydrogenation (hy-DRAH-juh-NAY-shun). A chemical process in which hydrogen is added to unsaturated fat molecules, breaking some double bonds and replacing them with single bonds. (16)

hydrolysis (hy-DRAH-luh-sus). The splitting of a compound into smaller parts by the addition of water. (15)

hydroxyl group. A chemical combination of hydrogen and oxygen that contains one atom of each element; part of the simple carbohydrate structure. (15)

hypotheses (HY-pah-thuh-seez). Testable predictions that explain certain observations. (5)

I

immersion freezing. A freezing method in food processing in which food is immersed directly in a nontoxic refrigerant to cause quick freezing. (28)

immiscible (ih-MISS-uh-bul). Incapable of mixing or blending. (9)

incomplete protein. A protein lacking one or more essential amino acids. (17)

incubation period. The time needed for bacteria to grow and ferment the milk during the process of culturing milk to make various products. (23)

independent variable. A factor that you change during an experiment. (5)

indicator. A substance contained on a paper that is used to test for acidity; color change of the paper indicates pH level. (10)

indigenous (in–DIJ–uh–nus). Found naturally in a particular environment, as the bacteria that help produce sauerkraut from cabbage. (22)

indirect-contact freezing. A freezing process in which food is placed on belts or trays and quickly chilled by a refrigerant circulating through a cold wall. (28)

inductive reasoning. A thinking process that draws a general conclusion from specific facts or experiences. (5)

inoculation. (23) Adding the starter culture to milk during the fermentation process that produces different cultured milk products.

insoluble (in-SAHL-yuh-bul). Incapable of dissolving. (3)

integrated pest management (IPM). Controlling pests with nonchemical deterrents. (2)

inversion. The process in which sucrose is hydrolyzed by acid and heat, rather than water, to produce sugars that are a combination of fructose and glucose. (15)

ion (EYE-ahn). Atoms that have either a positive or negative charge after an exchange of electrons. (8)

ionic bond. The bond formed by the transfer of electrons between atoms. (8)

ionic compound. Compounds that result when metals and nonmetals bond ionically with each other. (8)

ionization (eye-uh-nuh-ZAY-shun). The process of forming ions in water solution. (10)

irradiation. A food preservation process in which food is exposed to a controlled amount of radiation for a specific time to destroy organisms that would cause spoilage. (28)

J

joule (JOOL). A unit of heat energy; one joule (J) is equal to 0.239 calorie. (11)

K

kilocalorie (kcalorie). A unit of 1,000 calories used in measuring food energy. (11)

kilogray (kGy). Equals 1,000 grays. (28)

L

lactic acid. A waste product formed when carbohydrates are not completely metabolized. (14)

lactose intolerance. The inability to digest milk due to the absence of the enzyme lactase in the intestines. (23)

latent heat. The heat required to create a phase change without a change in temperature. (9)

lipids. A family of chemical compounds that are a main component in every living cell; includes triglycerides, phospholipids, and sterols. (16)

lipoproteins (lip-oh-PRO-teens). Large, complex molecules of lipids and protein that carry lipids in the blood. (16)

lyophilization (lye-ahf-uh-luh-ZAY-shun). A freeze-drying process that combines freezing and drying to preserve foods. First, a food is frozen; then it's treated to remove the solvent from dispersed or dissolved solids. In most food, this means removing water. (28)

M

macromolecules (mak-ro-MAHL-uh-kyools). Large molecules containing many atoms. (17)

major minerals. Minerals needed by the body in amounts of 0.1 g or more daily. (18)

mass number. The total protons and neutrons in an atom's nucleus. (8)

mass percent. The calculation used to describe the concentration of a solute in a solution. (20)

mastication. (13) The chewing process that prepares food for digestion.

matter. Anything that has mass and takes up space. (7)

medium. A substance through which something is transmitted or carried. (9)

megadoses. Excessively large amounts, as with vitamins. (18)

melting point. The temperature at which a substance changes from a solid to a liquid. (9)

membranes. Thin layers of tissue that wall in the cytoplasm of cells. (14)

meniscus (muh-NIS-kus). The bottom of the curve a liquid forms in a container. (3)

metabolic rate (met-uh-BAHL-ik). How fast the chemical processes of metabolism take place. (14)

metabolism. The process by which living cells use nutrients in many chemical reactions that provide energy for vital processes and activities. (14)

metric system. A common language used all around the world for measurement. (4)

micelles (my-SELLS). An aggregation or cluster of molecules often found in colloidal dispersions. (23)

microbes (MY-krohbs). Another name for microorganisms. (22)

microorganisms (my-kro-OR-gun-iz-ums). Single cells of microscopic size that cannot be seen by the human eye, but can be studied through a microscope. (22)

microwaves. Electromagnetic waves. (11)

milk solids. The solids remaining when water is removed to make dry milk; includes protein, carbohydrate, fat, minerals, and vitamins that were dissolved in the liquid portion of the milk. (23)

minerals. Inorganic elements needed by the body. (18)

mixture. A combination of two or more substances in which each substance keeps at least some of its original properties. (7)

modified-atmosphere packaging (MAP). A food packaging process that creates a specific gaseous environment for a food product in order to lengthen shelf life. (28)

molarity (moe-LAR-uh-tee). The number of moles of solute per liter of solution. (10)

mole. The unit of measure used to count atoms or molecules. (8)

molecule. The smallest unit of a molecular compound. (7)

monosodium glutamate (mahn-uh-SO-dee-um GLOOT-uh-mayt) or MSG. A salt that interacts with other ingredients to enhance salty and sour tastes. (6)

mouthfeel. How a food feels in the mouth. (6)

N

neutral. A water solution that contains an equal number of hydrogen and hydroxide ions. (10)

neutralization. A chemical reaction in which hydrogen ions from an acid react with hydroxide ions from a base to produce water molecules. (10)

neutron. An uncharged atomic particle found in the nucleus of an atom. (8)

nucleus (NOO-klee-us). The dense core of an atom. (8)

nutrient dense. Describes a food that provides a relatively high quantity of nutrients in comparison to a low-to-moderate number of kilocalories. (12)

nutrients. Substances that are found in food and needed by the body to function, grow, repair itself, and produce energy. (12)

nutrification. A process that adds nutrients to a food with a low nutrient-per-calorie ratio so that the food can replace a nutritionally balanced meal. (24)

nutrition. The science of how the body uses food. (12)

O

obesity. Having excessive fat and a body weight that is 20 percent or more above a healthful range. (11)

olfactory (ohl-FAK-tuh-ree). Identifies the organs related to the sense of smell. (6)

organic compound. A compound that contains the element carbon. (7)

osmosis (ahz-MOH-sus). The movement of fluid through a semipermeable cell membrane to create an equal concentration of solute on both sides of the membrane. (14)

osteomalacia (ahs-tee-oh-muh-LAY-shuh). Adult rickets, a vitamin D deficiency disease. (18)

osteoporosis (ahs-tee-oh-pore-OH-sus). A loss in bone density due to a prolonged deficiency of calcium. (18)

oxidation. The chemical reactions in which elements combine with oxygen. (12)

P

pancreatic juice. An enzyme-rich fluid that reduces food to smaller molecules during digestion in the small intestine. (13)

papain (puh-PAY-in). Three enzymes, which come from the papaya fruit, diluted with salt to make a dry powder. (19)

parasites. Organisms that grow and feed on other organisms. (25)

pasteurization. The process of using heat treatment to destroy bacteria, as in milk. (22)

pathogenic. Disease carrying. (25)

pellagra (puh-LAY-gruh). A disease that causes skin eruption, digestive and nervous disturbances, and eventual mental decline; caused by niacin deficiency. (18)

peptide bonds. Bonds between the nitrogen of one amino acid and the carbon of a second amino acid. (17)

periodic table. A chart that arranges elements by atomic number into rows and columns according to similarities in their properties. (8)

peristalsis (PEHR-uh-STAHL-sus). Waves of muscular contractions that force food through the digestive tract. (13)

pesticides. Chemicals used to kill insects, animal pests, and weeds. (25)

pH scale. A mathematical scale in which the concentration of hydrogen ions in a solution is expressed as a number from 0 to 14 to indicate acidity. (10)

phase. The physical state in which matter can exist, as in solid, liquid, or gas. (7)

phospholipid (fahs-foh-LIH-pud). A compound similar to a triglyceride but with a phosphorus-containing acid in place of one of the fatty acids. (20)

photosynthesis (foe-toe-SIN-thuh-sus). The process in which plants use the sun's energy to convert carbon dioxide and water into carbohydrate and oxygen. (15)

physical change. An alteration of a substance that does not change its chemical composition. (8)

physical property. A characteristic that can be observed or measured without changing the substance into something else. (7)

phytochemicals. Compounds that come from plant sources and may help improve health and reduce the risk of some diseases. (18)

plaque (PLAK). A mound of lipid material mixed with calcium and smooth muscle cells; plaque buildup in the arteries can lead to heart disease. (16)

polar molecule. A molecule with a clear division of opposite electrical charges. (9)

polymer (PAH-luh-mur). A large molecule formed when small molecules of the same kind chain together. (15)

polypeptide. A single protein molecule containing ten or more amino acids linked in peptide chains. (17)

pouch canning. (27) A food canning process similar to retort canning except that flexible tin packages replace cans and jars.

precipitate (prih-SIP-uh-tayt). To cause a solid substance to separate from a solution. (23)

precision. How close several measurements are to the same value. (4)

precursor (pri-KUR-sur). A compound that can be changed into a vitamin in the body. (18)

pressure processing. A food processing method in which containers of food are heated under pressure in a pressure canner. (27)

product. The elements or compounds formed during a chemical reaction. (8)

property. A feature that helps identify a substance. (7)

protons. Particles that have a positive electrical charge. The number of protons in the nucleus of an atom determines the atomic number of an element. (8)

pure substance. A substance made of only one kind of material and that has definite properties. (7)

pyloric sphincter. A circular muscle that controls food's rate of movement from the stomach to the small intestine. (13)

Q

quick breads. Bread products that do not need time to rise; they are usually made with baking powder or baking soda. (21)

R

radiation. The transfer of energy in the form of infrared rays. (11, 28)

rancid. The term that describes the unpleasant flavors that develop as fats oxidize in food. (16)

raw-pack. A canning method in which uncooked food is placed in a container that is filled with boiling water or juice and closed with a lid and a ring band; also referred to as cold-pack. (27)

reactant (ree-AK-tunt). The elements or compounds present at the start of a chemical reaction. (8)

Recommended Dietary Allowances (RDAs). The adequate amount of a specific nutrient needed by most healthy people. (12)

rehydration. Replacing water that was previously removed from a food. (26)

residue. The matter remaining after a chemical or physical process, as in the residue remaining on harvested foods. (25)

respiration. The metabolic process that releases energy from glucose when the glucose molecule in a cell is taken apart. (12)

restoration. The process in which nutrients that are lost in processing are returned to a food. (24)

retort canning. Pressure processing carried out by commercial food canners. (27)

retorts. Huge pressure canners used commercially. (27)

retrogradation (reh-tro-gray-DAY-shun). A condition in a food gel in which the amylose molecules shift and orient themselves in crystalline regions, forming a somewhat gritty texture in the mixture and possibly causing it to leak water; a pudding over time is an example.(15)

rickets. A vitamin D deficiency disease that causes soft, weak bones. (18)

S

saccharide (SAK-uh-ride). A sugar or a substance made from sugar. (15)

saliva. A mixture of water, mucus, salts, and the digestive enzyme amylase; secreted by the salivary glands in the mouth. (13)

saturated. A solution that contains all the solute that can be dissolved at a given temperature. (20)

saturated fat. A fat in which most of the fatty acids are saturated. In other words, the fatty acids contain all the hydrogen atoms their molecular structure can hold. (16)

scurvy. A circulatory disease, now known to result from vitamin C deficiency; causes bleeding gums, weakened blood vessels, and extreme fatigue. (12)

semipermeable (se-mee-PUR-mee-uh-bul). Allowing varying amounts of certain substances to pass through, as with a membrane. (14)

sensory characteristics. The qualities of a food identified by the senses. (6)

sensory evaluation. Scientifically testing food, using the human senses of sight, smell, taste, touch, and hearing. (6)

sensory evaluation panels. Groups of people who evaluate the sensory characteristics of food samples. (6)

shelf life. The time a food can be stored before deteriorating. (23)

single-acting baking powder. A leavening agent that releases carbon dioxide as soon as liquid is added to it; rarely sold in the United States. (21)

single bond. A covalent bond in which each atom donates only one electron to form the bond. (16)

smoke point. The temperature at which a fat produces smoke. (16)

solidification point. The temperature at which a melted fat regains its original firmness. (16)

solute (SAHL-yoot). The substance that is dissolved in a solution. (9)

solution. A homogeneous mixture in which one substance is dissolved in another. (7)

solvent (SAHL-vunt). The substance that dissolves another substance in a solution. (9)

specific heat. The amount of heat needed to raise the temperature of one gram of a substance 1°C. (11)

spores. Microorganisms in a dormant, inactive, or resting state. (25)

stabilizer. A substance that keeps a compound, mixture, or solution from changing its form or chemical nature. (24)

steam blanching. A pretreatment method used when dehydrating food in order to stop enzyme action; food is placed in a perforated basket over boiling water and blanched by steam. A set of two pans that stack for steaming may also be used. (26)

sublimation (sub-luh-MAY-shun). A phase change from the solid phase directly into the gas phase, skipping the liquid phase. (9)

substrate. The substance on which an enzyme works (the reactant). (19)

sulfiting. A pretreatment method used in food dehydration to prevent browning. Food is soaked in a solution of water and sodium metabisulfite or sodium bisulfite. (26)

sulfuring. A pretreatment method used in food dehydration to prevent browning. Food is placed on large trays that are stacked together and covered. Burning sulfur creates fumes that circulate around the food. The fumes contain sulfur dioxide gas. (26)

supersaturated (SOO-pur-SA-chuh-ray-tud). When a solution contains more dissolved solute than it would normally hold at that temperature. (15)

surface tension. An inward force or pull that tends to minimize the surface area of a liquid. (9)

sustainable farming. The process of producing food by natural methods that fit with local needs and conditions. (2)

syneresis (suh-NEHR-uh-sus). Water that leaks from a gel as it ages, as in pudding. (15)

syrup blanching. A pretreatment method that stop enzyme action in fruit to be dehydrated; cut fruit is soaked in a hot solution of sugar, corn syrup, and water, then drained and dried. (26)

T

taring. A procedure used to measure the mass of a substance but not its container. First mass the empty container; then subtract that value from the mass of the substance and the container together. (3)

taste blind. The inability to distinguish between the flavors of some foods. (6)

taste buds. Sensory organs located on various parts of the tongue. (6)

theory. An explanation based on a body of knowledge gained from many observations and supported by the results of many experiments. (5)

titration (tie-TRAY-shun). A common method used in the laboratory to determine the concentration of an acid or base. (10)

trace minerals. Minerals the body needs in daily amounts of 0.01 g or less. (18)

triglycerides (try-GLI-suh-ryds). The largest class of lipids, including nearly all of the fats and oils people typically eat. (16)

Tyndall effect. The ability to see the path of light as it passes through a colloidal dispersion. (20)

U

unsaturated. A solution that contains less solute than can be dissolved in it at a given temperature. (20)

unsaturated fat. A fat in which most of the fatty acids are saturated. That is, they're missing hydrogen atoms. (16)

V

variable. A factor that can change in an experiment. (5)

viscosity (vis-KAH-suh-tee). The resistance to flow. (15)

vitamins. Complex organic substances vital to life. (18)

volatile (VALL-uh-til). Easily changed into vapor. (6)

voluntary activities. Conscious and deliberate actions that burn energy in the body. (14)

W

water-bath processing. A process in canning food in which containers of food are heated in boiling water in a canning kettle. (27)

water-soluble vitamins. Vitamins that dissolve in water. (18)

whey. The protein found in the liquid that remains after fat and casein have been removed from milk. (23)

Credits

Cover Design: Pudik Graphics

Cover Image: Randall Sutter Photography

Interior Design: Design Associates, Inc.

Allpax Products Inc., 443

Ann's Portrait Designs, 8, 17, 18, 26, 36, 37, 164, 166, 167, 183, 207, 223, 233, 243, 249, 264, 267, 276, 279, 302, 303, 304, 312, 313, 317, 326, 330, 331, 333, 335, 336, 348, 369, 370, 373, 374, 390, 405, 407, 415, 423, 425, 426, 427, 432, 438, 463

Articulate Graphics, 6, 9, 14, 15, 50, 51, 54, 56, 65, 67, 69, 80, 93, 104, 105, 106, 108, 116, 117, 118, 121, 122, 130, 131, 133, 134, 136, 137, 139, 146, 147, 149, 151, 161, 162, 163, 178, 194, 195, 196, 197, 199, 207, 208, 209, 211, 220, 221, 224, 226, 240, 241, 242, 252, 258, 259, 260, 262, 276, 277, 296, 297, 298, 301, 311, 314, 317, 318, 319, 327, 342, 346, 347, 361, 373, 375, 439, 440, 444, 445, 448, 459, 461

Artesville, 14, 21, 31, 169, 176, 221, 239, 248, 270, 276, 288, 299, 346, 347, 354, 375, 425, 426, 442

ASEP Tech USA, 447

B.N.W. Industries/Belt-o-matic Dryers & Coolers, 430

James L. Ballard, 196

Roger Bean, 246

Keith Berry, 268

Culligan® Mark 100 Softener, © 2000 Culligan International Company, 138

Eating Disorders Awareness & Prevention Inc., Seattle, WA, 168

Envision, 353
  B.W. Holtman, 292
  Steven Needham, 24

Cheryl Fenton, 442

David R. Frazier Photolibrary, 22-23, 31, 88, 100-101, 225, 368, 391, 410, 411, 455

Tim Fuller, 41, 265

Food Pix, 17, 31, 64, 89, 123, 231, 261, 294, 302, 340, 342, 344, 362, 365, 390
  Susan Marie Anderson, 115
  David Bis, 223
  Thomas Firak, 313
  Benjamin Fisk, 447
  Gentl & Hyers/FPX, 264
  Brian Hagiwara, 243, 269
  Lisa Keenan, 10
  Susan Kinast, 245, 264
  Scott Lanza, 92
  Paul Poplis, 25

FPG International, 27
  Ken Chernus, 431
  Jeff Heger, 371
  Steve Hix, 31
  Spencer Jones, 315
  Keystone View Co. 177
  Jim Pickerell, 231, 354
  Ken Reid, 174
  Martin Rogers, 132
  Mark Scott, 176
  Allan H. Shoemaker, 287
  Chip Simons, 360

Stephen Simpson, 140

Adam Smith, 138

Telegraph Color Library, 34, 158, 364

Arthur Tilley, 37, 167

CT Tracy, 308

VCG, 11, 55, 452

Marshall Greenberg, 333, 366, 414

Dan Grossman, 185

Fundamental Photographs,
  Kristen Brochman, 9, 105
  Robert Mathena, 153
  Peticolas/Megna, 462
  Richard Megna, 90, 102, 120, 144, 150, 311
  Kip Peticolas, 314
  Paul S. Silverman, 145, 378

Index Stock,
  Gene Coleman, 292

International Stock,
  Julian Cotton, 40
  Willie Holdman, 460
  Charlie Westermann, 76

L' Equip®, 430

MDS Nordion,
  Peter Kunstadt, 459

Joe Mallon, 280

Ted Mishima, 7, 16, 43, 51, 53, 56, 62, 97, 107, 137, 148, 150, 165, 185, 187, 201, 215, 221, 235, 262, 281, 296, 297, 310, 315, 326, 328, 337, 363, 365, 375, 387, 389, 392, 393, 419, 424, 438, 445, 447, 449, 454

NASA, 457

Precision Weighing Balances, 52, 62

Peace Corps.,
  Andy Sprowl—Agribusiness Volunteer, Ecuador, 38

Index

A

Absolute zero, 160, 468
Absorption of nutrients, 199–200
Accuracy in measurement, 61, 468
Acesulfame-K, 388
Acetic acid, 147, 240, 346
Acetic-acid bacteria, 343, 347
Acidosis, 152
Acid rain, 156
Acids, 145
    in body, 152–153
    defining, 146, 468
    effect of, on protein, 271
    in foods, 152
    neutralization of, 147–148
    properties of, 148
    strength of, 146–147
Actin, 262
Activation energy, 295, 468
Active site, 297, 468
Additives, 462
    See also Food additives
Adenosine diphosphate (ADP), 207
Adenosine triphosphate (ATP), 207, 286, 468
Adipose tissue, 240, 248, 468
Adulterated food, 409, 468
Aeration, 244
Aerobic reaction, 342, 468
Age, influence on metabolism, 211
Aggregate, 314, 468
Agitation, 231–232
Agitation retorts, 444, 468
Agricultural engineer, 132
Agricultural protection, 39, 41
Agriculture
    biotechnology in, 22–23, 27, 32, 33
    environmental impacts of food science on, 39, 41
    genetically altered plants in, 292
    sustainable farming and, 41
Agriculture, U.S. Department of, 409, 411
Air as leavening agent, 325
Air pressure, 134–135
Albumen, 261, 468
Algal polysaccharides, 225
Alimentary canal, 193, 468
Alkalosis, 152
Alpha-D-glucose, 224

American Dietetic Association (ADA), 200
    fiber recommendation of, 228
Amine group, 257, 258, 468
Amino acids, 257, 259, 468
    essential, 268–269
Ammonia, 213
Ammonium bicarbonate, 330
Amylase, 194
Amylopectin, 224, 233, 468
Amylose, 224, 468
Anabolism, 206, 468
Anaerobic reaction, 343, 468
Anaerobic respiration, 343, 468
    See also Fermentation
Angel food cake, 337
Animal parasites, 408
Anorexia nervosa, 168, 468
Antacids, 203
Anthocyanins, 149
Antibodies, 268, 468
Antioxidants, 280, 283, 303, 385
Appeal, additives for, 386–387
Appearance, 91
Appert, Nicolas, 437
Aquifers, 129
Ascorbic acid, 303
    See also Vitamin C
Aseptic packaging, 464
Aseptic processing, 445–447, 468
Aspartame, 388
Atherosclerosis, 251, 385, 468
Atomic mass, 117–118, 118, 468
Atomic number, 116, 468
Atoms, 105–106, 116, 468
    structure of, 116–117
ATP. *See* Adenosine triphosphate (ATP)

B

Bacon, Francis, 25
Bacteria
    acetic-acid, 343, 347
    carbon-dioxide, 343
    in foodborne illness, 402–403
    increasing threats from, 420
    lactic-acid, 343, 344, 346–347, 369
    proteolytic, 344
    sources of, 344

Joule, 160, 474

K

Kerr, Alexander, 438
Ketones, 213, 263
Kidney stones, 263, 283
Kilocalories, 160, 165, 167, 183, 213, 474
    in food, 214–215
    See also Calories
Kilogray, 461, 474

L

Labeling, ingredient, 383
Laboratory
    equipment in, 49–52
    language in, 127
    safety, 53–54, 56
Laboratory panels, in sensory evaluation, 95
Lab technologist, 55
Lactase, 226, 362
Lactic acid, 215, 347, 369, 474
    as buffers, 389
Lactic-acid bacteria, 343, 344, 346–347, 369
Lactic acid fermentation, 357
Lactobacillus, 331
Lactobacillus bulgaricus, 344
Lactose, 221–222, 222, 226, 286, 348, 362, 369
    in milk, 363
Lactose intolerance, 362, 474
Lard, 247
Large intestine, absorption in, 199–200
Latent heat, 133, 474
Lavoisier, Antoine, 176–178
Law of conservation of mass, 122
LDLs. *See* Low-density lipoproteins (LDLs)
Leavened bread, 331
Leavened cakes, making, 335, 337
Leavened products, making, 331
Leavening agents, 325–337
    air as, 325
    steam as, 326
Leavenings
    as buffers, 389
    carbon dioxide in, 326–328, 330–331
Legumes, 225, 263–264
Length, measuring, 65
Lids, 438
Liebig, Justus von, 26
Light, effect of, on flavor, 253
Light activity, 212
Light cream, 367
Light whipping cream, 367
Lime juice, 145

Linoleic acid, 242
Lipids, 239–253
    categories of, 239–240
    defined, 239, 474
    See also Fats
Lipoproteins, 251–252, 268, 474
Liquid, 104
    measuring volume of, 70–71
    in yeast products, 334–335
Listeria, irradiation in reducing levels of, 398
Listeria monocytogenes, 407–408
Litmus, 149
Liver as source of cholesterol, 251
Low-density lipoproteins (LDLs), 252
Low-fat options, 249, 251, 255
Lye, 145, 149
Lymphatic system, 199
Lyophilization, 456, 474

M

Macromolecules, 257, 474
Magnesium, 286–287
    in milk, 362
Major minerals, 285–287, 474
Malnutrition, protein-energy, 272
Maltose, 194, 222, 348
Manganese, 287
Maple syrup, 387
Mason, John, 438
Mass, 52, 118, 120
    measuring, 67
Mass number, 117, 474
Mass percent, 310, 474
Mastication, 193, 474
Matter
    classification of, 104–107
    defined, 103, 474
    heat energy and, 160–161
    properties of, 103–104
Mayonnaise as emulsion, 318
Meal ready-to-eat (MRE), 450
Measurement, 61–69
    accuracy of, 61
    Celsius temperature scale, 67–69
    gaining skills and confidence in, 72
    metric prefixes, 63
    in metric system, 62–63, 65, 67
    precision of, 61
    recording metric data, 64–65
Measuring equipment, calibrating, 52–53
Meat, 183, 262–263
    irradiation of, 398
    tenderizing, 304

Meat tenderizers, 307
Medication, taking, 202
Medium, 130, 474
Megadoses, 280, 283, 474
Melting, 123, 131
Melting point, 131, 474
Melting range of triglycerides, 243
Membranes, 208, 474
Meniscus, 50, 474
Metabolic process, 205–209, 211–215
Metabolic rate, 208, 474
    math for, 212
Metabolism, 205–215, 299
    basal, 211, 217
    chemical balance during, 207–208
    defined, 205, 474
    energy for, 206
    function of, 205
    influences on, 208–209, 211
    set-point theory and, 216
    weight management and, 211–215
Metric system, 62–68, 474
Micelles, 360, 474
Microbes, 343, 474
    See also Microorganisms
Microbiologists, 27, 316
Microorganisms, 343, 474
    in foodborne illness, 402–403
    See also Bacteria; Molds; Yeast
Microvilli, 199
Microwave cooking, 163, 164
Microwaves, 163, 474
Milk, 183, 286, 359–376
    B vitamins in, 363
    calcium in, 362, 363
    carbohydrates in, 362
    cobalt in, 362
    complex nature of, 359
    composition of, 359–362
    concentrated, 366–367
    cooking with, 372–374
    copper in, 362
    dry, 367
    as emulsions, 318
    essential amino acids in, 363
    fat in, 361–362
    fluid, 366
    fortification of vitamin D in, 365
    heating, 373–374
    iodine in, 362
    iron in, 362
    lactose in, 363
    magnesium in, 362

minerals in, 362
molybdenum in, 362
niacin in, 362, 363
nickel in, 362
nutritional value of, 363
processing, 364–365
protein in, 360
riboflavin in, 362, 363
salt in, 362
separating, 375
as source of *Salmonella*, 406
storing, 372
thiamin in, 362, 363
trace elements in, 362
ultrahigh-temperature, 366
value of, 376
vitamin A in, 362, 363, 365
vitamins in, 362
Milk products
    cooking with, 372–374
    cultured, 377
    storing, 372
    types of, 366–370
Milk solids, 367, 374, 474
Minerals, 178, 195
    defined, 285, 475
    Dietary Reference Intakes (DRIs) for selected, 285
    function of, 285
    major, 285–287
    in milk, 362
    in solutions, 313
    trace, 287
Mixtures, 107, 475
    heterogeneous, 107
    homogeneous, 107, 139
    separating, 110–111
Moderate activity, 212
Modified-atmosphere packaging, 463–464, 475
Molarity, 150, 475
    calculating, 150
Molar mass, 120
Molasses, 230, 387
Mold fermentation, 350
Molds in foodborne illness, 402
Mole, 118, 475
Molecular motion, 160
Molecule, 106, 118, 120, 475
    polar, 129
Molybdenum, 287
    in milk, 362
Monocalcium phosphate, 328
Monoglycerides, 245

in milk, 360
structural, 268
structure of, 257–261
Proteolytic bacteria, 344
Protons, 116, 117, 476
Psychology and food choice, 88–89
Public health and safety, 39
Pure substances, 105–106, 476
Pyloric sphincter, 197, 476

Q

Quality control specialists, 28
Quick breads, 335, 476

R

Radiation, 162
defined, 458, 476
Radiation sterilization, 462
Rancid, 247, 476
Rancidity, 393
Raw-pack method, 438, 476
RDA. See Recommended Dietary Allowance (RDA)
R & D specialist (research and development), 29
Reactants, 122, 296, 476
Reaction, rates of, 163
Reasoning skills, 76
Recipe, 26
Recommended Dietary Allowance (RDA), 179–180, 182, 476
for protein, 269–270
Rectum, 199
Rehydration, 432–433, 476
Reliability, judging, 84
Religious traditions and food choices, 89
Report form, 78
Research chemist, 345
Residue, 410, 476
Respiration, 177, 342, 477
Restoration, 384, 477
Retinol, 281
Retort canning, 443, 477
Retorts, 443, 477
Retrogradation, 233, 477
Reversible chemical reactions, 123
Reversible physical changes, 123
Riboflavin, 278–279, 349
in milk, 362, 363
Rice, 183
in Japan, 236
Rickets, 275, 283, 393, 477
adult, 283
Ripening, 232
Ripening processes, 369–370

Room drying, 429

S

Saccharides, 221, 230, 477
Saccharin, 387–388, 390
Saccharomyces cerevisiae, 334
Safety
benefits of pasteurization, 364
food additives and, 393
frying, 248
laboratory, 53–54, 56
See also Food safety
Safety goggles, 53
Saliva,194, 477
Salivary glands, 194
Salmonella, 421, 461
irradiation and, 462
irradiation in reducing levels of, 398
Salts, 153
in milk, 362
in yeast products, 335
See also Ionic compounds; Sodium
Sandwich, nutrition anatomy of, 178
Saturated, 310, 477
Saturated fat, 241, 477
Saturation, 310
Sauerkraut, 347
Science
in real life, 31
See also Food science; Nutrition science
Scientific method, 61–69, 75–82
analyzing data, 77
conducting research, 79–80, 82
developing and revising theories, 78
experimentation, 76–77
forming hypothesis, 75–76
reporting results, 78
Scurvy, 176, 275, 477
Sedentary activity, 211–212
Selenium, 287
Semipermeable, 208, 477
Semisoft cheeses, 370
Senior protein biologist, 266
Sensory appeal, 96
Sensory characteristics, 91, 477
Sensory evaluation, 90–91, 95–96, 477
Sensory evaluation panels, 95, 477
Sensory scientist, 94
Serving, size of, 64
Set-point theory, 216
Shelf life, 364, 477
Shortened cakes, 335
Simple carbohydrates, 219, 221–223, 228, 230